Lecture Notes in Mechanical Engineering

Lecture Notes in Mechanical Engineering (LNME) publishes the latest developments in Mechanical Engineering—quickly, informally and with high quality. Original research reported in proceedings and post-proceedings represents the core of LNME. Volumes published in LNME embrace all aspects, subfields and new challenges of mechanical engineering. Topics in the series include:

- Engineering Design
- Machinery and Machine Elements
- Mechanical Structures and Stress Analysis
- Automotive Engineering
- Engine Technology
- Aerospace Technology and Astronautics
- Nanotechnology and Microengineering
- Control, Robotics, Mechatronics
- MEMS
- Theoretical and Applied Mechanics
- Dynamical Systems, Control
- Fluid Mechanics
- Engineering Thermodynamics, Heat and Mass Transfer
- Manufacturing
- Precision Engineering, Instrumentation, Measurement
- Materials Engineering
- Tribology and Surface Technology

To submit a proposal or request further information, please contact the Springer Editor of your location:

China: Ms. Ella Zhang at ella.zhang@springer.com
India: Priya Vyas at priya.vyas@springer.com
Rest of Asia, Australia, New Zealand: Swati Meherishi at
swati.meherishi@springer.com
All other countries: Dr. Leontina Di Cecco at Leontina.dicecco@springer.com

To submit a proposal for a monograph, please check our Springer Tracts in Mechanical Engineering at https://link.springer.com/bookseries/11693 or contact Leontina.dicecco@springer.com

Indexed by SCOPUS. All books published in the series are submitted for consideration in Web of Science.

More information about this series at https://link.springer.com/bookseries/11236

Kyoung-Yun Kim · Leslie Monplaisir ·
Jeremy Rickli
Editors

Flexible Automation and Intelligent Manufacturing: The Human-Data-Technology Nexus

Proceedings of FAIM 2022, June 19–23,
2022, Detroit, Michigan, USA, Volume 1

 Springer

Editors
Kyoung-Yun Kim
Industrial and Systems Engineering
Wayne State University
Detroit, MI, USA

Leslie Monplaisir
Industrial and Systems Engineering
Wayne State University
Detroit, MI, USA

Jeremy Rickli
Industrial and Systems Engineering
Wayne State University
Detroit, MI, USA

ISSN 2195-4356   ISSN 2195-4364   (electronic)
Lecture Notes in Mechanical Engineering
ISBN 978-3-031-18325-6   ISBN 978-3-031-18326-3   (eBook)
https://doi.org/10.1007/978-3-031-18326-3

This Springer imprint is published by the registered company Springer Nature Switzerland AG
The registered company address is: Gewerbestrasse 11, 6330 Cham, Switzerland

Preface

This volume of Lecture Notes in Mechanical Engineering (LNME) is one of two volumes including papers selected from the 31st International Conference on Flexible Automation and Intelligent Manufacturing (FAIM 2022), held in Detroit, Michigan, USA, from June 19 to 23, 2022. The FAIM 2022 conference was organized by Wayne State University, a Carnegie R1 Doctoral University, Member of the University Research Corridor of Michigan and located in midtown Detroit.

Flexible Automation and Intelligent Manufacturing (FAIM) is a renowned international forum for academia and industry to disseminate novel research, theories, and practices relevant to automation and manufacturing. For over 30 years, the FAIM conference has provided a strong and continuous presence in the international manufacturing scene, addressing both technology and management aspects via scientific conference sessions, workshops, tutorials, and industry tours. Since 1991, FAIM has been hosted in prestigious universities on both sides of the Atlantic and, in recent years, in Asia. The conference attracts hundreds of global leaders in automation and manufacturing research who attend program sessions where rigorously peer-reviewed papers are presented during the multiple-day conference. The conference links researchers and industry practitioners in a continuous effort to bridge the gap between research and implementation.

FAIM 2022 received 258 contributions from over 30 countries and over 120 institutions around the world. After a two-stage double-blind review, the technical program committee accepted 160 papers from 28 countries and affiliated with 91 institutions. 119 papers have been included in two LNME volumes, and 31 extended papers are published as fast-track articles in the journal of Robotics and Computer-Integrated Manufacturing and International Journal of Advanced Manufacturing Technology. A selection of these LNME articles will be invited to submit substantially extended versions to special issues in the International Journal of Computer-Integrated Manufacturing and Machines journal. We appreciate the authors for their contributions and would like to acknowledge the FAIM steering committee, advisory board committee, honorary chairs, the scientific committee members, and manuscript reviewers for their significant efforts, continuous support, sharing their expertise, and conducting manuscript reviews. Manuscript reviewers came from various locations around the world, representing 43 countries and affiliated with more than 210 institutions. With such effort and dignity, we were able to maintain the high standards of the papers included in the FAIM program.

Special thanks to FAIM 2022's honorary keynote speakers; Joseph Beaman (*Professor* Earnest F. Gloyna Regents Chair in Engineering, The University of Texas at Austin, USA), Brench Boden (Digital Enterprise Lead, Air Force Research Lab, Air Force Manufacturing Technology Program), Ryan Jarvis (Vice President & General Manager, Siemens Digital Industries), and Mario Santillo (Robotics Research Leader, Ford Motor Company). We would also like to acknowledge FAIM 2022's Industry Day speaker; Mark Dolsen (President, TRQSS, Inc.), Tom Hoffman (Portfolio Development Executive and Director of Business, Siemens), Lindsay Klee (Innovation, Commercialization & Technology Transfer Leader Innovation, Commercialization & Technology Transfer Leader, Wayne State University), Angie Lafferty (General Manager, Engineering & Skilled Trades, TRQSS, Inc.), Benjamin Messick (Director of Technology, ATCO Industries), Raymond Monroe (Executive Vice President, Steel Founders' Society of America SFSA), Sam Phillips (Digital Enterprise and Services at Siemens Digital Industries USA, Siemens), and Wendy Serra (Director of Advanced Manufacturing Services, Jacobs Engineering).

The book "Flexible Automation and Intelligent Manufacturing: The Human-Data-Technology Nexus - Proceedings of FAIM 2022" has been organized in two LNME volumes. The present volume—Volume 1—has been published via the open-access route, while Volume 2 has been published via the subscription route. The theme of FAIM 2022 was "Human-Data-Technology Nexus in Intelligent Manufacturing and Next Generation Automation." In both volumes, the papers have been organized into four thematic pillars in automation and intelligence streams, underpinning the conference's main theme: manufacturing processes, machine tools, manufacturing systems, and enabling technologies.

Manufacturing Processes: This thematic pillar encompasses new and innovative process such as additive, laser-based deposition, and hybrid processes; innovative materials in manufacturing; precision engineering; and processes at the micro and nanoscales.

Machine Tools: The machine tools thematic pillar focuses on research and development of manufacturing machine technologies. Specific topics in this area include but are not limited to numerical control and mechatronics; intelligent machine tools; process and condition monitoring; and computer-aided manufacturing.

Manufacturing Systems: Manufacturing systems are a thematic pillar that covers a broad range of manufacturing and automation topics. Topic areas with manufacturing systems include Industry 4.0; sustainable/green manufacturing; lean and post-lean manufacturing; human–robot collaboration; quality control and inspection; logistics and supply chain engineering; education and training; and more.

Enabling Technologies: Enabling technologies are a thematic pillar with topic areas that impact the three aforementioned pillars. Topics within this pillar cover applied artificial intelligence; machine learning; virtual/augmented reality; digital twins; manufacturing networks and security; and ontologies and information modeling.

We appreciate the partnership with Springer, ConfTool, and our sponsors for their fantastic support during the preparation of FAIM 2022. Thank you very much to the FAIM 2022 organizing team, whose hard work was critical to the success of the FAIM 2022 conference.

August 2022   Kyoung-Yun Kim
                                                                                     Leslie Monplaisir
                                                                                      Jeremy Rickli

FAIM 2022 Organization

Organizing Committee

General Chairs

Kyoung-Yun Kim	Wayne State University, USA
Leslie Monplaisir	Wayne State University, USA

Technical Program Chairs

Jeremy Rickli	Wayne State University, USA
Saravanan Venkatachalam	Wayne State University, USA
Murat Yildirim	Wayne State University, USA

Industrial Program Chairs

Ratna Chinnam	Wayne State University, USA
Qingyu Yang	Wayne State University, USA

Communication and Media Chairs

Sara Masoud	Wayne State University, USA
Yanchao Liu	Wayne State University, USA

Conference Managers

Mark Garrison	Wayne State University, USA
Cameron Morris	Wayne State University, USA

Assistant Conference Managers

Shengyu Liu	Wayne State University, USA
Zhenyu Zhou	Wayne State University, USA

Steering Committee – FAIM 2022

Frank Chen	The University of Texas at San Antonio, USA
Munir Ahmad	Khwaja Fareed University of Engineering & Information Technology, Pakistan
George-Christopher Vosniakos	National Technical University of Athens, Greece
Kyoung-Yun Kim	Wayne State University, USA

Advisory Board - Program Committee

Esther Alvarez de los Mozos	Universidad de Deusto, Spain
Americo Azevedo	INESC Porto, Portugal
F. Frank Chen	The University of Texas-San Antonio, USA
Paul Eric Dossou	ICAM Paris-Senart, France
Farnaz Ganjeizadeh	California State University, East Bay, USA
Dong-Won Kim	Chonbuk National University, South Korea
Stephen T. Newman	University of Bath, UK
Chike F. Oduoza	University of Wolverhampton, UK
Margherita Peruzzini	University of Modena and Reggio Emilia, Italy
Alan Ryan	University of Limerick, Ireland
Francisco Silva	Polytechnic of Porto, Portugal
Dusan Sormaz	Ohio University, USA
Leo De Vin	Karlstadt University, Sweden
George-Christopher Vosniakos	National Technical University of Athens, Greece
Lihui Wang	KTH Royal Institute of Technology, Sweden
Yi-Chi Wang	Feng Chia University, Taiwan

Honorary Chairs

Munir Ahmad (FAIM Founding Chair)	Khwaja Fareed University of Engineering & Information Technology, Pakistan
F. Frank Chen	The University of Texas at San Antonio, USA
William G. Sullivan (FAIM Founding Chair)	Virginia Polytechnic Institute & State University, USA

Scientific Committee

Rita Ambu	University of Cagliari, Italy
Jan Christian Aurich	Technische Universität Kaiserslautern, Germany
Fazleena Badurdeen	University of Kentucky, USA
Amarnath Banerjee	Texas A&M University, USA
António Bastos	Universidade de Aveiro, Portugal
Sara Behdad	University of Florida, USA

Panorios Benardos	National Technical University of Athens, Greece
Giovanni Berselli	University of Genoa, Italy
Satish Bukkapatnam	Texas A&M University, USA
Michele Cali	University of Catania, Italy
Osiris Canciglieri	Pontifical Catholic University of Parana, Brazil
Gaetano Cascini	Politecnico di Milano, Italy
Eric Coatanea	Tampere University, Finland
Ming Dong	Shanghai Jiao Tong University, PR China
Stephen Ekwaro-Osire	Texas Tech University, USA
Mohamed El Mansor	ENSAM, France
Zhaoyan Fan	Oregon State University, USA
Irene Fassi	Consiglio Nazionale delle Ricerche, Italy
Luís Pinto Ferreira	Instituto Superior de Engenharia do Porto, Portugal
Joerg Franke	Friedrich-Alexander University of Erlangen-Nuremberg, Germany
Liang Gao	Huazhong University of Science and Technology, China
Andrea Ghiotti	University of Padova, Italy
Karl Haapala	Oregon State University, USA
Witold Habrat	Rzeszow University, Poland
George Q. Huang	The University of Hong Kong, Hong Kong
Jingwei Huang	Old Dominion University, USA
Prashanth Konda Gokuldoss	Tallinn University of Technology, Estonia
Gül E. Kremer	Iowa State University, USA
Ismail Lazoglu	Koc University, Turkey
José Lima	Research Centre in Digitalization and Intelligent Robotics (CeDRI), Portugal
Angelos Markopoulos	National Technical University of Athens, Greece
Andrea Matta	Politechnico di Milano, Italy
Dimitris Mourtzis	University of Patras, Greece
Aydin Nassehi	University of Bristol, UK
Dimitris Nathanael	National Technical University of Athens, Greece
Pedro Neto	University of Coimbra, Portugal
Maria Teresa Pereira	Polytechnic of Porto, Portugal
Stavros Ponis	National Technical University of Athens, Greece
Duc Truong Pham	University of Birmingham, UK
R. M. Chandima Ratnayake	Professor, University of Stavanger, Norway
José Carlos Sá	Polytechnic of Porto, Portugal
Raul Duarte Salgueiral Gomes Campilho	ISEP, Portugal
Gilberto Santos	Polytechnic Institute Cávado Ave, Portugal
Panagiotis Stavropoulos	University of Patras, Greece

Hung- da Wan   The University of Texas at San Antonio, USA
Yong Zeng   Concordia University, Canada
Rhongho Jang   Wayne State University, USA
Zude Zhou   Wuhan University of Technology, China

Contents

Contents

Enabling Technologies

Manufacturing Processes

Die-Less Forming of Fiber-Reinforced Plastic Composites

Jan-Erik Rath$^{(\boxtimes)}$ ⓘ, Robert Graupner ⓘ, and Thorsten Schüppstuhl ⓘ

Institute of Aircraft Production Technology, Hamburg University of Technology, Denickestr. 17, 21073 Hamburg, Germany
jan-erik.rath@tuhh.de

Abstract. Fiber-reinforced plastics (FRP) are increasingly popular in light weight applications such as aircraft manufacturing. However, most production processes of thin-walled FRP parts to date involve the use of expensive forming tools. This especially hinders cost-effective production of small series as well as individual parts and prototypes. In this paper, we develop new possible alternatives of highly automated and die-less production processes based on a short review of current approaches on flexible thin-walled FRP production. All proposed processes involve robot guided standard tools, similar to incremental sheet metal forming, for local forming of the base materials. These include woven glass fiber fabrics which are locally impregnated with thermoset resin and cured using UV-light, woven commingled yarns made out of glass fibers and thermoplastic fibers which are locally heated and pressed, as well as pre-consolidated thermoplastic organo sheets which require selective heating for forming. General applicability of the processes is investigated and validated in practical experiments.

Keywords: Automation · Composite · Die-less forming · Fiber-reinforced plastic · Incremental forming

1 Introduction

Due to their high structural strength-to-weight-ratio, fiber-reinforced plastics (FRP) are exceedingly popular in aircraft manufacturing, medical and other lightweight applications. Basic manufacturing steps of FRP are shaping of the structure, impregnation of reinforcement fibers with resin, consolidation and curing/solidification. While thermoset polymer matrices are cured through chemical crosslinking, thermoplastic polymers can be softened and reshaped with the aid of heat [1]. Most FRP-production processes such as injection molding or thermoforming rely on solid molding tools, whose design and fabrication is time-consuming and costly. As a consequence, these processes are not economical for prototyping and small batch production, which is in conflict with the general trends of higher variant diversity and increasing cost pressure. Therefore, a reduction of the tooling effort up to die-less forming would be beneficial [2, 3].

For metal sheets, incremental sheet forming (ISF) is a flexible and die-less process for the production of individual parts and small series. Simple forming tools, usually with

© The Author(s) 2023
K.-Y. Kim et al. (Eds.): FAIM 2022, LNME, pp. 3–14, 2023.
https://doi.org/10.1007/978-3-031-18326-3_1

hemispherical tool ends or rotating balls, moved by CNC-machines or robots form the sheet. The workpiece is clamped by blank holders, so that the tools progressively introduce strains into the material. Different process variants include singe-point incremental forming (SPIF) with just one standard tool and double-sided incremental forming (DSIF) using two tools, one on each side of the sheet. Heat-assisted ISF has been demonstrated for the forming of thermoplastic sheets at research level [4, 5].

Due to the benefits of ISF, its application to FRP forming would be desirable. However, with high tensile strength and limited strain of the reinforcement fibers, deformation mechanisms of endless-FRP are significantly different to metal or pure thermoplastic sheets, and direct application of ISF is not possible [6, 7]. Instead, the main draping mechanisms of bending and shear of the fibers and the weave must be realized for forming. Thus, local deformation of the fabric in the current forming spot is only possible through fiber movement in adjacent regions of the weave. However, already generated final part geometries need to be maintained. Therefore, the development and evaluation of new processes for the die-less forming and production of FRP is focused in the joint research project iFish – incremental Fiber shaping.

The remainder of this paper is structured as follows: First, we review existing approaches on flexible and die-less forming of thin-walled, shell-shaped FRP components. Afterwards, we list requirements for die-less forming processes and conduct a functional analysis of material types and matching processing functions for a flexible production process similar to ISF. In basic experiments, we investigate the general applicability of the process ideas and discuss the results.

2   Related Work

Flexible forming can be classified into processes with rapid tooling, with flexible molds and without molds. Production of temporary molds for prototyping is often a manual process conducted by experts, although other rapid tooling processes such as additive manufacturing or the forming of metal molds by ISF have been used [7, 8].

A flexible molding method is multi-point forming, which uses a geometrically adjustable tool consisting of an array of pins whose lengths can be changed independently. Diaphragms can be used to smoothen the mold surface [9, 10]. For long endless-FRP parts with variable cross section, a patented concept describes the local forming of a pre-impregnated and consolidated solid thermoplastic FRP sheet (organo sheet) by heating the whole width but only a certain length of the sheet and subsequently forming the area in a modular press. As neighboring sections of the laminate are solid, strains are introduced and later released when the process is repeated in the respective sections [11]. Local curing of thermoset prepreg with pressure and heat was demonstrated by Cedeno-Campos et al. [12], however not forming the material.

Processes that do not require a mold at all include additive manufacturing. Extrusion-based processes such as Fused Deposition Modeling (FDM) have been used for FRP fabrication with short and long fibers, giving the possibility to arrange the fibers in the load path direction. However, several limitations prevent industrialization of fiber-3D-printing, including limited homogeneity and fiber volume content of the material, low productivity and challenging processing [3, 13]. Miller et al. [6] conceptualized and tested a flexible roll-forming process where thermoplastic FRP were passed through an array of individually controllable rollers, creating long, singly curved shapes with variable cross-section. Localized heating was accomplished by induction [14].

Few approaches were made to develop ISF-like processes for FRP production: Fiorotto et al. [7] applied a metal diaphragm and a nylon film to a woven thermoset prepreg via a vacuum bag in order to maintain the deformation during SPIF. Al-Obaidi et al. [15] sandwiched unidirectional basalt fiber laminates with polyamide matrix between aluminum sheets, globally heated them by hot air and applied conventional SPIF for forming using an additional steel sheet between the FRP-aluminum sandwich and the tool. The authors also used the same setup to form woven glass fibers with polyamide matrix sandwiched between metal sheets. Defects occurred, especially with higher wall angles [16]. A similar approach was made by Ambrogio et al. [17] who, however, only used short-fiber-reinforced polyamide and formed it together with an aluminum sheet by SPIF. Ikari et al. [18] used a round tool tip in an ISF-like setup to incrementally form short-fiber-reinforced organo sheets. Forming was conducted by successively moving the tool in x-y-direction, locally heating the thermoplastic by an infrared spot heater and pressing the tool onto the sheet to the target z-coordinate. Resulting shape accuracy and part quality need to be improved.

To conclude, no genuinely die-less process for the forming and production of doubly-curved, shell-shaped woven FRP without a metal sheet as support exists.

3 Process Analysis for Die-Less Forming

3.1 Materials

The aim of this project is to investigate and develop die-less forming and production of woven FRP without the use of metal support sheets. Woven fabric is targeted because of its high tailorable reinforcing effect as well as relatively good drapability [19]. As matrix material, thermoset as well as thermoplastic polymers are considered, which can be initially separate from the reinforcement fibers or already included in a semi-finished product. Separate matrix can either be initially liquid (thermoset) or solid (thermoplastic). Similarly, pre-impregnated fabrics are either formable (thermoset), semi-formable (thermoplastic semi-pregs, not considered here) or rigid (thermoplastic organo sheets) at room temperature. A special semi-finished and formable product are weaves of commingled yarns with reinforcement and thermoplastic fibers. Table 1 gives an overview of the described matrix materials and product types.

3.2 Requirements and Functional Analysis

The die-less forming processes shall make use of just two robot guided standard tools similar to DSIF, which form and solidify small areas of the FRP while moving along tool paths determined according to a forming strategy published elsewhere [20]. The paths follow the required fiber orientations from an already rigid starting point until the edge of the fabric. The forming sequence is based upon the required shear distribution as shear is the main draping mechanism for surfaces with double-curvature. Starting with

Table 1. Classification of matrix materials and product types

		Product type		
		Separate	Commingled	Pre-impregnated
Matrix material	Thermoset	A: liquid matrix		D: formable
	Thermoplastic	B: solid matrix	C: comm. weave with matrix fibers	E: pre-consolidated solid organo sheet

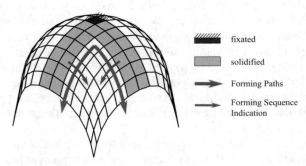

fixated

solidified

→ Forming Paths

→ Forming Sequence Indication

Fig. 1. Forming paths & forming sequence for a hemisphere

the forming paths requiring least shear, the desired geometry is formed and the FRP part produced adding a solidified path next to another, as indicated in Fig. 1.

Thus, depending on the materials used, the following functions must be fulfilled for forming and producing a FRP with just two robot guided standard tools:

1. Fixation of the fabric. According to the forming strategy, clamping of the fabric or laminate is primarily required in the starting point of the first forming paths, which is very likely in the midst of the fabric. For stability and realization of the clamping through mechanical elements, the FRP must be already cured/solidified in the starting point. Highly flexible dry fabrics (used in A & B in Table 1) require extra attention and possibly additional fixations on the edges in order to prevent unpredictable and undesirable draping before and during the forming process. A tacky pre-impregnated thermoset (D) could additionally stick onto itself when unwillingly folded. Thermoplastic fibers are usually less flexible than reinforcing fibers, so that commingled weaves (C) are more stable and less likely to deform extremely undesirably when fixated.

2. Acquiring formability of solid matrix material. Separate (B) or pre-consolidated (E) solid and non-deformable thermoplastic matrix materials need to be heated above melting temperature. This can be realized by multiple different heat sources including electrical resistance heating with contacting elements or heat guns, gas torches, infrared heating, laser light sources as well as ultrasonic elements or induction heating. Heating of separate matrix (B) is easier than of a flat or even already deformed organo sheet (E), especially when considering the need for targeted heating of a delimited area changing in size after each forming step. This is an essential basis for organo sheet forming without metal support sheets, as already formed areas need to be cold and solid while others need to be warm and drapable. While the first forming paths require heating of a major part of the sheet, areas to be heated are getting smaller during the process.

3. Local forming. For highest flexibility, two standard forming tools, one on each side of the fabric, should be guided by two individual industrial robots in a setup similar to the concept sketch in Fig. 2. Following the developed forming strategy, shear is introduced into the fabric only by out-of-plane deformation inducing compressive stresses in-plane in between the formed rigid areas. Thus, the forming tools do not need to introduce tension into the fabric directly by pulling a fixated point. In order to enable a time-efficient continuous processing while minimizing friction and undesired movement of fibers or matrix material, rotating ball or roller tool tips are desirable. The radius of the balls or rollers depends on the minimal edge radius to be formed. Flat tool tips or inflatable bladders can be considered as well, especially for processes requiring the application of pressure onto the composite in a comparably bigger area. The forming path would then consist of adding one individual pressed area after another, moving the tools to the next forming point while not in contact with the laminate. Regarding accessibility with industrial robots, rotationally symmetric tools are advantageous.

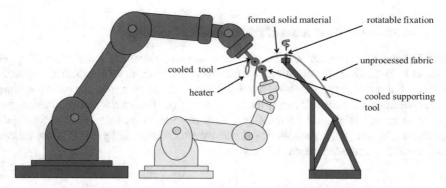

Fig. 2. Concept sketch for robotic die-less forming of thermoplastic co-weave

4. a) Impregnation. Unlike thermosets, thermoplastic polymers must be first heated above melting temperature in order to impregnate dry fiber fabrics. While impregnation immediately starts after heating woven commingled yarns (C), separate thermoplastic (B) or thermoset (A) polymers need to be handled and transferred to the reinforcement fabric which can be challenging. The former has higher viscosity and is therefore likely to deform fibers or fiber bundles rather than impregnating each individual fiber. Additionally, it easily solidifies if the required temperature is not maintained, which could be prevented by pre-heating the fabric.

4. b) Consolidation. Subsequent consolidation removes voids out of the matrix by applying pressure on the composite. The required pressing force between the tool ends must be generated by the handling devices. With higher matrix viscosity, higher pressure is required to fully impregnate the fabric and consolidate the FRP [21]. Especially in the case of thermoplastic matrices (B & C), this pressure needs to be maintained for a certain time. Not only prepregs, but even pre-consolidated organo sheets (E) need a certain but comparably lower pressure in order not to de-consolidate when heated above melting temperature for a specific time [22].

5. Curing/Solidification. Each forming path follows the final part geometry and curing/solidification while forming ensures that the acquired geometry is maintained. Thermoplastics (B, C, E) solidify during cooling, preferably under pressure to maintain consolidation, which can be supported by air jets or otherwise cooled tool ends. Thermoset polymers (A & D) need to build chemical crosslinks in order to cure. Depending on the resin, this curing process can for example be initiated thermally or by UV-radiation. The latter process, called photopolymerization, is significantly faster, allows easier resin handling, processing at atmospheric conditions and produces less styrene emissions. Selective curing is possible using either a laser or a conventional UV-light source in combination with local impregnation. However, in contrast to glass fibers, carbon and aramid fibers block UV-light, so that thermo-curing in a second processing step is necessary [23].

3.3   Allocation and Evaluation of the Processing Functions

Table 2 shows an allocation of the described processing functions to the material types of Table 1. In addition, feasibility of the matches is rated from *1* (feasible) to *3* (hardly feasible). If a material type does not require a certain processing step, the corresponding cell in the allocation matrix is empty and counted with *0*. The sum for each material type represents its unfeasibility. As a result, the three material types with least points, liquid separate thermoset (A), organo sheet (E) and woven commingled yarns (C), are further considered and evaluated for die-less forming.

Table 2. Matching material types and processing functions

		Fixation	Formability	Forming	Impregnation	Consolidation	Curing/Solidification	Total unfeasibility
		1	2	3	4 a)	4 b)	5	
Liquid separate thermoset	A	2		2	*1*	2	*1*	8
Solid separate thermoplastic	B	2	*1*	2	3	2	*1*	11
Commingled weave	C	*1*		*1*	*1*	2	*1*	6
Thermoset prepreg	D	3		3		2	3	11
Organo sheet	E	*1*	2	2		*1*	2	8

4 Preliminary Practical Experiments

4.1 Liquid Separate Thermoset Processing

Setup. Feasibility of local thermoset impregnation and UV-curing while forming a hemisphere according to the forming strategy was investigated using glass fiber twill weave with an areal weight of 160 g/m^2 and a size of 30 × 30 cm^2. Its center was fixated to a pole using a tack and the remainder of the fabric hanging freely. A handheld 2D quarter circle tool with a radius of 12 cm and a thickness of 1 cm was coated with release agent and manually positioned under the fabric. The so defined forming path was impregnated with UV-curable 3D-printing resin (PrimaCreator, Sweden) using a brush. Subsequently, a handheld laser with a wavelength of 405 nm and a power of ~600 mW was moved along the forming path, curing each spot for approximately 4 s. Afterwards, the tool was moved to the next position according to the forming strategy. In a further setup, in order to investigate die-less forming, the 2D tool was replaced by a rotatable metal ball as a standard "1D" ISF-tool, which was manually movable in fixed increments along a hemispherical surface around the fixation as the highest point of the hemisphere. In a forming path, the tool was moved to the desired point and the fabric manually impregnated and cured. Both setups are shown in Fig. 3.

Results. Both tools enabled the generation of hemispheres as depicted in Fig. 4. However, the surfaces show imprints of the respective tools so that overall part quality is not yet satisfactory. With a curing time of 4 s per forming point, sufficient part stability was achieved but surfaces were still tacky. Light conductivity led to undesired curing of already impregnated areas surrounding the laser spot, which was problematic especially when the metal ball was contacting the respective resin, leading to adhesion.

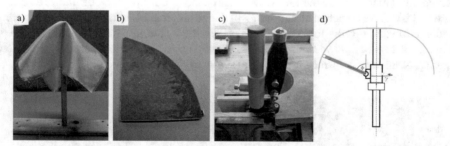

Fig. 3. a) Fixated glass fiber twill weave fabric, b) 2D quarter circle tool, c) Setup with rotatable metal ball ISF tool, d) Working principle of the metal ball tool setup

4.2 Organo Sheet Forming

Setup. Two-layered twill weave carbon fiber organo sheets (INEOS Styrolution, Germany) with styrene-acrylonitrile matrix, an areal weight of 245 g/m^2, 45% fiber volume

Fig. 4. a) Forming sequence and results of thermoset resin UV-curing using a 2D-tool, b) Result of thermoset resin UV-curing using a metal ball tool

content and a thickness of 0,6 mm were used to investigate the die-less forming process. The organo sheet sized 25×25 cm^2 was clamped at one or more rigid edges not requiring fiber movement in the respective forming step. Thus, also the starting point of the forming paths was fixed due to the rigidity of the organo sheet. Localized heating was acquired by using one infrared heater on either side of the sheet and masking areas not to be heated with reflective aluminum foil, shown in Fig. 5a). Temperature was measured using a thermography camera. After reaching 180 °C, the heating was disabled and the corresponding path was formed. Thereby, a handheld rotatable ball used in the setup in Fig. 3c) followed the forming path and pressed the sheet onto the 2D quarter circle tool of Fig. 3b) positioned under the sheet. While following the path, an air jet was directed onto the organo sheet exiting the forming zone for instant cooling.

Results. Shielding with aluminum foil enabled a clearly localized heating of the desired area as demonstrated in Fig. 5b), where only the upper side of the sheet was heated and temperature measured on the lower side. With double-sided heating, the target temperature of 180 °C was reached after around 40 s. The air jet enabled rapid localized cooling and solidification of the formed path. However, the resulting hemispherical frustum as depicted in Fig. 5c) showed large wrinkles mainly due to inhomogeneous and insufficient heating of already deformed areas by the stationary infrared heaters.

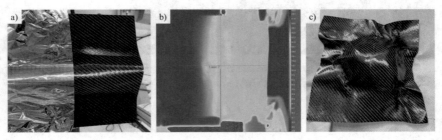

Fig. 5. a) Clamped organo sheet with aluminum foil shield, b) Resulting temperature distribution during infrared heating, c) Generated hemispherical frustum

4.3 Woven Commingled Yarn Processing

Setup. Glass fiber/polypropylene commingled twill weave (COMFIL, Denmark) with an areal weight of 700 g/m^2 and 60% reinforcement weight fraction was used to investigate the feasibility of local forming, impregnation and consolidation. The fabric of size 30×30 cm^2 was fixated to a pole, the 2D quarter circle tool was protected with release film and placed under the fabric to support the respective forming path. Local heating of the forming area to a temperature of ~190 °C, measured with a pyrometer, was accomplished by a hot air gun set to ~500 °C within 6 s. Moving along the forming path, the heated fabric was pressed onto the 2D-tool using a handheld roller protected by separating foil. The roller as well as the fabric exiting it were cooled by an air jet. The setup is depicted in Fig. 6.

Results. The performed process produced rigid material in each forming step and generated the desired hemispherical geometry as shown in Fig. 7a). Although the roller tool initially created relatively smooth surface in each forming path, the finally resulting surface quality of the whole part is inferior and individual forming paths are visible. As heating with the hot air gun is not as localized as required, material surrounding the forming path which already has been processed is heated and melted again, but not consolidated by the tools.

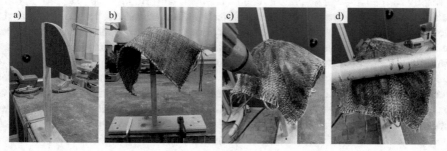

Fig. 6. a) Pole & 2D quarter circle tool, b) Commingled weave fixed to the pole, c) Heating with hot air gun, d) Forming, impregnation & consolidation with a roller

5 Discussion and Outlook

Die-less forming with two simple standard tools, each individually guided by a handling device such as an industrial robot, can be a solution to reduce cost, time and effort of prototype and small batch FRP production. Through a systematic functional analysis, material types and processing options for die-less forming were described and compared. Out of these options, local thermoset photopolymerization, organo sheet forming and local commingled weave processing have most potential and were investigated in preliminary experiments using 2D quarter circle tools. Furthermore, a rotating ball tool was used in local thermoset photopolymerization, die-lessly forming the material.

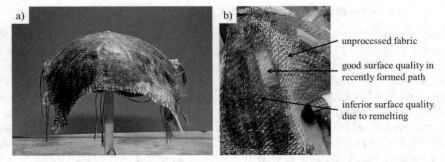

unprocessed fabric

good surface quality in
recently formed path

inferior surface quality
due to remelting

Fig. 7. a) Generated hemisphere, b) Surface quality in an intermediate forming step

Although UV-curing of thermoset resin showed good results and curing was relatively quick, processing was intricate. High resin viscosity led to easy wetting of the used tools, resulting in adhesion when the resin was cured. Due to toxicity of the resin and danger of the laser radiation, high safety measures are required. Furthermore, the process is limited to glass fibers. As curing takes place under atmospheric conditions, the process is simpler but part quality comparably poor. Consolidated organo sheets are advantageous in terms of stability while clamped as well as final part quality when carefully formed and not de-consolidated. However, uniform local heating proved difficult especially when the targeted area is already deformed and not perpendicular to the incoming infrared radiation. Other heating options such as movable laser or inductive heaters were considered but dismissed as too expensive and impractical. In addition, multiple repeated heating and/or a long time above melting temperature can lead to deconsolidation and decreased mechanical properties of the material. For commingled weave processing, heating is required for impregnation only in a small area the size of the tools. However, different heating options such as induction heating need to be further investigated as the heat-affected zone of the hot air gun used in the preliminary test was too big. Achieving a sufficient level of impregnation and consolidation in the FRP is one of the most challenging tasks in die-less forming. This is especially problematic with commingled weave processing as the pressed area might be too small and consolidation time too short. Thus, it may be reasonable to first create a semi-consolidated part with the desired geometry and outsource possible further consolidation to a downstream process step with different, comparably larger tools. Nevertheless, the processing of thermoplastic co-weaves is the overall most feasible option and will be further pursued and investigated in the project. This includes studies on heating methods, the influence of the tool type on part quality and optimal processing parameters.

Acknowledgements. Research was funded by the German Federal Ministry for Economic Affairs and Climate Action under the Program LuFo VI-1 iFish with project partners CompriseTec GmbH and carat robotic innovation GmbH.

References

1. Mallick, P.: Fiber-Reinforced Composites: Materials, Manufacturing and Design. CRC Press, Boca Raton (2008)
2. Bannister, M.: Challenges for composites into the next millennium — a reinforcement perspective. Compos. A Appl. Sci. Manuf. (2001). https://doi.org/10.1016/S1359-835X(01)00008-2
3. Zindani, D., Kumar, K.: An insight into additive manufacturing of fiber reinforced polymer composite. Int. J. Lightweight Mater. Manuf. **2**, 267–278 (2019)
4. Zhu, H., Ou, H., Popov, A.: Incremental sheet forming of thermoplastics: a review. Int. J. Adv. Manuf. Technol. **111**(1–2), 565–587 (2020). https://doi.org/10.1007/s00170-020-06056-5
5. Trzepieciński, T.: Recent developments and trends in sheet metal forming. Metals **10**, 779 (2020)
6. Miller, A.K., Gur, M., Peled, A., Payne, A., Menzel, E.: Die-less forming of thermoplastic-matrix, continuous-fiber composites. J. Compos. Mater. (1990). https://doi.org/10.1177/002199839002400401
7. Fiorotto, M., Sorgente, M., Lucchetta, G.: Preliminary studies on single point incremental forming for composite materials. Int. J. Mater. Form. (2010). https://doi.org/10.1007/s12289-010-0926-6
8. Sudbury, T.Z., Springfield, R., Kunc, V., Duty, C.: An assessment of additive manufactured molds for hand-laid fiber reinforced composites. Int. J. Adv. Manuf. Technol. **90**(5–8), 1659–1664 (2016). https://doi.org/10.1007/s00170-016-9464-9
9. Kaufman, S.G., Spletzer, B.L., Guess, T.L.: Freeform fabrication of polymer-matrix composite structures. In: Proceedings of International Conference on Robotics and Automation, pp. 317–322 (1997). https://doi.org/10.1109/ROBOT.1997.620057
10. Walczyk, D.F., Hosford, J.F., Papazian, J.M.: Using reconfigurable tooling and surface heating for incremental forming of composite aircraft parts. J. Manuf. Sci. Eng. (2003). https://doi.org/10.1115/1.1561456
11. Hauwiller, P.B., Strong, B.: Incremental Forming of Thermoplastc Composites. USA Patent 5,026,514, 25 June 1991
12. Cedeno-Campos, V.M., Jaramillo, P.A., Fernyhough, C.M., Fairclough, J.P.A.: Towards mould free composites manufacturing of thermoset prepregs. Incremental curing with localised pressure-heat. Procedia CIRP **85**, 237–242 (2019). https://doi.org/10.1016/j.procir.2019.09.020
13. Kállai, Z., Dammann, M., Schüppstuhl, T.: Operation and experimental evaluation of a 12-axis robot-based setup used for 3D-printing. In: ISR 2020. 52th International Symposium on Robotics. VDE Verlag, Berlin (2020)
14. Ramani, K., Miller, A.K., Cutkosky, M.R.: A new approach to the forming of thermoplastic-matrix continuous-fiber composites - part 1: process and machine. J. Thermoplast. Compos. Mater. (1992). https://doi.org/10.1177/089270579200500301
15. AL-Obaidi, A., Graf, A., Kräusel, V., Trautmann, M.: Heat supported single point incremental forming of hybrid laminates for orthopedic applications. Procedia Manuf. (2019). https://doi.org/10.1016/j.promfg.2019.02.101
16. AL-Obaidi, A., Kunke, A., Kräusel, V.: Hot single-point incremental forming of glass-fiber-reinforced polymer (PA6GF47) supported by hot air. J. Manuf. Process. (2019). https://doi.org/10.1016/j.jmapro.2019.04.036
17. Ambrogio, G., Conte, R., Gagliardi, F., de Napoli, L., Filice, L., Russo, P.: A new approach for forming polymeric composite structures. Compos. Struct. (2018). https://doi.org/10.1016/j.compstruct.2018.07.106

18. Ikari, T., Tanaka, H., Asakawa, N.: Development of a Novel Shell Shaping Method with CFRTP: Forming Experiment Using Localized Heating in Processing Point. MSF (2016). https://doi.org/10.4028/www.scientific.net/MSF.874.40
19. Flemming, M., Ziegmann, G., Roth, S.: Faserverbundbauweisen: Fertigungsverfahren mit duroplastischer Matrix. Springer, Heidelberg (1996). https://doi.org/10.1007/978-3-642-614 32-3
20. Rath, J.-E., Schwieger, L.-S., Schüppstuhl, T.: Robotic die-less forming strategy for fiber-reinforced plastic composites production. Procedia CIRP (2022). https://doi.org/10.1016/j.procir.2022.05.145
21. Martin, I., Del Saenz Castillo, D., Fernandez, A., Güemes, A.: Advanced thermoplastic composite manufacturing by in-situ consolidation: a review. J. Compos. Sci. (2020). https://doi.org/10.3390/jcs4040149
22. Ye, L., Chen, Z.-R., Lu, M., Hou, M.: De-consolidation and re-consolidation in CF/PPS thermoplastic matrix composites. Compos. A Appl. Sci. Manuf. (2005). https://doi.org/10.1016/j.compositesa.2004.12.006
23. Endruweit, A., Johnson, M.S., Long, A.C.: Curing of composite components by ultraviolet radiation: a review. Polym. Compos. (2006). https://doi.org/10.1002/pc.20166

Assessment of High Porosity Lattice Structures for Lightweight Applications

Rita Ambu[1]([⊠]) [iD] and Michele Calì[2] [iD]

[1] Department of Mechanical, Chemical and Materials Engineering, University of Cagliari, Via Marengo 2, 09123 Cagliari, Italy
rita.ambu@unica.it
[2] Department of Electric, Electronics and Computer Engineering, University of Catania, Viale A. Doria 6, 95125 Catania, Italy
michele.cali@unict.it

Abstract. Additive manufacturing (AM) methods have a growing application in different fields such as aeronautical, automotive, biomedical, and there is a huge interest towards the extension of their use. In this paper, lattice structures for AM are analysed with regards to stiffness and printability in order to verify the suitability for applications where the main requirement of efficiency in terms of stiffness has to be balanced with other needs such as weight saving, ease of manufacturing and recycling of the material. At this aim, lattice structures with high porosity unit cells and large cell size made of a recyclable material were considered with a geometrical configuration allowing 3D printing without any supports. The lattice structures considered were based on body-centred cubic (BCC) and face centred cubic (FCC) unit cell combined with cubic cell. Finally, a multi-morphology lattice structure obtained by mixing different unit cells is also proposed. The lattice structures were modelled and structurally analysed by means of finite element method (FEM), manufactured with a Fusion deposition modelling (FDM) printer and evaluated in relation to printability and dimensional accuracy. The results show that the proposed structure with mixed cells is potentially advantageous in terms of weight saving in relation to the mechanical properties.

Keywords: Lattice structure · Additive manufacturing · Supportless 3D printing · Geometrical configuration · High porosity

1 Introduction

The relevance of Additive manufacturing (AM) techniques has grown over the years since these methods have shown to be potentially advantageous in different fields of application. These include aerospace industry, automotive and biomedical field where AM is used in surgery for preoperative planning, implants or medical devices with many benefits since it allows to design devices customized according to the patients' needs [1]. Generally, a main goal in the design of additive manufacturing is to reduce the overall weight of a component satisfying at the same time structural requirements. This process is promoted by the capability of AM to produce complex geometries providing designers

K.-Y. Kim et al. (Eds.): FAIM 2022, LNME, pp. 15–26, 2023.
https://doi.org/10.1007/978-3-031-18326-3_2

in this way with a greater freedom. Besides appropriate geometries, lightening of a component can be obtained by means of porous or lattice structures. These structures can be obtained with different methods and can take up part of the geometry of the component or can be the structure itself.

In particular, topology optimization and generative design usually generate geometries of the components with non-functional areas lightened by a porous, stochastic distribution of the material according to loads and constraints acting on the part [2]. On the other hand, non-stochastic, or regular, structures can be modelled with different procedures. In particular, these structures can be built in Computer Aided Design (CAD) environment by replicating a unit cell along the three Cartesian directions, to obtain lattice-based geometries. Regular architectures can also be obtained by means of the Implicit Function Modelling (IFM) which is used, in particular, to obtain structures based on triply periodic minimal surfaces (TPMS) such as Gyroids, Diamond, Schwarz P-cell and others reported in literature [3]. The application of IFM requires a more computational effort with respect to the CAD method; however, complex geometrical configurations with local variable relative density and cell size can be generated, while it is more difficult to obtain geometries with local variable properties with CAD method.

Lattice structures have been considered in different areas for their lightness, structural properties and, in particular, for their energy absorption capability which is of interest in aerospace, automotive and marine structural components. Different geometries have been proposed including prisms, octet-truss, and similar [4–6].

In this paper lattice structures were considered and evaluated in terms of stiffness and printability [7, 8] for general purpose applications where the main requirement of efficiency in terms of stiffness and energy absorption is asked to be balanced with other needs such as lightness, material saving, ease of manufacturing and recycling. An example of a potential application is relative to packaging for breakable or sensitive components where requirements include lightness and recycling features for an optimal performance. AM lattice structures potentially offer the opportunity to accomplish these requirements as well as the possibility of creating personalized packaging which can be of interest for the final consumer.

Two different unit cells were considered. The first unit cell is obtained as a combination of a simple cubic (SC) unit cell and a body-centred cubic unit cell (BCC), while the other is a combination of the first unit cell and the face centred cubic (FCC) unit cell. Most of the research reported in literature analyses structures made with the basic unit cells BCC and FCC manufactured with metallic materials [9, 10], while others studies analyse the effect, in terms of structural performance, of the addition of struts with different orientations to these basic unit cells [11]. The application of structures made with unit cells similar to those considered in this study is mainly relative to the design of bone scaffolds made with biocompatible metal alloys [12, 13] where architectures with small cell size and tailored properties are required for an optimal performance.

High porosity structures with a large cell size made of a thermoplastic material were analysed, in order to verify the feasibility of these structures regarding stiffness and efficiency in manufacturing, considered that these are characterized by a reduced strut size with respect to cell size. Generally, the stiffness of a lattice structure depends on the geometrical arrangement and size of the struts [14, 15] and increases with larger strut size. It is also related to manufacturing, so that appropriate printing parameters have to be chosen in relation to the geometry and size of the lattice structure [16, 17]. As for manufacturing, the geometries considered in this study do not require supports for printing, which is advantageous for material saving. The models were structurally assessed with finite element analysis (FEA) by calculating the effective compressive and shear stiffness. Based on the numerical results obtained, a multi-morphology structure was finally proposed, modelled with both types of unit cells under study. This kind of structure is characterized by a variable volume fraction which can be also useful. The proposed structure was also evaluated by means of FE analysis, showing a good balance between stiffness and lightness. The lattice structures analyzed were 3D printed with Fusion deposition modelling (FDM) by using Polyethylene Terephthalate Glycol (PET-G), a thermoplastic polyester material with recycling capabilities. The 3D printed models were evaluated, in particular, in relation to the dimensional accuracy with respect to the CAD models. The results show that the lattice structures considered maintain adequate characteristics as for stiffness and manufacturing and can be useful when a main concern is lightness and material saving.

2 Modelling and Assessment of Lattice Structures

2.1 Modelling of Lattice Structures

The lattice unit cells were obtained by using a commercial parametric CAD modeller and the geometrical configurations are reported in Fig. 1. A simple cubic (SC) unit cell was combined with the unit cell body-centred cubic (BCC) to obtain the unit cell reported in Fig. 1a, that can be referred as SC-BCC. Then, the combination between the unit cell SC-BCC and the unit cell face-centred cubic (FCC) was considered to obtain the resulting unit cell reported in Fig. 1b, that can be referred as SC-BCC-FCC.

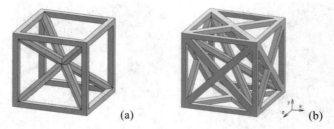

(a)   (b)

Fig. 1. Lattice unit cells: (a) SC-BCC; (b) SC-BCC-FCC.

A lattice structure is characterized by a value of the volume fraction given by V/V_0, namely, this parameter is defined as the ratio of the volume of the struts (V) to the equivalent volume occupied by the porous structure (V_0). The parameter can also be expressed in terms of porosity (P) by considering the relationship: $P = (V_0-V)/V_0$, which is generally expressed in terms of percentage.

The analysis performed was aimed to structurally evaluate high porosity structures. Therefore, a volume fraction of 0.10 equivalent to a porosity of 90% was considered.

In this way, it was possible to evaluate the effect of the geometry on the structural performance regardless of this parameter, which contributes to determine the stiffness of the structure. Lattice unit cells were designed assuming a cell size of 20 mm. A three-dimensional periodic array of the unit cell was then carried out along three mutually perpendicular directions to obtain the final architectures, each made of 4×4×4 cells. The number of cells of the models chosen for simulation allows to minimize the error in elastic modulus evaluation as shown in [18] where the variation of the stiffness of lattice models with different geometries by varying the number of cells was analysed. An evaluation of the mass of the two structures showed that SC-BCC unit cell allowed to produce a structure with a weight of about 42% lower with respect to that made of SC-BCC-FCC unit cells.

2.2 Numerical Assessment

The lattice structures were structurally analyzed by means of Finite Elements (FE) method. Each model was imported in a FE commercial software and meshed by using four nodes tetrahedral elements. Convergence testing was carried out in order to minimize the influence of mesh density on the results. The material chosen was Polyethylene Terephthalate Glycol (PETG), a thermoplastic resin whose mechanical properties are: $E = 2100$ MPa, $v = 0.3$.

First of all, each model was subjected to compression load. Uniaxial compression tests were performed by applying a uniform displacement, within the material elastic limit, to the top surface of the structure corresponding to 0.1% of compressive strain while the lower surface was fully constrained.

Figure 2 reports an iso-colour representation of Von Mises stress distribution, expressed in MPa, relative to the analyzed structures.

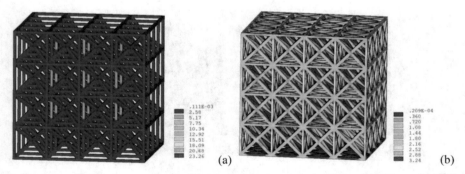

Fig. 2. Iso-colour representation of Von Mises stresses of lattice structures for compression loading: (a) SC-BCC; (b) SC-BCC-FCC

From the figure it can be observed that the lattice structure SC-BCC-FCC has a more homogeneous stress distribution with respect to the SC-BCC geometry. In fact, the SC-BCC structure is mainly affected by local stress concentration at the unit cell node of BCC lattice component which is induced by small radii corners of nodes. Researches on the failure mechanism of the BCC lattice structure showed that the stress concentration is the main cause [19] and a reduction of this effect can be obtained with appropriate design [20]. This effect is attenuated instead in the other structure for the presence of the FCC geometry.

For each model, the effective elastic modulus (E_{eff}) was evaluated based on Hooke's law. The reaction force was calculated by the FE solver while the homogenized stress was obtained by dividing the total reaction force by the total area of the loading plane. Since the applied strain is known, E_{eff} can be calculated.

The models were then subjected to shear load in order to evaluate the effective shear modulus (G_{eff}). With reference to the coordinate system reported in Fig. 1 with the origin considered coincident with the barycentre of each model, the shear modulus was evaluated by the application of a uniform displacement to the nodes on the outermost lateral face (+x) in the y direction, while the opposite face (−x) was fully constrained. The nodes on the top face (+y) of each model and those on the bottom face were also constrained in the x direction. The parameter of interest, analogously to the procedure previously described for E_{eff}, was evaluated starting from the reaction force calculated by the FE solver. The equivalent applied strain was 0.1%.

The graph reported in Fig. 3 depicts the effective compressive modulus and the effective shear modulus for the lattice structures analyzed.

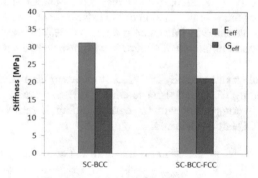

Fig. 3. Stiffness of lattice structures.

The lattice geometry SC-BCC-FCC shows higher values, both for E_{eff} and G_{eff}, compared to the other structure considered. Both structures show acceptable values of stiffness; however, SC-BCC-FCC structure can be considered as more efficient in terms of mechanical performance, also taking into account that, as previously observed, this geometrical configuration has a more homogeneous stress distribution with respect to the other considered.

However, as observed at the end of the previous paragraph, SC-BCC lattice structures, for the same volume fraction, are lighter than structures made of SC-BCC-FCC unit cells. Therefore, with the aim to reduce the overall weight, keeping at the same time adequate mechanical properties, a different structure was considered. This structure was obtained by combining both types of cells previously analysed so as to produce a multi-morphology lattice structure. In [21] it was shown that multi-morphology architectures are also advantageous since they can lead to improved energy absorption. In particular, the structure was obtained by introducing SC-BCC unit cells on the inside, while the unit cells located externally were modelled as SC-BCC-FCC cells. Figure 4 reports the lattice structure obtained by using the proposed approach.

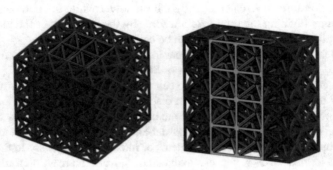

Fig. 4. Lattice structure made of mixed lattice cells.

In order to obtain a consistent solid model, an analogous size of the vertical struts for the two types of cells is required, to allow their superposition. This leaded to a slight increase, about 4%, of the volume fraction of the exterior component of the model. This structure, which was obtained by introducing $2 \times 2 \times 4$ (25%) lighter SC-BCC unit cells, allowed to reach a weight reduction of about 7% with respect to a structure with uniform unit cell type.

The hybrid model was structurally analyzed by means of FE analysis and the stiffness for compressive loading (E_{eff}) and shear loading (G_{eff}) was evaluated. Figure 5 shows the results obtained in comparison with the corresponding, in terms of volume fraction, uniform SC-BCC-FCC lattice structure.

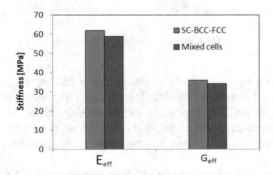

Fig. 5. Comparison of stiffness between uniform and mixed cells lattice structure.

The results obtained show a decrease of the stiffness in the hybrid structure, as expected, but with a limited difference between the values, assessing that this kind of approach is of interest for applications where weight reduction is a main requirement.

3 Manufacturing and Dimensional Assessment of Lattice Structures

3.1 Manufacturing with FDM Printer

The designed geometries were printed from a stereolithography (STL) file with the commercial Delta-type Anycubic Chiron printer using the PET-G material and the mold parameters shown in Table 1.

The FDM technique adopted allowed an accurate printing of the lattice structures without support, avoiding the delicate phase of supports removal, also with saving of material and time.

Table 1. Main printing parameters adopted.

Parameter	Working value	Variation range	Parameter	Working value	Variation range
Layer thickness [mm]	0.2	0.1–0.5	Print velocity [mm/s]	20	5–210
Initial thickness [mm]	0.3	0.1–0.6	Filling velocity [mm/s]	50	5–210
Perimeter threads	2	1- ∞	Outer wall print velocity [mm/s]	20	5–210
Horizontal expansion %	0	0–100	Lower surface print vel. [mm/s]	30	5–210
N° upper layers	3	1–∞	Movement velocity [mm/s]	100	5–210
N° lower layers	3	1–∞	Lower layers print vel. [mm/s]	25	5–210
Fill density	20	10–30	Print acceleration [mm/s^2]	1000	0–1000
Fill configuration	Zig Zag	–	Feedback distance [mm]	6	0–300
Print temperature [°C]	230	180–240	Feedback velocity [mm/s]	40	30–60
Print bed temperature [°C]	70	20–100	Fan speed %	100	0–100
Flow %	100	0–100	Print bed adhesion type	Brim	–
Initial layer flow %	105	0–100	Brim line number	3	1–∞

Figure 6 shows the printed lattice structures SC-BCC, SC- BCC-FCC, and the structure with mixed cells previously discussed.

Fig. 6. Lattice structures obtained with 3D Anycubic Chiron printer: SC-BCC (at left), SC-BCC-FCC (at center), Mixed-cells lattice (at right).

The lattice structures were manufactured by enabling the control of the acceleration and variability of the feedback of the head (nozzle). Z Hop was also been enabled during print retraction and cooling. These settings, together with the arrangement of the printing parameters used, make possible to limit only to the "Layer thickness" and the "Feedback velocity", the variations of the printing parameters necessary during prints without support with PET-G material. Among the printing parameters adopted, these, in fact, were the most critical ones for the accuracy of the printed models as the environmental conditions of temperature and humidity varied.

Using the printing parameters shown in Table 1, it was possible to optimize the values of the "Layer thickness" and the "Feedback velocity" as temperature and humidity changed (Fig. 7). The values shown in Fig. 7 allow to print the structures with considerable accuracy (see 3.2 subsection), without suffering appreciable flexural effects due to overhang.

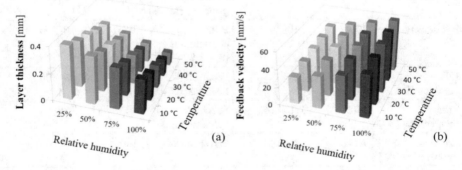

Fig.7. Printing parameters: (a) Layer thickness; (b) Feedback velocity.

The dimensions of the unit cells (20 × 20 × 20 mm) and the geometric layout adopted in the printed lattice structures ensure that the sections of the structure always have inclinations less than or equal to 45°. In particular, the inclined struts inside the cube are inclined respectively at angles of 35. 26° (Fig. 8a) and 45° (Fig. 8b).

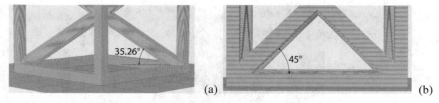

Fig.8. Inclined struts inside the unit cell: (a) 35.26° inclined struts; (b) 45° inclined struts.

3.2　Dimensional Assessment of 3D Printed Cellular Structures

Three special Azure Kinect DK sensors, equipped with advanced artificial intelligence system, were used to detect and quantify with high accuracy printing errors on the above-mentioned lattice structures (Fig. 9). Each Kinect sensor contains a depth sensor, camera and orientation sensor and, thus, is a compact "all-in-one" device usable with multiple modes by compiling the appropriate subroutines in Matlab environment. The most complete and accurate acquisitions (accuracy of 0.01 mm) were obtained by arranging 3 DK sensors at 120° around each printed structure (Fig. 9b).

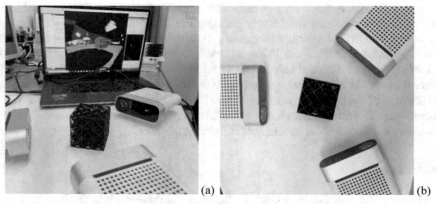

Fig. 9. (a) Azure Kinetic DK sensors acquisition system setup; (b) acquisition layout.

The acquisition with the three special Azure Kinect DK sensors has made possible to evaluate the printing errors for all the lattice structures manufactured. Figure 10 shows the comparison, in terms of dimensional deviation, between the proposed lattice with mixed cells structure printed with Anycubic Chiron printer using the PET-G and the designed STL model relative to the mixed-cells structure.

In detail, it was found that the maximum error occurs on the struts of the upper zone, which were printed last. In particular, the maximum positive errors (dimensions greater than the design ones) of magnitude equal to 0.22 mm occurred on the horizontal struts of the ends of the cubic structure, while the negative ones (dimensions smaller than the design ones) of magnitude equal to −0.12 mm occurred on the struts inclined by 45° in the central area of the cube (Fig. 10).

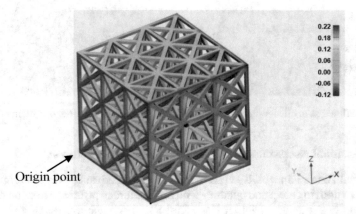

Fig. 10. Dimensional deviation between the printed structure and the STL model.

4 Conclusions

High porosity lattice structures for AM were analyzed in terms of stiffness and printability. Appropriate geometrical configurations of the unit cells with low strut size in relation to cell size have proven a satisfactory performance in terms of stiffness and accuracy in printing. The hybrid structure proposed can further reduce the overall weight with a limited decrease of stiffness. With the approach suggested in this study, potentially, the different type of unit cells should be easily arranged as required and the percentage of lighter unit cells can be varied, according to the final geometry of the part, considered that libraries of the basic unit cells could also be made available. As for manufacturing, support-less 3D printing has shown to be advantageous in terms of material as well preserving the dimensional accuracy of the produced parts.

References

1. Praveena, B.A., Lokesh, N., Buradi, A., Santhosh, N., Praveena, B.L.,Vignesh, R.: A comprehensive review of emerging additive manufacturing (3D printing technology): methods, materials, applications, challenges, trends and future potential. Materials Today: Proceedings **52**, 99–110 (2022)
2. Vlah, D., Žavbi, R., Vukašinović, N.: Evaluation of topology optimization and generative design tools as support for conceptual design. In: International Design Conference-DESIGN 2020, pp. 451–460. Cambridge University Press, England (2020)
3. Gabbrielli, R., Turner, I.G., Bowen, C.R.: Development of modelling methods for materials to be used as bone substitutes. Key Engineering Materials 361–363, Daculsi, G., Layrolle, P., Eds.: Scientific.net: Zurich, Switzerland, pp. 903–906 (2008)
4. Niu, J., Choo, H.L., Sun, W.: Finite element analysis and experimental study of plastic lattice structure manufactured by selective laser sintering. Proceedings of the Institution of Mechanical Engineers, part L. Journal of materials: Design and Applications **231**(1–2), 171–178 (2017)
5. Dong, L., Deshpande, V., Wadley, H.: Mechanical response of Ti–6Al–4V octet-truss lattice structures. Int. J. Solids Struct. **60–61**, 107–124 (2015)

6. Horn, T.J., Harrysson, O.L.A., Marcellin-Little, D.J., West, H.A., Lescelles, B.D.X., Aman, R.: Flexural properties of Ti6AlV4 rhombic dodecahedron open cellular structures fabricated with electron beam melting. Addit. Manuf. **1–4**, 2–11 (2014)
7. Tao, W., Leu, M.C.: Design of lattice structure for additive manufacturing. In: International Symposium on Flexible Automation (ISFA) IEEE, pp. 325–332 (2016)
8. Plocher, J., Panesar, A.: Review on design and structural optimization in additive manufacturing: towards next-generation lightweight structures. Mater. Des. **183**, 108164 (2019)
9. Ren, X.L., Xiao, L., Hao, Z.: Multi-property cellular material design approach based on the mechanical behaviour analysis of the reinforced lattice structure. Mater. Des. **174**, 107785 (2019)
10. Lee, K.-W., Lee, S.-H., Noh, K.-H., Park, J.-Y., Cho, Y.-J., Kim, S.-H.: Theoretical and numerical analysis of the mechanical responses of BCC and FCC lattice structures. J. Mech. Sci. Technol. **33**(5), 2259–2266 (2019)
11. Fadeel, A., Mian, A., Al Rifaie, M., Srinivasan, R.: Effect of vertical strut arrangements on compression characteristics of 3D printed polymer lattice structures: experimental and computational study. J. Mater. Eng. Perform. **28**(2), 709–716 (2018)
12. Wang, L., Kang, J., Sun, C., Li, D., Cao, Y., Jin, Z.: Mapping porous microstructures to yield desired mechanical properties for application in 3D printed bone scaffolds and orthopaedic implants. Mater. Des. **13**, 62–68 (2017)
13. Jiang, C.P., Wibisono, A.T., Pasang, T.: Selective laser melting of stainless 316L with face-centred-cubic based lattice structures to produce rib implants. Materials **14**, 5962 (2021)
14. Kladovasilakis, N., Tsongas, K., Kostavelis, I., Tzovaras, D., Tzetzis, D.: Effective mechanical properties of additive manufactured strut-lattice structures: experimental and finite element study. Adv. Eng. Mater. **24**, 2100879 (2022)
15. Nazir, A., Arshad, A.B., Hsu, C.P., Jeng, J.Y.: Effect of fillets on mechanical properties of lattice structures fabricated using multi-jet fusion technology. Materials **14**(9), 2194 (2021)
16. Tang, C., Liu, J., Yang, Y., Liu, Y., Jiang, S., Hao, W.: Effect of process parameters on mechanical properties of 3D printed PLA lattice structures. Composites Part C: Open Access **3**, 100076 (2020)
17. Asadi-Eydivand, M., Solati-Hashjin, M., Farzad, A., Osman, N.A.A.: Effect of technical parameters on porous structure and strength of 3D printed calcium sulfate prototypes. Robotics Computer-Integrated Manufacturing **37**, 57–67 (2016)
18. Maskery, I., Aremu, A.U., Parry, L., Wildman, R.D., Tuck, C.J., Ashcroft, I.A.: Effective design and simulation of surface-based lattice structures featuring volume fraction and cell type grading. Mater. Des. **155**, 220–232 (2018)
19. Li, P.: Constitutive and failure behaviour in selective laser melted stainless steel for microlattice structures. Material Science Eng.: A **622**, 114–120 (2015)
20. Bai, L., Yi, C., Chen, X., Sun, Y., Zhang, J.: Effective design of the graded strut of bcc lattice structure for improving mechanical properties. Materials **12**, 2192 (2019)
21. Alberdi, R., et al.: Multi-morphology lattices lead to improved plastic energy absorption. Mater. Des. **194**, 108883 (2020)

Machine Tools

Development of a Sensor Integrated Machining Vice Towards a Non-invasive Milling Monitoring System

Panagiotis Stavropoulos[✉], Dimitris Manitaras, Christos Papaioannou, Thanassis Souflas, and Harry Bikas

Laboratory for Manufacturing Systems and Automation (LMS), Department of Mechanical Engineering and Aeronautics, University of Patras, 26504 Rio Patras, Greece
`pstavr@lms.mech.upatras.gr`

Abstract. The future of manufacturing processes is the fully autonomous operation of machine tools. The reliable autonomous operation of machine tools calls for the integration of inline quality control systems that will be able to assess in real time the process status and ensure that the machine tool, process and workpiece are complying with the manufacturing tolerances and requirements. Sensor integrated tooling for machining processes can significantly contribute towards this goal as they can facilitate monitoring close to the actual process. However, most of the solutions proposed so far are highly expensive or very complex to integrate and operate in an industrial environment. To this end, this paper proposes an approach for a sensor integrated vise using low-cost industrial sensors that can easily be integrated in existing machine tools in a non-invasive fashion. The development and dynamic analysis of the system is presented, along with an experimental verification against a lab-scale, high accuracy sensing setup

Keywords: Machining monitoring · Sensor integrated tooling · Digital machining · Modal analysis

1 Introduction

Digitalization of manufacturing processes is one of the key strategic research pathways for the manufacturing industry, since it can enable future factories to operate autonomously or with minimal human intervention with automated decision-making algorithms that ensure reliable, safe and productive operation. Initiatives, such as Industry 4.0 have showcased the importance of the digital transformation of manufacturing and provide roadmaps for implementation of technologies and concepts related to this topic. To this end, both academia and industry invest heavily on developing technologies that can enable this digital transformation [1].

Key enabling technologies for this digital transformation are the inline quality control systems that constantly monitor the manufacturing process, detect deviations from the optimal operation and apply corrective measures. Especially when machining is concerned, the stability of the process is one of the most crucial factors that affect its

K.-Y. Kim et al. (Eds.): FAIM 2022, LNME, pp. 29–37, 2023.
https://doi.org/10.1007/978-3-031-18326-3_3

quality. The presence of chatter, which is a self-excited vibration, is detrimental for the tool life, surface quality of the workpiece and safe operation of the machine tool [2]. This phenomenon has been studied for years and several technologies have been developed that try to suppress chatter, based on inline monitoring and control systems. However, the industrial implementation of such systems is still limited and one of the key reasons is the fact that sensing systems for machining require complex integration of sensors and data acquisition devices in machine tools. End users are reluctant to integrate an invasive system in their machine tool, which would require modifications in the machine tool frame, spindle housing, etc. On top of that, high investment costs for the acquisition and integration of the sensing system adds an additional barrier for smaller companies to adopt such technologies. A promising solution to this issue that can foster the digitalization of machining processes is sensor integrated tooling. The concept of sensor integrated tooling is based on installation of sensors in replaceable components of the machining system (e.g. tool holders, fixtures, vices), which provides a non-invasive monitoring approach and can be transferred seamlessly between different machine tools.

2   State of the Art

Since sensor integrated tooling is a very promising solution for monitoring of the machining process, academia and industry have put a lot of effort into the development of such systems. The most common approach that has been followed is the integration of sensors in the tool holder or cutting tool, since it enables to reach as close to the machining process as possible. Xie et al. [3] have integrated a vibration sensor and six capacitance sensors, along with the required electronics for wireless transmission in a ring structure that has been installed externally on the tool holder, in order to estimate the wear level of the cutting tool. Bleicher et al. [4] have integrated a capacitive MEMS sensor in the internal structure of the tool holder that has been used to measure vibrations and transmit the data wirelessly through Bluetooth. Rizal et al. [5] have integrated a rotating dynamometer, based on strain gauges, in the outer geometry of the tool holder to measure the cutting force during milling. Totis et al. [6] have integrated triaxial piezo-electric force transducers between the insert cartridge and the body of the milling head, in order to measure the individual cutting forces on each cutting edge. In a similar fashion, Luo et al. [7] have integrated polyvinylidene fluoride (PVDF) sensors between the insert cartridge and the body of the milling head to measure the cutting forces. Both approaches enable measurement very close to the cutting zone; however, their applicability is limited in large, indexable milling cutters. In order to address this issue, Cen et al. [8] used PVDF sensors that were adhered on the shank of the end mill. The strain signals generated by the sensors were used to calculate the cutting forces at the end mill tip, by treating it as a cantilever beam and using the Euler-Bernoulli beam theory. A significant challenge regarding wireless sensory devices is related to the fact that they need to be charged, thus introducing downtime to the equipment. To tackle this, Osta-sevicius et al. [9] have developed a self-powered wireless sensor integrated tool holder for tool wear monitoring, which uses the tool holder vibrations to excite a piezoelectric transducer. The transducer charges a capacitor that powers the low-power electronics of the tool holder. The integration of sensors on fixtures has also been investigated. Liu et al. [10] integrated PVDF thin-film sensors into fixtures to monitor the cutting forces on thin-wall aircraft structural parts. Rezvani et al. [11] have replaced the jaws of a vice with sensor integrated plates with strain gauge and PZT sensors to monitor the clamping force and cutting forces during milling. Apart from academic approaches there are also commercial sensor integrated tooling systems, such as the iTENDO tool holder from Schunk [12] or the spike tool holder from promicron [13].

Although the aforementioned approaches are very promising and can offer high quality measurement are based on expensive sensing elements and complex integrations that significantly increase their implementation cost. All those aspects lead in complex or very expensive systems or both, that are not economically viable for SMEs and small manufacturers that want to digitalize their machining processes.

To this end, this paper proposes a simple and low-cost solution for sensor integrated tooling, through the integration of a MEMS accelerometer on a machining vice. In the rest of the paper, the dynamic analysis of the monitoring setup and its experimental validation are presented. Moreover, the sensor integrated vice is validated in a case study of chatter detection, compared to an expensive, lab-scale setup to prove its performance. Finally, the results and conclusions of this study are presented.

3 Approach

For the development of the sensing setup a commercial, general-purpose milling vice (Vertex VA-6) has been equipped with a low-cost MEMS accelerometer from Micromega (IAC-CM-U). The first step of the development is the modelling of the dynamic behavior of the sensor integrated vice, in order to determine the best integration approach and to ensure that the dynamic behavior of the system will not impact the measurement quality due to the operating conditions during the milling process. An adapter plate was manufactured and integrated on the back of the steady jaw of the vice, on which the sensor was be installed (Fig. 1).

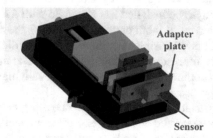

Adapter plate

Sensor

Fig. 1. Integration approach for the sensory vice

The first step of the approach is the modal analysis of the system that can give a first estimation of its dynamic response. Table 1 shows the analysis results. The modal analysis has been setup in ANSYS and validated experimentally with an impact hammer test. The experimental equipment used for the impact test are a Kistler 9724A5000 impact hammer and a Kistler 8762A10 tri-axial accelerometer. Labview was used to calculate the Frequency Response Functions (Fig. 2). As it can be observed, there is a general agreement between simulation and experimental validation.

Table 1. Modal analysis results

Mode #	Frequency [Hz]	Effective mass ratio [%]		
		X-axis	Y-axis	Z-axis
1	1171	34.76	0.00	0.00
2	2157	59.18	96.94	0.00
3	3301	2.48	0.00	0.00
4	3850	0.17	0.00	0.00
5	3937	0.00	0.00	0.76
6	4714	0.05	0.00	0.00
7	4832	0.00	0.08	2.94
8	4962	0.00	0.01	0.34

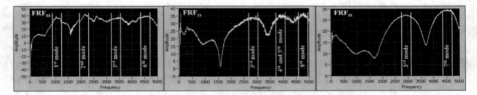

Fig. 2. Experimentally obtained frequency response functions

The first natural frequency of the sensory vice occurs at 1171 Hz. The target milling machine where it is going to be installed has a spindle with a maximum rotating speed of 3600RPM. Even with a multi tooth cutter (e.g. 6 teeth), the maximum tooth passing frequency that will be observed during machining in this specific machine tool is 360 Hz. Therefore, there is no risk for resonance phenomena. For other machine tools, capable of high-speed machining, another vice should be selected with higher natural frequencies; however, the overall approach is still the same.

The next step is to quantify the effect of the operating conditions in the dynamic behavior of the sensor integrated vice. A very efficient method of determining the dynamic behavior of a system in a wide frequency range is the harmonic analysis. The intermittent cutting during milling introduces dynamic loads that have a harmonic nature. As a result, it is possible to quantify the dynamic behavior of the sensor integrated vice during machining through a harmonic analysis. The harmonic analysis has been setup with a harmonic force acting on the workpiece. The cutting forces that were used in the analysis were $F_x = 600N$, $F_y = 280N$ and $F_z = 550N$, which are typical values for roughing of hardened steel [14]. The vibration amplitudes at the sensor position have been measured during the analysis. The results of the harmonic analysis are presented in Fig. 3. The simulation results showcase the stiffness of the integration point of the sensor. As a result, no unwanted compliance will be introduced during the operation of the system, interfering with the process.

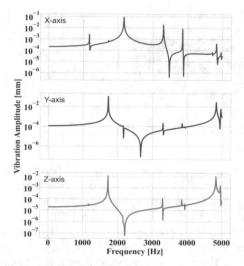

Fig. 3. Vibration amplitudes at the sensor location

4   Case Study on Chatter Detection

In order to test the actual suitability of the sensor integrated vice for monitoring of the milling process, a case study on chatter detection was conducted. The whole monitoring system was comprised of the sensory vice and a Labjack T7 data acquisition system, which fed the vibration data in real time to a personal computer (Fig. 4).

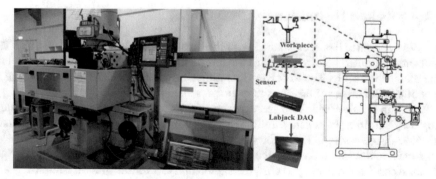

Fig. 4. Monitoring system setup

An indicative vibration signal in the feed axis, as well as the plot of is Fast Fourier Transform (FFT) is presented in Fig. 5. In general, a good signal quality can be observed with a slight noise level ranging the whole frequency spectrum.

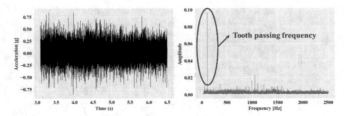

Fig. 5. Vibration signal (left) and its FFT (right) using the sensory milling vice

The case study that has been selected to validate the performance of the sensory vice as an enabler for a milling monitoring system was chatter detection. The chatter detection system is based on a proprietary development of the Laboratory for Manufacturing Systems and Automation and is described in detail in [15]. For the sake of completeness, a short description is given here. The chatter detection algorithm is based on vibration signals in the feed and cross-feed axes, which are decomposed with Variational Mode Decomposition (VMD). From the decomposed signal, the modes that are related to chatter are analyzed and chatter related features are extracted from them in the time and frequency domains. The features are fed to a Support Vector Machine (SVM) classifier to detect chatter status. In [15], the system was developed and tested using a highly expensive, lab-scale setup, comprised of a Kistler 8762A10 ceramic shear accelerometer and a National Instruments PXI-4472 sound and vibration module for data acquisition. In order to validate the performance of the sensory milling vice, the system was retrained and tested with data coming from the proposed system. Using the stability lobe diagrams of the machine tool, tool holder and cutting tool system, generated in [15], the process parameters that led to stable and chatter machining were selected for the machining experiments.

5   Results and Discussion

The SVM classifier has been trained with the data from the sensory milling vice. The results are presented below, as well as the results that have been achieved with the lab-scale monitoring setup. The Receiver Operating Characteristic (ROC) curves, as well as the confusion matrices of the classifiers are presented (Fig. 6 and Fig. 7).

As it can be observed from the experimental results, the chatter detection system using the data from the sensory milling vice has similar chatter detection performance with the lab-scale setup. This shows that the sensory milling vice can be considered as a robust and reliable data source, compared to an expensive lab-scale monitoring system. Apart from the reduced cost, the significantly lower integration complexity and increased durability of the sensory milling vice can render it as a promising sensor integrated tooling solution for smaller machine shops and SMEs.

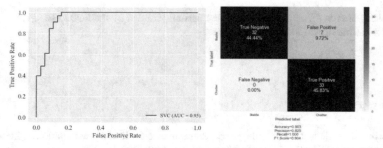

Fig. 6. ROC curve (left) and confusion matrix (right) of the sensory milling vice

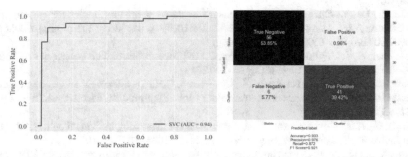

Fig. 7. ROC curve (left) and confusion matrix (right) of the lab-scale setup

6 Conclusions

The scope of this study was the development of a low-cost sensor integrated tooling system for milling, suitable for industrial implementation. The system was based on low-cost monitoring equipment (sensor and DAQ system) and a commercial, general-purpose milling vice. Based on the results derived from this study the following conclusions can be drawn:

- The simulation of the dynamic behavior of the sensor integrated tooling system is a crucial element of the development phase, since it can enable the correct selection of the sensor integration approach
- The experimental modal analysis of the sensor integrated vice has validated the predicted dynamic behavior
- The performance of the sensor integrated vice as a data source for chatter detection has been validated against an expensive, lab-scale monitoring setup

Future work should include the integration of an edge computer to eliminate the need for a personal computer and provide a plug-and-play solution. Moreover, advanced signal processing algorithms should be employed to eliminate the noise existing in the signal. Finally, closing the loop with the machine tool is another important aspect to enable real-time process control and chatter suppression.

Acknowledgements. This research has been partially funded by the H2020 EU Project DIMO-FAC – Digital Intelligent MOdular FACtories.

References

1. Charbonneau-Genest, M., Gamache, S.: Prerequisites for the implementation of industry 4.0 in manufacturing SMEs. Procedia Manufact. **51**, 1215–1220 (2020)
2. Munoa, J., et al.: Chatter suppression techniques in metal cutting. CIRP Ann. **65**(2), 785–808 (2016)
3. Xie, Z., Lu, Y., Chen, X.: A multi-sensor integrated smart tool holder for cutting process monitoring. Int. J. Adv. Manufact. Technol. **110**(3–4), 853–864 (2020). https://doi.org/10.1007/s00170-020-05905-7
4. Bleicher, F., Schörghofer, P., Habersohn, C.: In-process control with a sensory tool holder to avoid chatter. J. Mach. Eng. **18**(3), 16–27 (2018)
5. Rizal, M., Ghani, J.A., Nuawi, M.Z., Che-Haron, C.H.: Development and testing of an integrated rotating dynamometer on tool holder for milling process. Mech. Syst. Sign. Process. **52–53**, 559–576 (2015)
6. Totis, G., et al.: Development of a dynamometer for measuring individual cutting edge forces in face milling. Mech. Syst. Signal Process. **24**(6), 1844–1857 (2010)
7. Luo, M., et al.: A wireless instrumented milling cutter system with embedded PVDF sensors. Mech. Syst. Signal Process. **110**, 556–568 (2018)
8. Cen, L., Melkote, S.N., Castle, J., Appelman, H.: A wireless force-sensing and model-based approach for enhancement of machining accuracy in robotic milling. IEEE/ASME Trans. Mech. **21**(5), 2227–2235 (2016)
9. Ostasevicius, V., Karpavicius, P., Jurenas, V., Cepenas, M., Cesnavicius, R., Eidukynas, D.: Development of universal wireless sensor node for tool condition monitoring in milling. Int. J. Adv. Manufact. Technol. **110**(3–4), 1015–1025 (2020). https://doi.org/10.1007/s00170-020-05812-x
10. Liu, D., Hu, Y., Zhang, D., Luo, H.: Milling force monitoring with thin-film sensors integrated into fixtures. Int. J. Adv. Manufact. Technol. **103**(1–4), 1519–1527 (2019). https://doi.org/10.1007/s00170-019-03666-6
11. Rezvani, S., Kim, C.-J., Park, S.S., Lee, J.: Simultaneous clamping and cutting force measurements with built-in sensors. Sensors. **20**, 3736 (2020)
12. Schunk iTENDO. https://schunk.com/de_en/news/highlights/reports/article/94625-feinfuehliger-werkzeughalter/. Accessed 15 Jan 2022
13. promicron spike. https://www.pro-micron.de/spike/?lang=en. Accessed 15 Jan 2022
14. Hoang, T.D., Nguyen, N.T., Tran, D.Q., Nguyen, V.T.: Cutting forces and surface rough-ness in face-milling of SKD61 hard steel. J. Mech. Eng. **65**, 375–385 (2019)
15. Souflas, A.: Generation of manufacturing process knowledge for process optimization: a case study on milling, Diploma Thesis, University of Patras, Greece (2021). http://hdl.handle.net/10889/15028

Effect of Ultrasonic Burnishing Parameters on Burnished-Surface Quality of Stainless Steel After Heat Treatment

Rizwan Ullah[✉] [iD], Eric Fangnon[iD], and Juha Huuki[iD]

Mechanical Engineering Department, Aalto University, Espoo, Finland
`rizwan.ullah@aalto.fi`

Abstract. Ultrasonic burnishing induces beneficial compressive stresses and high surface quality in components with contact as a functional requirement. It was observed in previous work that some burnishing parameters can hinder burnishability of stainless steels. In this research tangential misalignment angles (TMA) for burnishing were varied considering as-supplied and heat-treated stainless steel. Properties such as surface hardness and surface roughness were measured after burnishing process. Electron Backscatter Diffraction was performed to characterize microstructure using Matlab (MTEX) to calculate average grain areas. By changing burnishing parameters, i.e., shaft rotational speed and burnishing tool diameter, it was observed that burnishing was less successful. Nevertheless, significant improvement in burnished surface quality was observed after heat-treatment process. In addition, grain size characterization revealed mean grain area reduction from 26 μm^2 for unburnished to 11 μm^2 and 3 μm^2 for burnished and heat-treated samples respectively. Most importantly this work reveals the enhanced possibility of burnishing stainless steels after heat-treatment with varying tangential misalignment angles.

Keywords: Ultrasonic burnishing · Heat treatment · Surface finish · Tangential misalignment

1 Introduction

Burnishing is a modern and effective solution to improve the surface properties in different applications that require an excellent surface finish and dimensional accuracy [2]. Conventional finishing treatments such as polishing, grinding, lapping, or honing are used in the mechanical industry. Burnishing is one finishing technique that serves as an alternative to traditional grinding processes. Burnishing methods which include often roller burnishing, ball burnishing, diamond burnishing are mostly used on rotating components that have high-quality requirements, such as automotive bearing parts, crankshafts or axles [3, 4]. Burnishing is also used to get close tolerance in areas like automobile, aircraft, defense, machine tool, hydraulic and pneumatic equipment, and home appliances [5].

© The Author(s) 2023
K.-Y. Kim et al. (Eds.): FAIM 2022, LNME, pp. 38–47, 2023.
https://doi.org/10.1007/978-3-031-18326-3_4

Ultrasonic burnishing is a modern method used for finishing metal surfaces by forging at very high frequency. The method is not so well known than roller, ball or diamond methods. Burnishing is a finishing process commonly applied to improve the surface integrity, i.e. surface roughness, hardness, residual stresses and microstructure of a mechanical component [6]. In ultrasonic burnishing, forging is done with a ball-shaped finishing head at very high ultrasonic frequency while workpiece rotate along fixed axis, as represented by a schematic in Fig. 1. A constant spring load is applied on burnishing tool to keep it in contact with the workpiece surface. The process does not remove any material. The method causes the plastic deformations on the part surface which creates residual stresses in the worked surface and improve surface quality [7]. Priyadarsini et al., recently showed an overview of past research on surface integrity in burnishing processes; ball burnishing, roller burnishing and low plasticity burnishing (LPB) [8]. The results of this meta-study reveal burnishing to be an effective technique for improving surface properties [6].

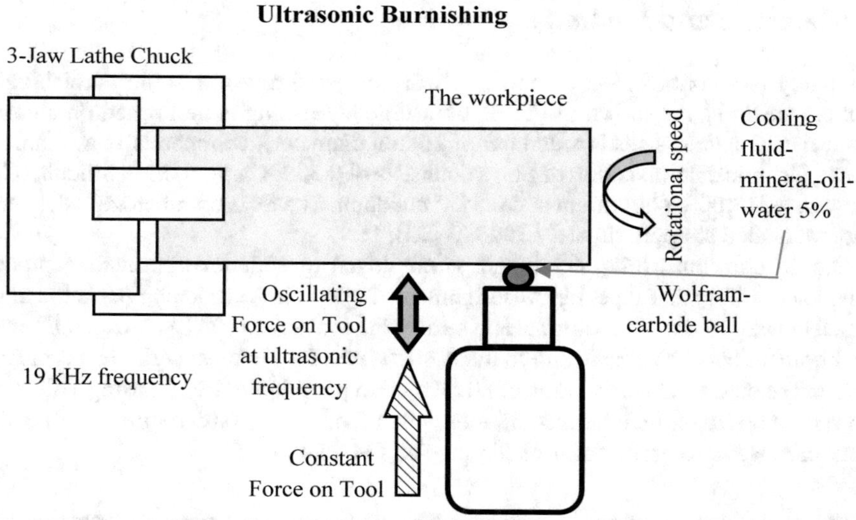

Ultrasonic Burnishing

Fig. 1. Schematic of ultrasonic burnishing process

Alshareef et al., investigated ball burnishing of AISI 8620 steel affects its microstructure, roughness, and residual stress. The author reported that burnishing improves the surface roughness of AISI 8620 steel by more than 60% when applied after turning. In addition, the results showed that the layer thickness of the workpiece was approximately 15 μm after turning and burnishing [9].

S. Ramesh investigated the effect of ball burnishing process on equal channel angular pressing of Mg-4Zn-1Si alloy. The microstructure analysis showed that grain size was 6.6 μm before and 3.3 μm after burnishing. The author stated that ball burnishing process leads to improved surface roughness and induces residual compressive stresses [10].

Zhen-yu, states that the surface burnishing process has a significant effect on the microstructure of copper, such as the grain size and the density of geometrically necessary dislocation, but has little effect on the micro texture [11].

Tangential misalignment is quite easy to control in turning operations or with planar surfaces, but with a double-curved-surfaces, the control of the angle can become challenging as burnishing tool head has round surface and shaft could have misalignments of its own [6]. Little research work has been conducted on tangential misalignment in ultrasonic burnishing. This necessities the importance to study tangential misalignment angle between the shaft and burnishing tool head. Many authors have examined burnishing methods like ball or roller effect on material integrity [7, 12–15]. The effect of ultrasonic burnishing on microstructural mechanism for cylindrical workpiece with different tangential alignments has been understudied. For this reason, the need to evaluate the effect of ultrasonic burnishing on the surface finish, crystallography and microstructure evolution of workpiece is recognized.

2   Materials and Methods

This study is a continuation of previous work on exploring ultrasonic burnishing of stainless steel [1]. As shown in Fig. 2, ultrasonic burnishing is performed on a corrosion resistant stainless steel (round bar of 20 mm diameter), commercially available as Stavax. Chemical composition of Stavax consists of 0.38% Carbon, 0.9% Silicon, 0.5% Manganese, 13.60% Chromium and 0.3% Vanadium. Stavax is commercially delivered as soft annealed to approximately 190 HB [15].

Hiqusa ultra burnishing equipment is employed to undertake burnishing process as depicted in Fig. 2. Tangential Misalignment Angle (TMA) or kappa (κ) is the angle formed between shaft-axis's normal and burnishing tool axis. A 0° TMA would mean no misalignment between burnishing tool and shaft-axis. In previous work [1], burnishing was performed with chuck rotation of 80 revolution per minute (RPM), using a diameter of 6 mm tungsten carbide burnishing tool with a feed rate of 0.05 mm/rev. Burnishing in this case was successful for limited range of TMA i.e., 0°–5°.

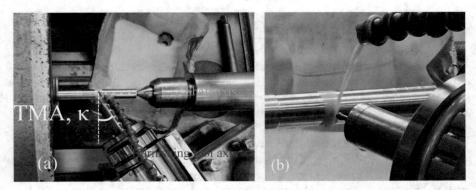

Fig. 2. Burnishing equipment: (a) tangential misalignment angle (TMA) κ, (b) semi-spherical burnishing tool head

However, in the present work, burnishing parameters were changed to 500 RPM and burnishing tool diameter of 4 mm. Tool feed and burnishing frequency of tool head were kept constant [1]. At this configuration, burnishing was only possible at 0° TMA while higher TMAs resulted in deteriorated surfaces. Additionally, as-delivered Stavax was heat treated using the procedure as prescribed in Table 1.

Table 1. Heat treatment procedure

Sequence	Operation	Time (min)	Temperature	Cooling
#1	Pre-heating	33	700 °C	–
#2	Austenitizing	33	1030 °C	–
#3	Tempering 1	120	250 °C	Air cooling to 33 °C for 20 min
#4	Tempering 2	120	250 °C	Air cooling to 33 °C for 20 min

Hard scales on the surface of the round bar as result of heat treatment were removed by turning, leaving the bar with the final diameter of 19.5 mm. Burnishing with varying TMAs (κ) was performed on heat-treated Stavax. The variations were made within the range of 0°–5° with 1° interval and of 10°–40° with 10° interval. At each TMA, burnishing was carried out for 5 mm distance on circumference of the shaft, which is referred here as burnished band.

2.1 Surface Roughness

Surface roughness of the burnished bands was measured on the cylindrical surface using MarSurf PS 10 apparatus which employed the stylus type measuring procedure and 2.5 mm cut off length.

2.2 Surface Hardness

Surface hardness (Rockwell C scale) was measured at cylindrical surface on the burnished bands using SwissMax 300 (Gnehm, Thalwil, Switzerland) equipment.

2.3 EBSD (Sample Preparation and Post Processing)

For crystallographic characterization, Electron Backscatter Diffraction (EBSD) was conducted with Zeiss Ultra 55, equipped with EBSD detectors (2005) and Oxford HKL Nordlys data acquisition. Sample preparation for EBSD consists of following steps:

- Grinding with SiC paper of FEPA grit sizes #1200 and #2500 for 3 and 5 min respectively.
- Automatic polishing with Tegramin-20 equipment employing diamond pastes of 9 μm, 3 μm (polishing time = 5 min each), following with 1 μm and 0.25 μm (polishing time = 10 min each). Samples were cleaned with ethanol and dried with hot air.

Results from EBSD were further processed with MTEX (version 5.7.0) toolbox package with Matlab (2021b). All the grains were constructed considering boundaries resulting from misorientation of 3° (deg). Mean grain area (MGA) was calculated with MTEX, using the data resulting from grain reconstruction.

Mainly two samples were considered for EBSD analysis. Their respective designation and description are shown in Table 2.

Table 2. Sample designation

Designation	Description
Sample #1	As delivered (Soft annealed) and Burnished
Sample #2	Heat treated (as in Table 1) and Burnished

Figure 3 highlights a schematic representation of EBSD scan areas for a burnished sample. Two indentation marks (Fig. 3) were made at distances of 60 μm and 110 μm away from the burnished edge. Burnished zone is expected to be up to 100 μm away from edge while the rest of the area i.e., after 100 μm and towards center of shaft, is considered as core material, with no effect of burnishing [1].

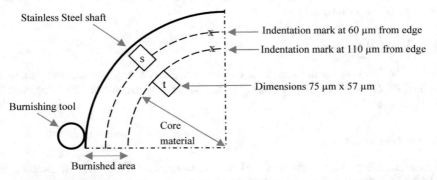

Fig. 3. Schematic of cross section of the burnished shaft, representing EBSD measurement locations

3   Results and Discussion

3.1   Surface Roughness

Minimum surface roughness, Ra value, in previous work [1], was achieved at 0° TMA with a value of 0.389 μm. In the present work, the measured surface roughness (Ra) at various TMAs after the heat-treatment, are reported in Fig. 4.

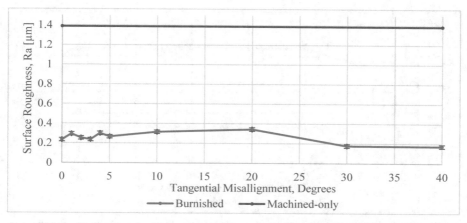

Fig. 4. Surface roughness Ra, measured at cylindrical surface on burnished bands

Results pointed out that surface roughness is less affected by varying tangential misalignment angle. This makes heat-treatment a very feasible choice for Stavax material with current set of burnishing parameters.

Unlike to previous work [1], surface roughness (Fig. 4) does not increase linearly but rather fluctuate with in 0.2–0.4 μm range. Unburnished-machined (turning only) surface had produced surface roughness of 1.387 μm. Ultrasonic burnishing has decreased surface roughness by 83% at 0° TMA and an even larger surface roughness reduction of 87.5% was observed at 40° TMA.

Using smaller tool diameter, surface roughness is decreased compared to previous results [1] by maximum of 55%.

3.2 Surface Hardness

Trend of surface hardness on Rockwell (HRC) scale, is depicted in Fig. 5. TMAs in range of 0°–5° [1], had shown lowest hardness value of approx. 30 HRC (converted from HV10 scale to HRC scale) at 2°, while it showed higher values 33–34 HRC (converted from HV10 scale to HRC scale) at 0°, 4° and 5°; thus, forming a concave shape of maximum peaks for various TMAs. In present work, a similar trend is seen at TMAs for sample#2 (Fig. 5). Magnitude of error bars indicate that concave effect may not be very significant. Nevertheless, hardness has increased by minimum of 1.5% at 2° TMA while maximum increase was calculated as 3.6% at 30° TMA, compared to hardness of the machined (turning) surface. Surface hardness of heat-treated Stavax after turning process has been recorded as 50.42 HRC. In Fig. 5, hardness is affected by TMA in linear increasing order up to 10°, but after which hardness remains constant at highest value of 52.2 HRC.

The observed difference of measured surface hardness of STAVAX in as-supplied [1] and heat-treated state from 26 HRC (converted from HV10 scale to HRC scale) to 50.42 HRC respectively, can be safely attributed to the microstructure transformations resulting from the heat treatment.

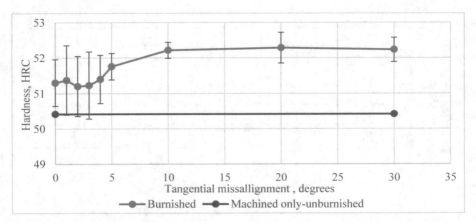

Fig. 5. Surface hardness measured on Rockwell HRC scale for various tangential misalignment angles

3.3 Grainsize Characterization

The EBSD misorientation maps of sample#1, taken at the core of the material and at burnished zone are shown in Fig. 6 (a), (b) respectively. Their corresponding inverse pole figure (IPF) is shown in Fig. 6 (d). These EBSD scans were captured at location marked as s and t in Fig. 3. In addition, EBSD misorientation map for sample#2 as shown in Fig. 6 (c), was taken at the core of material.

Figure 6 (a) shows a relatively evenly distributed ferritic grains across the map with grain areas ranging from 10–150 μm^2 with the MGA of about 26 μm^2. The burnished area (s) manifests regions of coarse grains with relatively reduced grain size compared to the core material, as depicted in Fig. 6 (b). The reduction of the MGA from 26 μm^2 for base material to 11 μm^2 for burnished zone, evidence the effect of burnishing on the grain areas near the surface.

EBSD of heat treated Stavax as shown in Fig. 6(c) manifests not only a further reduction in average grain area to 3 μm^2 but also a change in morphology of grains to needle-like structure evidencing a transition from ferritic to martensitic microstructure.

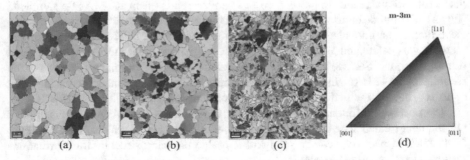

Fig. 6. EBSD maps for burnished Stavax (a) t, core material (b) s, burnished area (c) for heat treated specimen at core material (d) inverse pole figure

Analyzing further, the effect of burnishing and heat treatment on zone s and zone t can be attained by quantifying the grain area distribution on the scanned maps. Figure 7 (a) and (b) show histograms of grain area distribution of scanned EBSD maps for zone s and zone t for sample#1 respectively. Similarly, Fig. 7 (c) depicts the grain area distribution of the core of sample#2. The results show that 17.6% of grains in zone t are above 50 μm^2 while only 13.5% are above 50 μm^2 for zone s while for sample#2, all needle-shaped grains are below 20 μm^2.

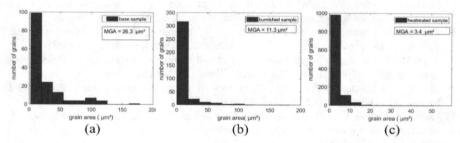

(a)   (b)   (c)

Fig. 7. Histogram depicting grain area distribution of burnished Stavax (a) t, core material (b) s, burnished area (c) for heat treated specimen at core material

Worth to note is the correlation of measured surface hardness and grain areas of respective zone t and zone s. Increased surface hardness from 50.42 HRC at zone t to higher surface hardness values at zone s corresponds to reduction in average grain area from 26 μm^2 to 11 μm^2.

4 Conclusions

Ultrasonic burnishing is a finishing process that improves surface properties including higher surface hardness and lower surface roughness. Resulted components have high wear resistant properties and last longer in service. This study finds out the positive influence of heat treatment of stainless steel on burnishing. Following conclusions can be made:

1. Stainless is better suited for burnishing after heat-treatment for specific burnishing parameters. Surface properties, i.e., roughness and hardness, are improved compared to soft annealed state.
2. Burnishing process after heat treatment results in surface properties which vary with tangential misalignment angle, however burnishing ability is not hindered, at any specific tangential misalignment angle.
3. Burnishing has resulted in smaller grain areas within 100 μm from the burnished edge compared to the rest of material, evidencing effect of burnishing on near surface grain sizes. For heat treated sample, average grain area is further reduced with needle shaped grains.

References

1. Huuki, J., Ullah, R., Laakso, S.: Process limitation of ultrasonic burnishing for commercially available martensitic stainless steel. Procedia Manufact. **51**, 885–889 (2020). https://doi.org/10.1016/j.promfg.2020.10.124
2. Nguyen, T.T., Cao, L.H., Nguyen, T.A., Dang, X.P.: Multi-response optimization of the roller burnishing process in terms of energy consumption and product quality. J. Clean. Prod. **245**, 119328 (2020)
3. Huuki, J., Laakso, S.V.: Surface improvement of shafts by the diamond burnishing and ultrasonic burnishing techniques. Int. J. Mach. Mach. Mater. **19**(3), 246–259 (2017)
4. Dzierwa, A., Markopoulos, A.P.: Influence of ball-burnishing process on surface topography parameters and tribological properties of hardened steel. Machines **7**(1), 11 (2019)
5. John, M., Stalin, R., et al.: An investigation of ball burnishing process on CNC lathe using finite element analysis. Simul. Model. Pract. Theory **62**, 88–101 (2016)
6. Huuki, J., Laakso, S.V.A., Ullah, R.: Effect of tangential misalignment in ultrasonic burnishing. Procedia Manuf. **38**, 1540–1546 (2019)
7. Llumà, J., Gómez-Gras, G., Jerez-Mesa, R., Rue-Mascarell, J., Travieso-Rodriguez, J.A.: Mechanical strengthening in S235JR steel sheets through vibration-assisted ball burnishing. Metals **10**(8), 1010 (2020)
8. Priyadarsini, C., Ramana, V.V., Prabha, K.A., Swetha, S.: A review on ball, roller, low plasticity burnishing process. Mater. Today Proc. **18**, 5087–5099 (2019)
9. Alshareef, A.J., Marinescu, I.D., Basudan, I.M., Alqahtani, B.M., Tharwan, M.Y.: Ball-burnishing factors affecting residual stress of AISI 8620 steel. Int. J. Adv. Manufact. Technol. **107**(3–4), 1387–1397 (2020). https://doi.org/10.1007/s00170-020-05119-x
10. Ramesh, S., Anne, G., Kumar, G., Jagadeesh, C., Nayaka, H.S.: Influence of ball burnishing process on equal channel angular pressed Mg-Zn-Si alloy on the evolution of microstructure and corrosion properties. SILICON **13**(5), 1549–1560 (2021)
11. Zhen-yu, Z., Qiu-yang, Z., Cong, D., Ju-yu, Y., Zhong-yu, P.: Effect of surface burnishing process with different strain paths on the copper microstructure. J. Manuf. Proces. **71**, 653–668 (2021)
12. Attabi, S., Himour, A., Laouar, L., Motallebzadeh, A.: Effect of ball burnishing on surface roughness and wear of AISI 316L SS. J. Bio Tribo-Corros. **7**(1), 1–11 (2021)
13. Revankar, G.D., Shetty, R., Rao, S.S., Gaitonde, V.N.: Analysis of surface roughness and hardness in ball burnishing of titanium alloy. Measurement **58**, 256–268 (2014)
14. El-Axir, M.H.: An investigation into roller burnishing. Int. J. Mach. Tools Manuf **40**(11), 1603–1617 (2000)
15. Uddeholm Finland. Uddeholm Stavax ESR. https://www.uddeholm.com/finland/fi/products/uddeholm-stavax-esr/. Accessed 12 Feb 2022

High Precision Fabrication of an Innovative Fiber-Optic Displacement Sensor

Zeina Elrawashdeh[1,2(✉)], Philippe Revel[3], Christine Prelle[3], and Frédéric Lamarque[3]

[1] ICAM, Grand Paris Sud, 77127 Lieusaint, France
zeina.elrawashdeh@utbm.Fr, zeina.elrawashdeh@icam.Fr
[2] University of Technology of Belfort and Montbélliard - ICB COMM Laboratory,
90400 Sevenans, France
[3] University of Technology of Compiègne, Centre Pierre Guillaumat, 60200 Compiègne, France

Abstract. This study presents the high precision fabrication technique, employed to manufacture a 3D conical grating, used as the reflector element, for a fiber-optic displacement sensor. To get high performance in terms of the surface quality, as well as a dimensional precision, the surface of the reflector must be a polished-mirror surface. To do so, a high precision turning machine along with aluminum alloy were the technical choices made. Two prototypes with different geometric dimensions, have been fabricated using the same machining strategy. Single crystal diamond tool was chosen, to obtain high surface roughness. The followed machining procedure was divided into two main parts; the first part achieves several cuts, to get the desired dimensions, and the last cut is deduced to get the desired nanometric roughness. Good results have been obtained, which validates the followed machining procedure.

Keywords: Machining · Cutting tool · Roughness

1 Introduction

With the development of modern industry, high precision fabrication techniques of sensors and instrumentation tools is gaining more attention and becoming an essential research topic in various domains. Several research studies presented different fabrication techniques of sensors and instrumentation tools, along with different levels of complexity.

The study presented by Jéssica Santos Stefano et al. [1], has illustrated the manufacturing techniques of low-cost disposable electrochemical sensors for the possibility of using consumables as a base material for their construction providing attractive characteristics, such as simplicity, sustainability, and applicability in a single device. The strategies used to manufacture such sensors are screen & stencil printing, laser-scribing, and pencil drawing. These techniques do not require complicated methods nor expensive equipment; however, the fabrication steps must be strongly considered because they influence the performance of the final device. In the field of medicine & healthcare, Sudhanshu Nahato et al. [2] presented a feasibility study towards fabricating

© The Author(s) 2023
K.-Y. Kim et al. (Eds.): FAIM 2022, LNME, pp. 48–55, 2023.
https://doi.org/10.1007/978-3-031-18326-3_5

custom-designed surgical instruments for knee and hip replacement using metal additive manufacturing. To establish the feasibility, several tests were conducted where the additive manufacturing-built materials were compared with the traditionally manufactured material. Another study presented in [3], illustrated the design of a simple and high-performance flexible strain sensor, based on gold nanoparticles and a polyimide substrate. The fabrication included drop-casting and high temperature annealing. The sensor revealed high linearity with low power consumption. It will be used for detection of human motion and subtle strain.

Several other technologies such as machining techniques are highly recommended, as they provide the desired accurate dimensions, where the accuracy of a workpiece can be improved by surface measurement and compensation machining. The study presented by Zao Zao et al. [4], introduced the design of an on-machine measurement device based on a chromatic confocal sensor, it can inspect workpiece surfaces with larger depths and slopes. The machine tool is equipped with three linear hydrostatic axes (X, Y and Z axes) and two rotational axes, the chromatic confocal sensor is installed on the rotational axis platform.

The objective of this study is to present the high precision machining technique employed to fabricate the reflector part of a 3D fiber-optic linear displacement sensor. This sensor is targeted to measure the linear displacement with high resolution, for an axis performing a simultaneous motion, of rotation and translation at the same time.

2 Sensor Principle

The sensor consists of two fiber-optic probes associated to a highly reflective surface. Each probe has one center emission fiber and four reception fibers placed around the emission fiber. The sensor performance when it is associated to a planar surface has been already analyzed [5, 6, 7]. In classical configuration, the emission fiber placed in the center emits light on a flat reflective surface. The light reflected by the surface is injected in the reception fibers and guided to a PIN photodiode. The voltage output of the sensor is a function of the mirror displacement (Fig. 1). The mirror displacement is millimetric, the emission fiber diameter is approximately 460 µm, the reception fiber diameter is 240 µm and the space between these two is 30 µm.

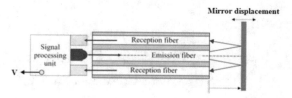

Fig. 1. Fiber-optic sensor.

When translating the flat mirror perpendicularly to the probe axis, the sensor response curve is as shown in Fig. 2.

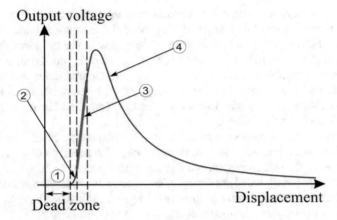

Fig. 2. Response curve of the fiber-optic displacement sensor.

As seen from Fig. 2, the sensor response curve consists of four zones. The first one is the dead zone, where the reception fibers cannot collect the reflected light because of the space between the emission and the reception fibers. Zones 2 and 4 are strongly non-linear with poor resolution. Zone 3 is the most interesting working zone because of its high sensitivity and linearity. On the other hand, this zone has a limited measurement range (less than 200 μm). For this reason, and in order to increase the measurement range the displacement direction of the flat mirror can be different from the normal vector orientation of its surface, resulting in the multiplication of the nominal range value by $(\sin \varepsilon)^{-1}$ factor, where ε is the inclination angle related to the grating axis [5]. This inclined mirror configuration, has been duplicated to increase more the linear measurement range of the sensor, in this configuration two fiber-optic probes will be needed to avoid the transition between two successive inclined step as shown in the following Fig. 3. The dimensions for the length, height and the angle for this grating is fixed with a MATLAB model, that evaluates the sensor performance as a function of these dimensions.

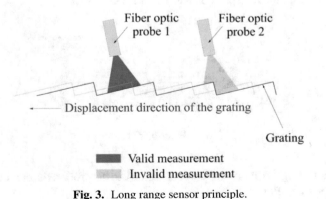

Fig. 3. Long range sensor principle.

The current study interests in the linear displacement measurement of an axis performing a rotational motion. To satisfy this criteria, the inclined mirror configuration is replaced with 3D cones assembled grating [8], which gives the result illustrated in the following figure (Fig. 4).

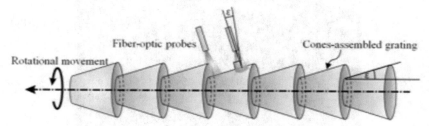

Fig. 4. 3D cones assembled grating configuration.

The following paragraphs will show the machining technique employed to fabricate the cones assembled grating.

3   High Precision Machining Technique

The manufacturing process has been done using a high precision turning machine along with single crystal diamond tool; and that allows to get sub-micrometric dimensional precision, along with a surface roughness of several nanometers. The performance of the surface quality are influenced by the quality of the cutting tool. However, the single crystal diamond tool can only be used in machining for certain materials; due to the chemical reaction, which occurs between the carbon of the diamond with the carbide substances (Fe, Ti, etc.). For that reason, aluminum alloy 2017 was chosen to make the 3D cones assembled grating. When a ductile material like the aluminum is machined with a diamond tool, the lubrification facilitates the cutting process, and used to eliminate the chip, so that it won't damage the machined surface. The geometric characteristics of the machined object define the relative movement between the tool and the sample to machine. Therefore, the high precision machining process of the cones assembled grating, which is the key element of the fiber-optic displacement sensor fabricated in this research work, was done in Roberval laboratory, because this lab is equipped with a high precision turning machine, which allows to have a micrometric precision for each step of the conical grating, in addition to a high surface quality (20–5 nm roughness).

3.1   The High Precision Turning Machine

The high precision turning machine is a prototype with two perpendicular displacement axes and one spindle with magnetic bearing. It has been designed at the end of 1980's by the European society of Propulsion (SEP) and was targeted to produce aspherical surfaces in various domains. It was transferred to Roberval laboratory in 1994 and fixed in an air-conditioned room at ±21 °C (Fig. 5), this machine was fixed on a concrete floor slab isolated from the main slab [9].

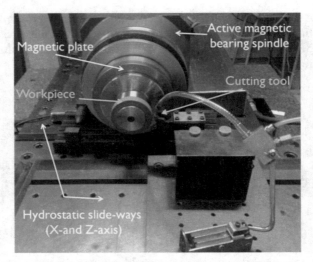

Fig. 5. The high precision turning machine [10].

3.2 The Machining Strategy of the Cones Assembled Grating

Firstly, the cones assembled grating have been geometrically modelled on MATLAB. The model aims to get the sensor performance, in terms of resolution and measurement range as a function of the cones' assembled grating dimensions mentioned in the following figure (Fig. 6).

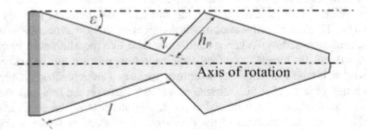

Fig. 6. The geometric dimensions of the conical step.

Where:

- l: the step length (μm) was fixed to 1573 μm.
- h_p: the step segment (μm) was fixed to 194 μm.
- ε: the step angle (°) was fixed to 4.62°.
- γ: the angle at the bottom of the step (°) to 130°.

Each step on the cones' assembled grating has been machined with several cuts at several different depths, to get the optimal geometric profile (Fig. 7).

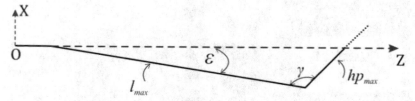

Fig. 7. Geometric profile of each step for machining.

Where:

- X: Axe of the tool holder slide of the machine.
- Z: Axe of the spindle holder slide of the machine.

For the prototype considered, 10 steps were fabricated for the cones assembled grating. Every step is machined with 7 successive cuts. The last finishing cut allow to get a high surface quality (Fig. 8). The depth of the first six cuts was fixed to 18 μm, 10 μm for the seventh cut with a speed ($V_a = 500$ μm/s). For the last cut, the depth was fixed to 5 μm, along with a speed of ($V_a = 50$ μm/s), to insure high surface roughness. Firstly, the tool follows the trajectory from point A to point B, then point B to point C.

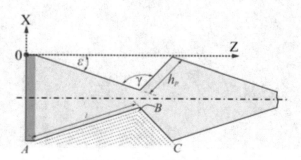

Fig. 8. Machining strategy.

The first step of the conical grating fabrication consisted of turning a cylinder, whose diameter is 55 mm, firstly, with a carbide tool in order for it to be centered on the spindle axis, and in a second time with a single-crystal diamond tool, with a radius of curvature ($R_c = 2$ mm), this tool allows to get a polished-mirror surface. The following figure shows the tool orientation with respect to the spindle, which is the axis of rotation. As seen from the following figure, the initial workpiece is a cylinder with a diameter of 55 mm; this cylinder has been straightened and turned (Fig. 9).

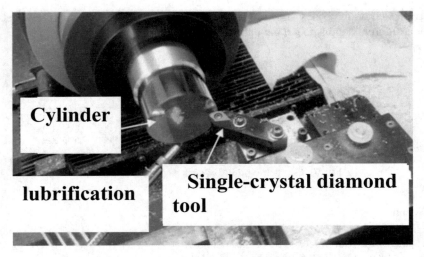

Fig. 9. Fabricated cones assembled grating.

The following figure presents the fabricated cones assembled grating on the cylinder (Fig. 10).

Fig. 10. Fabricated cones assembled grating.

The roughness obtained on the cones' assembled grating is nanometric and high micrometric dimensional precision has been also obtained.

4 Conclusion

This research paper presents the machining strategy used to fabricate the element reflector for a fiber-optic linear displacement sensor. This sensor will be used to measure the displacement of an axis in rotation, the reflector element is a conical grating made from alluminium alloy.

References

1. Stéphano, J.S., et al.: Different approaches for fabrication of low-cost electrochemical sensors. Current opinion in electrochemisty **32**, 100893 (2022)
2. Nahata, S., Ozdoganlar, O.B.: Feasibility of metal additive manufacturing for fabricating custom surgical instrumentation for hip and knee implants. In: 47 th SME North American Manufacturing Research Conference, Penn State Behrend Erie, Pennsylvnia (2019)
3. Xu, X.-L., Li, S.-X., Yang, Y., Sun, X.-C., Xia, H.: High-performance strain sensor for detection of human motion and subtle strain by facile fabrication. Measurement **189**, 110658 (2022)
4. Chen, Z., et al.: Development of an on-machine measurement system for ultra-precision machine tools using a chromatic confocal sensor. Precision Eng. **74**, 232-241 (2022)
5. Prelle, C., Lamarque, F., Revel, P.: Reflective optical sensor for long-range and high-resolution displacements. Sensors and Actuators **127**(1), 139–146 (2006)
6. Khiat, A., Lamarque, F., Prelle, C., Pouille, P., Leester-Schadel, M., Büttgenbach, S.: Two-dimension fiber optic sensor for high-resolution and long range linear measurements. Sensors and Actuators **158**(1), 43–50 (2010)
7. Girão, P.M.B.S., Postolache, O.A., Faria, J.A.B., Pereira, J.M.C.D.: An overview and a contribution to the optical measurement of linear displacement. IEEE Sens. J. 1(4), 322–331 (2001)
8. EL Rawashdeh, Z., Revel, P., Prelle, C., Letort, P., Lamarque, F.: Fiber-optic sensor for long range displacement measurement of a rotating spindle. In: 2016 11th France-Japan & 9thEurope-Asia Congress on Mechatronics (MECATRONICS) /17th International Conference on Research and Education in Mechatronics (REM)
9. Jouini, N.: Etude de l'intégrité de surfaces obtenues par usinage de haute précision d'un acier à roulement 100Cr6; application à la tenue en fatigue de contact, Mémoire de thèse, UTC (2011)
10. Revel, P., Jouini, N., Thoquenne, G., Lefebvre, F.: High precision hard turning of AISI bearing steel. Precis. Eng. **43**, 24–33 (2016)

3D Printing of Hydrogel-Based Seed Planter for In-Space Seed Nursery

Yanhua Huang⬤, Li Yu⬤, Liangkui Jiang, Xiaolei Shi⬤, and Hantang Qin$^{(\boxtimes)}$⬤

Iowa State University, Ames, IA 50010, USA
qin@iastate.edu

Abstract. Interest in manufacturing parts using 3D printing became popular across academic and industrial sectors because of its improved reliability and accessibility. With the necessity of self-sustentation, growing plant in space is one of the most popular topics. Carboxymethyl cellulose (CMC) is one of the best candidates for sprouting substrate with 3D printing fabrication as it is non-toxic, biodegradable, and suitable for extrusion-based 3D printing. Soybeans were placed into the designed and printed CMC gel with different orientations. Without visible light, soybeans with hilum facing side had the highest water absorption average comparing those facing up or down. Hydrogel weight dominated the water absorption efficiency. These findings signified that bean orientation affects the sprouting process. This study demonstrates the substrate geometry and seed orientation impacts on germination of soybeans, proposed guidelines for optimizing the sprouting process for high-level edible plants and promoting innovated in-space seed nursery approach.

Keywords: 3D Printing · Hydrogel · Soybean · Sprouting · In-Space

1 Introduction

In the last three decades, 3D printing, as one of the additive manufacturing processes, has been defined, developed, and improved [1]. Unlike traditional subtractive manufacturing methods, 3D printing works by building pieces layer by layer based on a computer-aided design (CAD) file. This method allows for more frequent geometry iterations than conventional manufacturing approaches such as casting or forging. Because of this, 3D printing-assisted product prototypes can have faster iterations and more adjustable dimensions, also known for improved customizing capability. A wide range of materials can be 3D printed, such as polymers [2], ceramics [3], metal alloys [4], and their composites [5]. Even more, these materials can blend with different feedstocks to improve the functionalities of the 3D printed products. These feedstocks include: solid particles [6], viscus semi-solids [7], liquids [8], and aqueous solutions [9]. Hydrogels have been explored and refined to fit the 3D printing approach as one of the polymer-liquid mixtures [2].

Because of the shear-thinning characteristics, hydrogels have a lower inclination to flow post extrusion from the nozzle under low-shear circumstances [2]. 3D printed

© The Author(s) 2023
K.-Y. Kim et al. (Eds.): FAIM 2022, LNME, pp. 56–63, 2023.
https://doi.org/10.1007/978-3-031-18326-3_6

hydrogels can assist wound healing [10], nutrient delivery [11], medication delivery [2, 12], and organ repair [13] applications. Aside from these characteristics, hydrogels and 3D printing are renowned for their porous structure and inhomogeneous material distributions. These characteristics lead to anisotropic mechanical properties in 3D printed objects, which promote the usage of 3D printed hydrogel in seed sprouting applications. The porous structure allows embedded seeds to breathe, while the anisotropic mechanical performance replicates solid planting conditions.

Soilless cultivation is one of the most common topics for studying plants using hydrogel sprouting. Soilless farming is the practice of cultivating plants without the use of soil. The system's complexity is divided into aeroponics, aquaponics, and hydroponics [14]. Only the plant and the growth substrate are present in aeroponics. The materials should have the following attributes to ensure that the plant grows in a healthy state:

- Maintain mechanical integrity of the connection and any suitable form changes during the growing cycle [15].
- Give the gemination nourishment or water (e.g., sprouting, rooting) [16].
- Prevent evaporation of water [17].
- Keep diseases like fungus, bacteria, and other pathogens [18].
- Create an atmosphere that is bioactive and non-toxic [19].

Since cellulose is the polymerizing repeating unit, carboxymethyl cellulose (CMC) chain carries a considerable amount of hydroxyl groups. They can preserve water and provide a bioactive environment while ensuring attachment from viscous polymer chains and hydrogen bonding. CMC is non-toxic to most plants when there are no halogen elements present. Pathogen isolation can be achieved using 3D printing by printing gels separately for different seeds. Moreover, CMC hydrogels possess reliable ability to retain 3D high-aspect ratio struc-tures, distinguishing itself from some alternative materials like methylcellulose and hydroxypropyl methylcellulose.

Growing edible plants and producing food in space is advantageous. They can absorb carbon dioxide and create oxygen gas on the one hand. It also provides food for astronauts and completes the carbon cycle. On the other hand, planting in space was more difficult than on Earth. Firstly, microgravity can cause soil particles to fly out of control. Second, the soil has a limited water retention capacity compared to hydrogel [20]. Hydrogel as a growing medium in space is both efficient and cost-effective. Finally, because of the reduced space requirements, hydrogel-based growing plants can be multilayer-packed, resulting in less weight and volume.

Bean orientation has been shown to affect light absorption and seedling emergence rate in previous studies [21]. It is still unknown if the direction has an effect on sprouting when no visible light is present. It is also uncertain how much water is required for soybeans to sprout. To tackle these two problems, more study is required. Different size 3D-printed hydrogels were created as growing media for sprouting soybeans in this study. Soybeans were planted in hydrogels with varying orientations for a 5-day sprouting observation using an infrared camera. The weights of printed hydrogel, dry beans, and sprouted bean weights were recorded, as well as the sprouting length, which was photographed and recorded for analysis.

2 Materials and Method

2.1 Materials

The soybeans were selected with a weight range from 0.26 to 0.29 g per bean, supplied by Well luck Co., Inc. (Jersey City, NJ). CMC-gel was formulated by swelling CMC 6000 Fine Powder curtesy of Ticalose (White Marsh, MD) with deionized water at 8%, 10%, and 12% w/w.

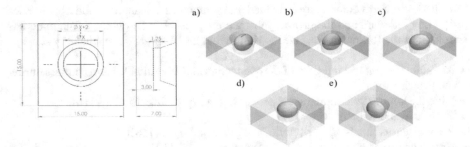

Fig. 1. Left: Printed models schematics with changing dimensions marked with X (unit: mm). Right: Bean orientations a) hilum faces up (HU), b) hilum faces down (HD), c) hilum faces up (HS), d) hilum faces side and corner (HSC), and d) hilum faces side and side (HSS).

2.2 Hydrogel Printing

Twenty-four hours post swelling, CMC-gel were packed into a 50 mL syringe and centrifuged at 2000 rpm for 4 min to remove trapped air bubbles. Models with different diameters (5 mm, 6 mm, and 7 mm, X marked dimension, the outwards drafting (X + 2) were expanded accordingly) were designed and exported as STL files, as shown in Fig. 1 left. STL files were sliced with Procusini.club web interface [22]. The CMC gel was printed using Procusini Dual 4.0 (Freising, Germany) with an optimized setting. After printing, printed gels were temporarily stored in the fridge at 4 °C.

2.3 Seed Planting

Each seed with no visible defects was selected and planted in the central cavity of the printed seed planter made of CMC-gel. To investigate the effect of seed orientation on sprouting performance, three seed orientations were chosen: hilum faces up (HU), hilum faces down (HD), and hilum face sides (HS), as shown in Fig. 1 right a. Furthermore, two seed orientations were chosen: hilum face sides and corner (HSC) and hilum face side and side (HSS), as shown in Fig. 1 right b. To eliminate the hole size effects in the orientation-controlled group, the HSC and HSS groups were seeded in 6 mm hydrogel. All seeds were planted facing one of the corners of the hydrogels to eliminate the orientation effects in the diameter-controlled groups.

3   Results and Discussion

3.1   Hydrogel Printing Quality Control and Mass Evaluation

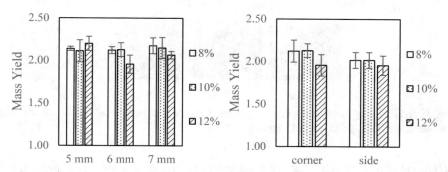

Fig. 2. Soybean weight yield on Day 5 normalized by Day 0.

Printed CMC-hydrogel weight was measured after the printing, as shown in Fig. 2. On one hand, with the same composition of the hydrogel, increasing diameter resulted in mass decreases due to the theoretical volume decrease. On the other hand, with the same diameter, reducing CMC weight percentage will result in a non-significant mass decrease due to the density of the formulated CMC hydrogel decrease. With a maximum of 3% error rate, the printing performance was stable and reliable under the sample size and manufacturing method [23].

Under the consideration of size difference from each soybean, the yielding masses at Day 5 were normalized by the dry mass of the soybeans at Day 0 shown in Fig. 3. And 4. In general, the average mass yields decrease when the CMC weight percentage increases. According to previous studies [2], increasing CMC weight percentage leads to the hydrogels' mechanical strength and viscosity of the increase. It prevents the beans from swelling, therefore reducing the water absorption. This also affected the volume changes.

3.2   Volume Evaluation

The soybean swelled significantly during the sprouting process due to the high-water content in the CMC-gel contact. In general, water absorption is the major reason for volume increases, as shown in Fig. 4. Changes in the orientation of downward planted soybeans (HD) were observed by examining infrared images. The soybeans changed to hilum facing sides (HS). In addition, HSC and HSS conditions were evaluated to seek difference when hilum faces to the corner or the side of the printed hydrogel from a top-down view.

To understand the relationship between sprouting and experimental controlled variables, soybean volumes at Day 0 and Day 5 was estimated using the elliptical sphere volume equation. The swelling volume change ratio is calculated by dividing soybean

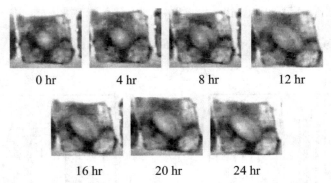

0 hr   4 hr   8 hr   12 hr

16 hr   20 hr   24 hr

Fig. 3. Infrared images show volume changes for the embedded soybean of first 24 h of swelling the size of the CMC-gel at 0 h is 15 mm square.

volume at Day 5 by Day 0. Results are shown in Fig. 4. The volume expansion rate decreases in average at the condition of HSC and CMC percentage increase. This is due to the improved mechanical strength provided with an increasing percentage of the CMC. This orientation requires the beam to swell against the share force instead of linear modules. On the other hand, when HSS, the improved mechanical strength reduces the shape deformation caused by swelling from the perpendicular direction of the hilum location, preventing the hydrogel from distortion, and providing less compression against the swelling process.

3.3   Soybean Density Evaluation

It is uncommon to evaluate the density changes of the sprouted seed. However, with limited space and hydrogel substrate, density changes are critical for seeds to grow in healthy conditions. This is important information providing guidance in choosing appropriate soil or sprouting substrate. Previously, evidence indicated soybean's density decreased overall during the sprouting stage [24]. This study evaluates the density change by the ratio between mass changes over volume changes shown in Fig. 5. When the density change ratio is larger than one, it indicates that the condition of density increases.

Soybeans' density, on average, decreased in the orientation control group except for 12% CMC for HSC condition and 8% CMC for HSS conditions. This density increase is evidence of water absorption shortage. Therefore, these conditions are not suitable for soybean sprouting. From another perspective, the density of the soybean sprouted in the 8% CMC hydrogel showed density decrease on average. Moreover, the 95% confidence interval upper limits of 5 mm and 7 mm gel are less than one, indicating they are the optimal condition for soybean sprouting across this study.

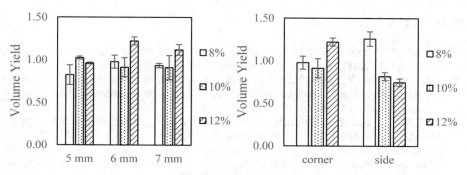

Fig. 4. Soybean volume yield on Day 5 normalized by Day 0.

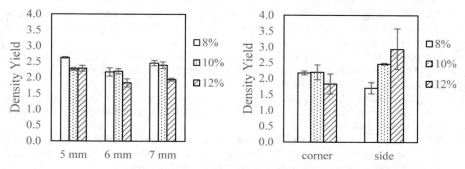

Fig. 5. Soybean density yield on Day 5 normalized by Day 0.

4 Conclusions

With the assistance of accessible and customizable 3D printing, high water retention, biodegradable, non-toxicity, and capacity of extrusion at ambient conditions, CMC was chosen to formulate hydrogel providing attachment, water, and isolation in the application of soybean sprouting. Compared with different orientations, HD soybeans rotate to HS providing sprouts with better growing freedom. HU orientation limited the water absorption by not allowing hilum to contact the hydrogel. Within the HSC and HSS conditions, only 12% gel with HSC and 8% gel with HSS condition yielded density increased soybean sprouting, demonstrating these conditions limited water availability to the soybean. Among different weight percentages of the CMC content in the hydrogel formula, 8% CMC hydrogel sprouted soybean yields the highest mass gain in average. Compared with different diameters of the central cavity and different planting orientations, both 5 mm and 7 mm gel printed with 8% CMC gel yields density decreased soybean in the 5-day sprouting process, which suggested the optimal sprouting conditions in this study.

These findings provide guidelines and further research suggestions regarding seed sprouting with limited resources and complex environments. Further in-space experiments will assist in analyzing the ultimate performance of sprouting seeds, using hydrogel as growing media.

References

1. Horvath, J.: A brief history of 3D printing. In: Mastering 3D Printing, pp. 3–10. Apress, Berkeley, CA (2014). https://doi.org/10.1007/978-1-4842-0025-4_1
2. Jiang, X., Huang, Y., Cheng, Y., et al.: Effects of lyophilization on the release profiles of 3D printed delivery systems fabricated with carboxymethyl cellulose hydrogel. Polym 13, 749 (2021). https://doi.org/10.3390/POLYM13050749
3. Li, J., Wu, C., Chu, P.K., Gelinsky, M.: 3D printing of hydrogels: rational design strategies and emerging biomedical applications. Mater. Sci. Eng. R Reports 140, 100543 (2020)
4. Karunakaran, R., Ortgies, S., Tamayol, A., et al.: Additive manufacturing of magnesium alloys. Bioact Mater. 5, 44–54 (2020). https://doi.org/10.1016/J.BIOACTMAT.2019.12.004
5. Huang, Y., Jiang, L., Li, B., et al.: Study effects of particle size in metal nanoink for electro-hydrodynamic inkjet printing through analysis of droplet impact behaviors. J. Manuf. Process 56, 1270–1276 (2020). https://doi.org/10.1016/J.JMAPRO.2020.04.021
6. Shen, W., Zhang, X., Jiang, X., et al.: Surface extraction from micro-computed tomography data for additive manufacturing. Procedia Manuf 53, 568–575 (2021). https://doi.org/10.1016/J.PROMFG.2021.06.057
7. Fafenrot, S., Grimmelsmann, N., Wortmann, M., Ehrmann, A.: Three-dimensional (3D) printing of polymer-metal hybrid materials by fused deposition modeling. Mater. 10, 1199 (2017). https://doi.org/10.3390/MA10101199
8. Jiang, L., Huang, Y., Zhang, X., Qin, H.: Electrohydrodynamic inkjet printing of Polydimethylsiloxane (PDMS). Procedia Manuf. 48, 90–94 (2020). https://doi.org/10.1016/J.PROMFG.2020.05.024
9. Qin, H., Dong, J., Lee, Y.S.: Fabrication and electrical characterization of multi-layer capacitive touch sensors on flexible substrates by additive e-jet printing. J. Manuf. Process 28, 479–485 (2017). https://doi.org/10.1016/J.JMAPRO.2017.04.015
10. Xu, Z., Hwang, D.G., Bartlett, M.D., et al.: Alter macrophage adhesion and modulate their response on hydrophobically modified hydrogels. Biochem Eng. J. 165 (2021). https://doi.org/10.1016/J.BEJ.2020.107821
11. Ma, L., Shi, Y., Siemianowski, O., et al.: Hydrogel-based transparent soils for root phenotyping in vivo. Proc Natl Acad Sci U S A 166, 11063–11068 (2019). https://doi.org/10.1073/pnas.1820334116
12. Cheng, Y., Qin, H., Acevedo, N.C., et al.: 3D printing of extended-release tablets of theophylline using hydroxypropyl methylcellulose (HPMC) hydrogels. Int. J. .Pharm 591 (2020). https://doi.org/10.1016/J.IJPHARM.2020.119983
13. Daly, A.C., Pitacco, P., Nulty, J., et al.: 3D printed microchannel networks to direct vascularisation during endochondral bone repair. Biomaterials 162, 34–46 (2018). https://doi.org/10.1016/J.BIOMATERIALS.2018.01.057
14. Kalossaka, L.M., Sena, G., Barter, L.M.C., Myant, C.: Review: 3D printing hydrogels for the fabrication of soilless cultivation substrates. Appl Mater Today 24, 101088 (2021). https://doi.org/10.1016/J.APMT.2021.101088
15. Sinnett, D., Morgan, G., Williams, M., Hutchings, T.R.: Soil penetration resistance and tree root development. Soil Use Manag 24, 273–280 (2008). https://doi.org/10.1111/J.1475-2743.2008.00164.X
16. Savvas, D., Gruda, N.: Application of soilless culture technologies in the modern greenhouse industry - a review. Eur. J. Hortic Sci. 83, 280–293 (2018). https://doi.org/10.17660/EJHS.2018/83.5.2
17. Gupta, S.C., Larson, W.E.: Estimating soil water retention characteristics from particle size distribution, organic matter percent, and bulk density. Water Resour. Res. 15, 1633–1635 (1979). https://doi.org/10.1029/WR015I006P01633

18. Fierer, N.: (2017) Embracing the unknown: disentangling the complexities of the soil micro-biome. Nat. Rev. Microbiol **1510**(15), 579–590 (2017). https://doi.org/10.1038/nrmicro.2017.87
19. Blok, C., De Kreij, C., Baas, R., Wever, G.: Analytical methods used in soilless cultivation. Soil Cult Theory Pract 245–289 (2008). https://doi.org/10.1016/B978-044452975-6.50009-5
20. Phuangsombut, K., Suttiwijitpukdee, N., Terdwongworakul, A.: Nondestructive classification of mung bean seeds by single kernel near-infrared spectroscopy (2017). 101142/S179354581650053X10
21. Parrish, D.J., Leopold, A.C.: Transient changes during soybean imbibition. Plant Physiol **59**, 1111–1115 (1977). https://doi.org/10.1104/PP.59.6.1111
22. Procusini Club. https://www.procusini.club/. Accessed 13 Nov 2021
23. Nuchitprasitchai, S., Roggemann, M., Pearce, J.M.: Factors effecting real-time optical monitoring of fused filament 3D printing. Prog Addit Manuf **2**, 133–149 (2017). https://doi.org/10.1007/S40964-017-0027-X/TABLES/4
24. Kuznetsov, O.A., Brown, C.S., Levine, H.G., et al.: Composition and physical properties of starch in microgravity-grown plants. Adv. Sp Res. **28**, 651–658 (2001). https://doi.org/10.1016/S0273-1177(01)00374-X

Modelling and Simulation of Automated Hydraulic Press Brake

Ilesanmi Daniyan[1]([⊠]) [ID], Khumbulani Mpofu[1] [ID], Bankole Oladapo[2] [ID], and Rufus Ajetomobi[3]

[1] Tshwane University of Technology, Pretoria 0001, South Africa
afolabiilesanmi@yahoo.com
[2] De Montfort University, Leicester LE1 9BH, UK
[3] Afe Babalola University, P.M.B. 5454, Ado Ekiti, Nigeria

Abstract. In this study, a reconfigurable hydraulic press brake was designed using Solidworks and simulated on a hydraulic Automation Studio Fluidsim. The designed press brake comprises of the frame balance, conveyor rollers and support, belt, chuck, six hydraulic cylinders assembled with bolts and nuts. The buckling force was determined analytically and compared with the Finite Element Analysis (FEA) simulation to prevent distortion of length and section. The Von mises stress theory was used to determine the stress, resultant load and displacement. The results obtained from the FEA simulation were compared with the mechanical properties of the hydraulic press brake. The maximum stress induced is significantly lower than the tensile strength of the hydraulic press brake. Hence, the stress induced due to bending cannot cause the cast alloy to yield. Also, the buckling force significantly exceeds the resultant force giving no chances for buckling. The designed hydraulic press brake is flexible enough to control using hydraulic cylinders and enhances sufficient strength and rigidity during clamping and loading conditions.

Keywords: Buckling force · Distortion · Hydraulic cylinder · Press brake

1 Introduction

According to Thomas *et al.* [1], a hydraulic press is a machine press which uses hydraulic cylinder to generate a compressive force to perform various pressing operations such as metal forging, punching, stamping, etc. The press provides an efficient means of pushing and pulling, rotating, thrusting and controlling load [2, 3]. Some hydraulic press applications include compression moulding, injection moulding, drawing, forging, blanking, coining, clamping, compacting, Forming, pad forming, potting, punching, and stacking, bending, stamping and trimming [4, 5]. The use of hydraulic cylinders for controls boasts of cost-effectiveness, high rate of production, positive response to changes, ease of control of parameters and primarily suitable when a heavy workpiece is to be machined [3, 6, 7]. Other advantages include tonnage adjustment and cycle time maximisation [8, 9]. According to Maneetham and Afzulpurkar [10], Hydraulic Servo

K.-Y. Kim et al. (Eds.): FAIM 2022, LNME, pp. 64–78, 2023.
https://doi.org/10.1007/978-3-031-18326-3_7

Systems (HSS) have been used in many modern industrial applications by their small size to power ratios and their ability to apply considerable force and torque.

On the other hand, by using a simulation model for hydraulic systems, the dynamic performance of these systems may be validated in the absence of actual hardware, which is accomplished via the use of specialised modelling and simulation tools [11–14]. In addition, the bending forces and moment can easily be predicted using simulation tools to determine the magnitude of stain, buckling and distortion [15–17]. This will enhance the use of hydraulic cylinders with sufficient clamping force that ensures adequate strength without distortion. This study aims to design a reconfigurable press brake assembly with hydraulic cylinders for holding a workpiece and adjusting the ram height during machining operations on a press brake. This is to enhance adequate clamping and precision during manufacturing operations. Despite productivity gains achieved through automation of design routines and manufacturing tasks, the authors Kumar et al. [18] and Ulah et al. [19] report that nearly 85% of all fixture processes and design plans are still performed manually, and detailed optimisation plans are rarely created. The interchangeability of parts is critical to the successful operation of any mass production facility because it allows for quick assembly and lower unit costs. Mass production methods demand fast and easy positioning for accurate operations [20, 21]. When designing jigs and fixtures, the strength of the clamp should be sufficient to hold the workpiece firmly in place and to withstand the strain of the cutting tool without springing [22, 23]. When producing large quantities of different materials on a large scale, a significant amount of time is spent setting up the device and clamping it [24, 25]. According to Pachbhai and Raut [26] as well as Daniyan et al. [27], hydraulic cylinders instead of manual adjustment are characterised by quick and automatic adjustment, greater accuracy, high productivity, consistent performance clamping force, and repeatable clamp location. Computer-aided design, modelling, and simulation tools have been used to improve the development of fixtures. For instance, Ruksar et al. [28] carried out the FEA and optimisation of machine fixtures, while Wang [29] applied a polynomial fit-based simulation method in a hydraulic actuator control system. Shrikant and Raut [30] employed computer-aided design for fixture development. It is necessary to reconfigure existing machines to have efficient work holding capacity to increase overall productivity, location accuracy, and surface finish quality of the finished product. The study aims to design the locating, supporting and clamping methods for a reconfigurable press brake using hydraulic controls.

2   Methodology

This paper proposes a six-cylinder automated hydraulic brake press. The press brake is constructed with a balanced frame, conveyor rollers and support, belt, chuck, and six hydraulic cylinders that are bolted and nutted together. Solidworks was used to design and model the fixture. According to Khurmi and Gupta [31], the maximum distortion energy theory for yielding is expressed as Eq. 1.

$$(\sigma t_1)^2 + (\sigma t_2)^2 - 2\sigma t_1 \times \sigma t_2 = \left(\frac{\sigma t_1}{FS}\right) \tag{1}$$

where: σt_1 is the maximum princial stress (N/m^2); σt_2 is the minimum principal stress (N/m^2), and σy_t is the stress at yield point (N/m^2); F.S is the factor of safety. The maximum and minimum principal stress calculated from Von mises stress analysis is given as 2.39365×10^5 N/m^2 and 5.44655×10^7 N/m^2 respectively. The volumetric parameters for the entire model are given as mass: 442.634 kg, volume: 0.056748 m^3, density: 7800 kg/m^3, and weight: 4337.82 N. Buckling is a possibility in the lower beam, which is the area where the fixture is loaded. The analytical results are compared to those obtained from the FEA simulation to determine the likelihood of buckling. With a length of 150 mm, the support for the tested section flexural rigidity equals 6.6×10^{-6}. Nm^{-2}. This is calculated as the modulus of elasticity and moment of inertia for the section under consideration. As a result, the buckling force is represented by Eq. 2.

$$F_b = \frac{El \cdot \pi^2}{L_c^2}. \tag{2}$$

F_b is the buckling force (N), EI is the flexural rigidity 6.6×10^6 Nm^{-2}, and L_c is the effective length (m). Since both ends are pinned, the effective length equals the actual length. Hence,

$$L_c = L \tag{3}$$

The model of the designed fixture assembly and the assembly drawing are shown in Fig. 1.

1	Hoisted Stand		2	
12	Frame balance (left)		1	
3	Frame balance (right)		1	
4	Conveyor roller support		1	
5	Conveyor roller support		21	
6	belt 1		1	
7	belt 2		1	
8	belt 3		1	
9	chuck		2	
10	stand		6	
11	Hyraulic Cylinder Assembly		6	

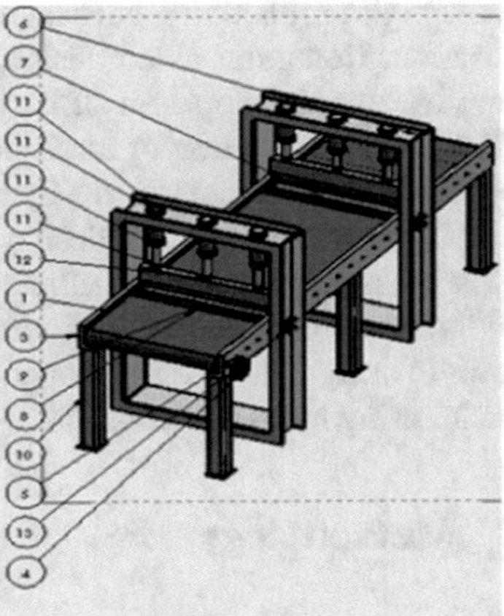

Fig. 1. The model assembly drawing of the fixture.

From Eq. 2, the buckling force is calculated as 2.8958×10^9 N. Due to the fact that the section only has to support a resultant load of 499.716 N, the buckling force exceeds the resultant force, thus, giving no chances for buckling. The design was based on the maximum tonnage of the press brake, which is determined by the material type, thickness, length, and method of bending and clamping. When performing Von Mises stress analysis, failure or yielding occurs at a point in a member where the distortion strain energy is most significant [31, 32]. Furthermore, according to the results of a simple tension test, the shear strain energy per unit volume in a bi-axial stress system reaches the limiting distortion energy at the yield point per unit volume at the yield point.

The area of the piston is expressed by Eq. 4.

$$A = \frac{\pi D^2}{4} \tag{4}$$

$$A = \frac{3.142D^2}{4} = 0.7854 \, d^2 m^2$$

where 'D' is the internal diameter of the piston-cylinder (m). The stress-induced is expressed by Eq. 5.

$$\sigma = \frac{F}{A} \tag{5}$$

'F' is the force applied (N), and 'A' is the piston cross-sectional area (m^2). Introducing the maximum stress given as 2.47903×10^5 N/m^2, reaction force 69.4426 N calculated from Von mises stress analysis and cross-sectional area calculated from Eq. 2 as 0.7854 d^{2}m^2 into Eq. 3; we have

$$2.47903 \times 10^5 = \frac{69.4426}{0.7854 \, d^2}$$
$$d = 0.0188 \, \text{m} \quad or \quad 18.8 \, \text{mm}$$

Using a safety factor of 2 and correcting to the nearest standard size, the piston diameter is calculated as 40 mm; therefore, the area is calculated as

$$0.7854 \times 0.04^2 = 1.26 \times 10^{-3} m^2$$

The piston will be subjected to shear stress; hence its thickness should be sufficient to resist failure by shearing. The minimum thickness of the piston required to resist shearing is given by Eq. 6.

$$t = \frac{pd}{2\sigma} \tag{6}$$

where: d is the internal diameter of the piston-cylinder is 0.04 m, σ is the maximum allowable stress (7.23826×10^8 N/m^2) and ρ is the pressure in the cylinder (2.47903×10^5 N/m^2), and thickness is calculated as 0.006849 m. Using a safety factor of 2, the thickness is calculated as 0.015 m to the nearest standard thickness. The volumetric properties of the hydraulic cylinder are as follow; mass: 0.618573kg; volume: 8.03342e-005 m^3; density: 7700 kg/m^3 and weight: 6.06202 N.

2.1 Computer Aided Modelling and Simulation

The modelling and simulation for the two components under investigation (hydraulic cylinder and assembly fixture) were carried out in the Solidworks 2018 environment.

The study type is to investigate the stress, displacement and strain of the hydraulic cylinder and fixture analysis. The the linear elastic isotropic model type and the Von mises failure criterion was used to determine the stresses induced in the component member, resultant loads and the corresponding displacements. The general standard static analysis of the finite element modelling was set up for the model analysis. From the material database, the mechanical properties of the materials selected for the hydraulic cylinder and assembly fixture (stainless steel 304 and cast alloy steel, ASTM A216) were selected. This was followed by the free body model of the components and the assignment of the loading conditions vis-à-vis the service requirements. Next is the dicretisation of the model. This is to mesh the developed models into finite elements and the application of the mesh control. The properties of stainless steel 304 employed for the design of the hydraulic cylinder are presented in Table 1 while Table 2 presents the mechanical properties of the cast alloy steel employed for the design of the fixture model.

Table 1. Mechanical properties of stainless steel 304 [33]

S/N	Parameter	Value
1	Yield strength	$2.40e + 008$ N/m^2
2	Tensile strength	$5.90 + 008$ N/m^2
3	Elastic modulus	$2.1e + 011$ N/m^2
4	Shear stress	$5.429e + 008$N/m^2
5	Poisson's ratio	0.28
6	Density	7700 kg/m^3
7	Shear modulus	$7.6e + 010$ N/m^2
8	Thermal expansion coefficient	17e-006 /C

Table 2. Properties of cast alloy steel (ASTM A216) for the fixture model [34].

S/N	Parameter	Value
1	Yield strength	$5.56e + 008$ N/m^2
2	Tensile strength	$7.30e + 008$ N/m^2
3	Elastic modulus	$2.11e + 011$ N/m^2
4	Shear stress	$3.36062e + 008$ N/m^2
5	Poisson's ratio	0.29
6	Density	7850 kg/m^3
7	Shear modulus	$8.2 + 010$ N/m^2
8	Thermal expansion coefficient	1.5e-005 /Kelvin

The linear elastic isotropic model type was selected and the Von mises failure criterion was employed for the failure analysis. A mesh size of 2 mm was employed in the Solidworks environment to mesh model into finite elements.

Using a mesh interval of 0.2 mm, it was observed that the computational time decreases with an increase in the mesh size up to 2.0 mm for the hydraulic cylinder. Further increase in the mesh size up to 2.6 mm resulted in a slight increase in the computational time. Hence, the mesh size of 2.0 mm which produced the least computational time (152 s) was selected for the hydraulic cylinder. For the fixture, it was observed that the computational time decreases with an increase in the mesh size up to 2.2 mm. Further increase in the mesh size up to 2.6 mm resulted in a slight increase in the computational time. Hence, the mesh size of 2.0 mm which produced the least computational time (367 s) was selected for the fixture (Fig. 2).

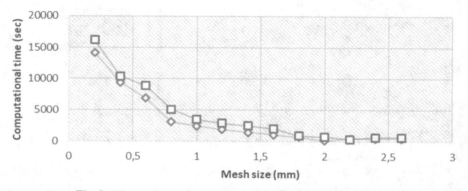

Fig. 2. The mesh time and the corresponding computational time.

2.2 Model Control of Triplet Cylinder

The category of the technical properties used to control the component variables can be assigned to other variables in the "read" or "write" mode for sending or receiving control signals. The driving force is an assignable variable to apply a driving force to the component. If there is not enough pressure, this force will drive the rod-piston assembly. The curve defining the external driving force is expressed in terms of the percentage of the cylinder position. From 0% to100%, the force is applied during the extension of the cylinder until the cylinder reaches the end of its stroke. Once the end of stroke is reached, the curve used will be in the −100% to 0% quadrant. Between 0% and 100%, if the read value is positive and there is not enough pressure to oppose, the rod-piston assembly will retract; inversely, but if the force is opposing, it will extend. Between −100% and 0%, if the value of the force is positive and there is not enough negative pressure, the piston-rod assembly will retract and extend if negative. This curve is of null value by default—external mass assignment variable of mass to allow the dynamic change of the mass during simulation. The default unit of the variable is the kilogram. The resistive force assignable variable is to apply a resistive force to the component. This force is resistive and will oppose the displacement of the piston-rod assembly. The curve that defines this force is expressed in terms of the percentage of the cylinder position. When the cylinder is extending, the curve is read in the 0–100% quadrant; inversely, the force will be read between −100 to 0% quadrant when the cylinder retracts. The value of this force can only be positive by convention. Figure 3 presents the mechanical working principles of the double-acting triplet cylinder.

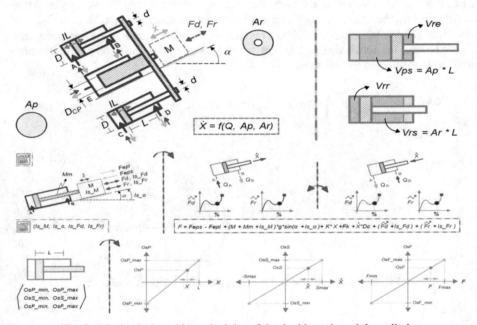

Fig. 3. Mechanical working principles of the double-acting triplet cylinder.

2.3   The Operational Model of the Triplet Cylinder

The hydraulic press brake utilises mechanically connected cylinders that operate in parallel. Linear actuators are devices that convert fluid energy to mechanical energy. As the name implies, the linear actuators will deliver the powers straight. In fluid power systems, linear actuators are often available with various components attached to the end of the rod. Mechanical linkage, levers or cables can be attached to the cylinder to transform the force in the type of movement wanted—technical modelling category of the properties that affect the components simulation model. The drop-down list options allow to edit other parameters or enable/disable the performance curve modelling. For the operating condition, the category of the properties relates to the components operating conditions, especially those that describe its operation limits. Most of these properties assess a faulty component and automatically trigger a failure, thus, activating the respective simulation option as "Automatic Failures". The maximum force that can be applied to the component, or the maximum force range that the directional valve command can apply in proportional operation mode can be selected. The maximum pressure supported by the component supposes the option Monitor Faulty Components is activated in the simulation options. In that case, a visual warning will be displayed next to the component to inform the user that the value is exceeded during the simulation. If the option "Automatic Failure Trigger" is activated in the troubleshooting branch, the user will trigger a failure when this maximum value is reached during simulation. The failure must first be declared and selected, which will only be triggered with this property. The maximum distance travelled by piston per unit time is the distance moved by the piston from one end of the cylinder to the other end. Suppose the option "Monitor Faulty Components" is activated in the simulation options, a visual warning will be displayed next to a component to inform user that the value is exceeded during simulation. If the option "Automatic Failure Trigger" is activated in the troubleshooting branch, the user will trigger a failure when this maximum value is reached during simulation. The failure must first be declared and selected. It will then only be triggered with this property.

Figure 4 shows the model design of the automatic and manually operated triple cylinders.

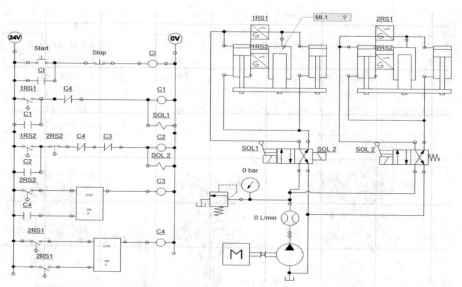

Fig. 4. Model design of the automatic and manually operated triple cylinders.

3   Results and Discussion

The result of the simulation of the hydraulic cylinder using Solidworks and the linear elastic isotropic model type and the Von mises failure criterion is presented in Tables 3 and 4 as well as Fig. 5. While Table 3 summarises the reaction forces and moments, Table

4 and Fig. 5 present the strain, stress and displacement analysis. The resultant force from Table is 69.4426 N with the highest reaction experienced along the vertical axis (Y-axis).

Table 3. Reaction forces and moments.

Selection set	Units	Sum X	Sum Y	Sum Z	Resultant
Reaction forces	N	−1.49012e-007	69.4426	2.44007e-006	69.4426
Reaction moment	N.m	0	0	0	0

It can be seen in Table 4 that the deformation per unit length is negligible, if not completely non-existent. It indicates that the clamping force is sufficient to prevent distortion in this particular instance. Furthermore, the stress-induced is minimal, and the cylinder will not yield to the applied force due to this stress.

Table 4. Strain, stress and displacement analysis for the hydraulic cylinder.

Name	Type	Min	Max
Strain 1	ESTRN: Equivalent strain	1.57209e-018 Element: 9335	8.29254e-007 Element: 22978
Stress 1	VON: von Mises stress	4.48733e-007 N/m^2 Node: 13269	247903 N/m^2 Node: 38330
Displacement 1	URES: Resultant displacement	0 mm Node: 158	0.000114313 mm Node: 37571

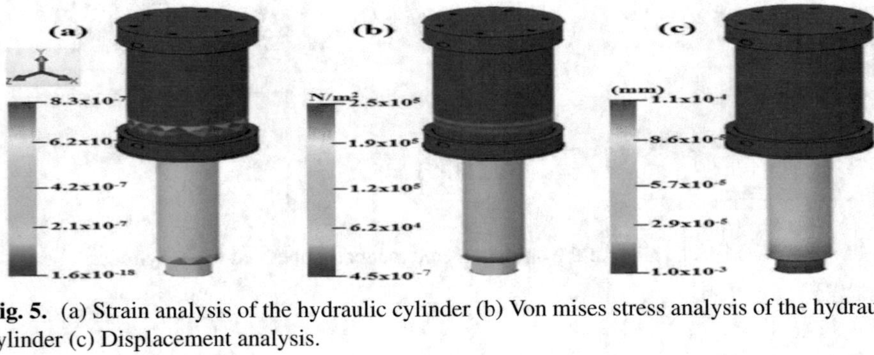

Fig. 5. (a) Strain analysis of the hydraulic cylinder (b) Von mises stress analysis of the hydraulic cylinder (c) Displacement analysis.

Figure 5 (b) shows the modelling result of the stress-induced in the hydraulic cylinder due to machining. The maximum stress induced is 2.47903×10^5 N/m^2 while the minimum is 4.48733×10^{-7} N/m^2. From Fig. 5(c), the maximum relative displacement of the cylinder from its mean position is 0.000114313 mm. Comparing the magnitude of the maximum stress induced in the cylinder to the yield strength of the material (2.40×10^8 N/m^2), then it can be concluded that the material is not likely to fail under the required service condition.

Table 5 presents the summary of the reaction forces and moments for the fixture. The resultant reaction force obtained from Solidworks simulation is 499.716 N. This force is insufficient to produce any bending as the resultant bending moment is zero.

Table 5. Reaction forces and moment.

Selection set	Units	Sum X	Sum Y	Sum Z	Resultant
Reaction Forces	N	9.51035	499.625	0.000166947	499.716
Reaction Moment	Nm	0	0	0	0

The summary of results of the simulations for the strain, stress and displacement analyses are presented in Table 6 and Fig. 6 for the fixture. The maximum and minimum strains were found to be 5.58621×10^{-7} and 2.26322×10^{-16} respectively. Both are negligibly insignificant. In this case, the fixture orientation does not change significantly while the bending operation is being performed.

Table 6. Strain, stress and displacement analysis for the fixture.

Name	Type	Min	Max
Strain1	ESTRN: Equivalent Strain	2.26322e-016 Element: 2491	5.58621e-007 Element: 26763
Stress1	VON: von Mises Stress	5.44655e-007 N/m^2 Node: 33371	239365 N/m^2 Node: 54832
Displacement1	URES: Resultant Displacement	0 mm Node: 315	8.77695 mm Node: 56061

From Fig. 6a the maximum strain is 5.5862×10^{-7} while the minimum is 2.2633×10^{-16}. For the entire fixture model, the maximum stress induced is 2.26322×10^5 N/m^2 while the minimum is 5.44655×10^{-7} N/m^2 (Fig. 6b). From Fig. 6c, the maximum relative displacement of the cylinder from its mean position is 8.77685 mm. As shown in Fig. 6c, the front beam has a larger displacement while the beam along the neutral plane has a smaller displacement. This is due to the fact that at the neutral plane, the beam is not under the influence of any stress either compressional or tensional stress. The Von Mises analysis also revealed that the maximum and minimum stress-induced are 2.394×10^3 N/m^2 and 5.447×10^{-7} N/m^2 respectively. The maximum stress induced is lower than the yield strength (5.56×10^8 N/m^2) of the cast alloy from which the fixture was designed (ASTM A216). As a result, the stress induced by bending may not cause the cast alloy to yield. The tensile strength is also sufficient to withstand bending forces without displacement or distortion as the maximum value of displacement is 8.77695 mm.

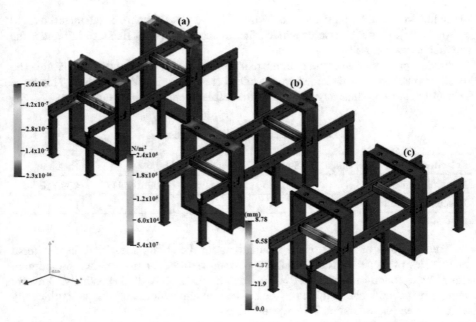

Fig. 6. (a) Strain analysis of the entire fixture (b) Stress analysis of the entire fixture (c) Displacement analysis of the entire fixture

Figure 7 shows the hydraulic circuit comprising a configurable 3/n way valve with three connections and negligible hydraulic resistance and the 4/n way valve with four connections. It also comprises a double-acting cylinder with a shock absorber at the stroke end. The connected pressure loads control the cylinder piston while the shock absorber can be adjusted using two adjustable screws. The piston of the cylinders contains a permanent solenoid that can be used to operate a proximity switch. The diameter of the piston is 20 mm with a maximum stroke length of 200 mm. The tank is a part of the pump unit and is integrated into it. To reduce the risk of damaging the component, the filter with negligible hydraulic resistance limits the amount of contamination in the fluid. The pump unit delivers a volumetric flow, with the operating pressure being limited by an internal pressure relief valve within the pump units housing. There are two tank connections on the pump. In addition, the relief valve is included in the circuit, which is closed in the normal position. Assume that the operating pressure has been reached at one of the end openings, the other opening opens when the pressure falls below the current level. The valve closes with the pilot pressure generated by the input pressure, resulting in the valve being closed again. It also has a pilot stage and the main stage; when the pilot stage is open. Thus, there is less volumetric flow through it than when the main stage is closed.

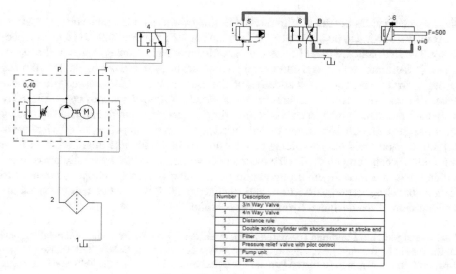

Number	Description
1	3/n Way Valve
1	4/n Way Valve
1	Distance rule
1	Double acting cylinder with shock adsorber at stroke end
1	Filter
1	Pressure relief valve with pilot control
1	Pump unit
2	Tank

Fig. 7. The hydraulic circuit

Figure 8 shows the variation in force as the piston position varies. The maximum force at a piston position of 46 mm is 0.58 kN. The magnitude of the force applied decreases with an increase in the piston position. At a maximum stroke length of 200 mm, the magnitude of the force becomes negligibly small. The simulation result from the AUTOMATION STUDIO Fluid sim is in agreement with the FEA simulation, which calculates the resultant reaction force as 0.499716 kN. This force is insufficient to produce any bending, strain or displacement, which confirms conclusively that the designed hydraulic brake press has sufficient strength to withstand bending stresses and forces without distortion.

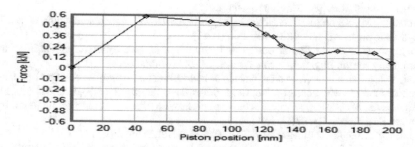

Fig. 8. Change in force with piston position.

4 Conclusion

A reconfigurable hydraulic press brake was designed using Solidworks and simulated on a hydraulic AUTOMATION STUDIO Fluidsim. The maximum strain, stress, and displacement values obtained from manual Solidworks simulation and Von mises stress analysis were found to be 8.29×10^{-7} N/m^2, 2.48×10^3 N/m^2 and 0.000114 mm respectively. The hydraulic cylinder boasts greater efficiency than manual means or the use of a jack. It facilitates quick adjustment and greater accuracy in equipment and workpiece setting as the ram retreats automatically, and the machine is quickly returned without any waste of time. The results indicate that the hydraulic cylinder actuator can sufficiently withstand the machining forces while providing sufficient strength and rigidity during machining operations. Hence, the reconfigured actuator system possesses efficient work holding capacity without sacrificing rigidity and stiffness. Results obtained from FEA simulation when compared with the mechanical properties of the hydraulic press brake indicate that the reconfigurable hydraulic press brake possesses adequate strength to prevent buckling, strain, distortion, and displacement. Future work can consider the development of the designed hydraulic press brake.

Acknowledgement. Funding: The authors disclosed receipt of the following financial support for the research: Technology Innovation Agency (TIA) South Africa, Gibela Rail Transport Consortium (GRTC), National Research Foundation (NRF grant 123575) and the Tshwane University of Technology (TUT)."

References

1. Thomas, A.T., Parameshwaran, R., Kumar, R.D., Mohanraja, S., Harishwaran, M.: An investigation on modelling and controller design of a hydraulic press. In: Third International Conference on Advances in Control and Optimisation of Dynamical Systems, pp. 719–725. Kanpur, India (2014)
2. Oladapo, B.I., Balogun, V.A., Adeoye, A.O.M., Ige, O., Afolabi, S.O.: Experimental analysis of electro-pneumatic optimisation of hot stamping machine control systems with on-delay timer. J. Appl. Res. Tec. **15**(4), 356–364 (2017)
3. Sumaila, M., Okonigbon, A., Ibhadodhe, A.O.A.: Design and manufacture of a 30-ton hydraulic press. AU J.T. **14**(3): 196–200 (2011)
4. Oladapo, B.I., et al.: Model design and simulation of automatic sorting machine using proximity sensor. Eng. Sci. Technol. Int. J. **19**, 1452–1456 (2016)
5. Lai, W.K., Rahmat, M.F., Abdul Wahab, N.: Modeling and controller design of pneumatic actuator system with control valve. Int. J. Smart Sens. Intell. Syst. **5**(3), 624–644 (2012)
6. Eltantawie, M.A.: Design, manufacture and simulation of a hydraulic bending press. Int. J. Mech. Eng. Robot. Res. **2**(1), 1–9 (2013)
7. Lange, K.: Handbook of Metal Forming. McGraw-Hill, New York (1975)
8. Adeoye, A.O.M., Aderoba, A.A., Oladapo, B.I.: Simulated design of a flow control valve for stroke speed adjustment of hydraulic power of robotic lifting device. Proc. Eng. **173**, 1499–1506 (2019)
9. Ferreira, A.J., Sun, P., Gracio, J.: Close loop control of a hydraulic press for springback analysis. J. Mater. Process. Technol. **177**, 377–381 (2006)
10. Maneetham, D., Afzulpurkar, N.: Modeling, simulation and control of high speed nonlinear hydraulic servo system. J. Autom. Mob. Robot. Intell. Syst. **4**(1), 94–103 (2010)
11. Prabhu, S.M., Wendlandt, J., Glass, J., Egel, T.: Multi-Domain modeling and simulation of an electro-hydraulic implement system. The MathWorks, Inc, pp. 1–8

12. Adeoye, A.O.M., Oladapo, B.I., Adekunle, A.A., Oladimeji, A.J., Kayode, J.F.: Design, simulation and implementation of a PID vector control for EHVPMSM for an automobile with hybrid technology. J. Mater. Res. Technol. **8**(1), 54–62 (2019)
13. Koivo, A.J., Thoma, M., Kocaoglan, E., AndradeCetto, J.: Modeling and control of excavator dynamics during digging operation. J. Aerosp. Eng. **9**, 10–18 (1996)
14. Krutz, J.E., Thompson, D.F., Krutz, G.W., Allemang, R.J.: Design of a hydraulic actuator test stand for non-linear analysis of hydraulic actuator systems. In: Proceedings of the Automation Technology for Off-road Equipment, pp. 169–183 (2002)
15. Farsi, M.A., Behrooz, A.: Bending force and spring-back in v-die-bending of perforated sheet-metal components. J. Brazil Soc. Mech. Sci. Eng. **33**(1), 45–51 (2011)
16. Florica, M.G., Gheorghe, A., Lucian, L., Vasile, A.C.: Spring back prediction of the bending process using finite element simulation. In: Proceeding 7th International Multi-disciplinary Conference, pp. 261–266, Baia Mare, Romania (2007)
17. Balogun, V.A., Oladapo, B.I., Adeoye, A.O.M., Kayode, J.F., Afolabi, S.O.: Hysteresis analysis of Thornton (IP6, IP12E and TH5V) magnetic materials through the use of Arduino microcontroller. J. Mats. Res. Tech. **7**(4), 443–449 (2018)
18. Kumar, V.V., Yadav, S.R., Liou, F.W., Balakrishnan, S.: A digital interface for the part designers and the fixture designers for a reconfigurable assembly system. Math Probl. Eng. **943702**, 1–13 (2013)
19. Saif, U., Guan, Z., Wang, B., Mirza, J., Huang, S.: A survey on assembly lines and its types. Front. Mech. Eng. **9**(2), 95–105 (2014). https://doi.org/10.1007/s11465-014-0302-1
20. Pachbhai, S.S., Raut, L.P.: A review on design of fixtures. Int. J. Eng. Res. Gener. Sci. **2**(2), 2091–2730 (2014)
21. Swami, A., Kondhalkar, G.E.: Design, development and analysis of hydraulic fixture for machining engine cylinder block on VMC. Int. Res. J. Eng. Technol. **3**(8), 463–469 (2016)
22. Warule, V.S., Sawant, S.M.: Design and analysis of milling fixture for HMC. J. Basic Appl. Eng. Res. **16**, 1362–1365 (2015)
23. Hui, L., Weifang, C., Shengjie, S.: Design and application of flexible fixture. Proc. CIRP **56**, 528–532 (2016)
24. Harini, T.V., Sastry, K.R., Narayana, Y.V.: Design and clamping force analysis of vacuum fixture to machine aerospace components. IOSR J. Mech. Civil Eng. **11**, 40–45 (2014)
25. Komal, B., Bhise, S.M.: Design and development of hydraulic fixture for VMC. Int. J. Res. Appl. Sci. Eng. Tech. **6**, 174–182 (2015)
26. Pachbhai, S.S., Raut, L.P.: Design and development of hydraulic fixture for machining hydraulic lift housing. Int. J. Mech. Eng. Robot. Res. **3**(3), 204–214 (2014)
27. Daniyan, I.A., Balogun, A.V., Adeodu, A.O., Oladapo, B.I., Peter, J.K., Mpofu, K.: Development and performance evaluation of a robot for lawn mowing. Proc. Manuf. **49**, 42–48 (2020)
28. Ruksar, S.D., Ram Babu, T.S., Rao, M.J.: Finite element analysis and optimisation of machine fixture. Int. J. Current Eng. Technol. **4**, 2156–2172 (2015)
29. Wang, J.: The application of a polynomial fit based simulation method in hydraulic actuator control system. In: IEEE (2012)
30. Shrikant, V.P., Raut, L.P.: Computer-aided fixture design: a review. Int. J. Adv. Eng. Res. Technol. **2**(1), 2348–8190 (2014)
31. Khurmi, R.S., Gupta, J.K.: A Textbook of Machine Design. Eurasia Publishing House Ltd., New Delhi (2005)
32. Daniyan, I.A., Adeodu, A.O., Oladapo, B.I., Daniyan, O.L., Ajetomobi, O.R.: Development of a reconfigurable fixture for low weight machining operations. Cogent Eng. **6**(1579455), 1–17 (2019)
33. Material Type Cost. https://web.mit.edu/course/3/3.11/www/modules/props.pdf. Accessed 2 May 2022

34. Daniyan, I., Mpofu, K., Fameso, F., Adeodu, A.: Numerical simulation and experimental validation of the welding operation of the railcar bogie frame to prevent distortion. Int. J. Adv. Manuf. Technol. **106**(11–12), 5213–5224 (2020). https://doi.org/10.1007/s00170-020-04988-6

Assessment of Reconfigurable Vibrating Screen Technology for the Mining Industries

Boitumelo Ramatsetse[1], Khumbulani Mpofu[2] , Ilesanmi Daniyan[2(✉)] ,
and Olasumbo Makinde[2]

[1] Educational Information and Engineering Technology, Wits School of Education,
Johannesburg 2193, South Africa
boitumelo.ramatsetse@wits.ac.za
[2] Tshwane University of Technology, Pretoria 0001, South Africa
afolabiilesanmi@yahoo.com

Abstract. Vibrating screens are very vital in the mineral processing industries for the beneficiation (separation) of mineral particles into different sizes. The breaking down of vibrating screens due to unforeseen contingencies have reduced the productivity of these machines thereby reducing the competitiveness, availability and reliability of these machines for the set production target made by the company. Also since human wants are insatiable, fluctuation in the mineral concentrates demands has been an inevitable scenario, thus, reducing the efficiency of the mineral beneficiation industries. During peak mineral concentrates demand, most of these industries do not have an option than to purchase another beneficiation screen in order to meet up with the continuously increasing production demand. A solution called the Reconfigurable Vibrating Screen (RVS) that can cover the gaps created by machine breakdown, and ensure that the variations in quantity of mineral concentrates needed by customers are met. In this paper, a state of configurations achieved by RVS as compared to the existing conventional vibrating screens was made. In addition to this, a market assessment of the proposed RVS and other existing screening technologies was performed. The index parameters used for this analysis are capacity, reliability, efficiency, versatility and cost. From the comparative analysis, it was observed that there are high advantages for using RVS for beneficiation operations in the mineral beneficiation in place of existing vibrating screens.

Keywords: Mineral processing · Production targets · Reconfigurability · Vibrating screen

1 Introduction

The need for adaptable mining systems with possibility of reconfiguration, to carter for ever changing production, demands is becoming more evident in the mining industries. In an environment where production targets are not constantly met, machine breakdowns due unforeseen circumstances and ever changing customer demands are of concern,

K.-Y. Kim et al. (Eds.): FAIM 2022, LNME, pp. 79–87, 2023.
https://doi.org/10.1007/978-3-031-18326-3_8

adaptable beneficiation technologies promises to be a solution to address such challenges. Small to medium mining companies who cannot afford the luxury of buying big machines when demand of their products need to be scaled is another eminent challenge to be addressed. Considering the beneficiation machines especially Vibrating Screens (VS) that are used for separation of crushed mineral particles obtained from primary, secondary and tertiary crushers into different sizes. The different types of VS used in the mining industries are horizontal screens, linear screens, gyratory screens, vibrating screens, circular motion screens, elliptical motion screens and resonance screens. There are a variety of mining and processing equipment that are currently designed and manufactured all over the world, but the aforementioned type of screens are built using dedicated manufacturing systems, which thus makes this machine restricted to particular type, screen structure, efficiency and production target; which in the long run results in the aforementioned problem. In view of this, the RVS is the newly designed and developed beneficiation machine, with an adjustable screen structure to meet up with variance in mineral volume products demanded by customers at varying time at the production site. The machine is also of utmost importance to meet the varying design specifications needed by different mining companies. Nevertheless, there is no doubt that the Reconfigurable Manufacturing System (RMS) concept has been proposed to meet changes and uncertainties of manufacturing environment and this objective would be achieved by reconfiguring hardware or software resources [1]. The major uncertainty of reducing the efficiency of the conventional vibrating screen is downtime. Meanwhile some designers and manufactures are seeking alternative ways to come up with new innovative screening methods. Currently, in the mining and material processing industries, screens are installed for operation for certain number of years, later, the same screen will be replaced with a bigger or alternatively a smaller structure simply because it is not able to respond to new production demands. Also, when there is high mineral particle demand, most of these industries are forced to buy another screen to be able to meet up with the fluctuating demand, thus increasing the company's operating cost and reducing the profits substantially. Due to the low efficiency from the conventional vibrating screen, this provide designers a leading edge for new innovative solutions that utilize the reconfigurability concept. Reconfigurability is a very valuable concept in responding to meet the highlighted challenges of mining and processing industry in the world. A RVS is an innovative solution that is designed to have a cost-efficient production, with a higher standardization level by modularizing the structural components.

This machine has been designed to ensure higher mineral concentrates productivity as well as meeting the company's exact production target, which presently has been infeasible due to downtime, unscalable machinery, which has been seen as inevitable scenario in the mining industries. According to Vorster and De la Garza [2], the different types of downtime costs that contribute to high maintenance costs spent on mining machines are associated resource impact costs, lack of readiness costs, service level impact costs and alternative method impact costs. Associated resource impact costs concern the effects of the failure on other components of the team. Lack of readiness costs are penalty costs assessed when an item that should be constantly available is not. Service level impact costs measure the decreased productivity of a fleet of equipment when a portion of that fleet has failed. Alternative method impact costs occur when a different method of production must be used due to the failure of a given component of the original production team. Different research has been done to show the negative impacts of down time on the mining operations and profit of the company. Edwards *et al.* [3] used regression analysis to predict the expected downtime cost rate for tracked hydraulic excavators in opencast mining industry. Their model shows that, machine weight is an excellent predictor of downtime cost. Vorster and Sears [4] proposed failure cost profiles which measure the expected cost per unit time in terms of the duration of the interval out of service. For a fixed repair time, the work featured the introduction of a cost related criterion that also takes into account the relative productivity of equipment that may be assigned to different kind of works. By doing so, the authors were able to decide replacement and task assignments for a fleet of similar equipment.

Operator-induced consequential costs were studied by Edwards *et al.* [3]. The work includes skill level of operators, fatigue, morale, and motivation. The authors suggest that operator's skill has the most important factor concerning the performance of the equipment. Considering operational poor practices, Pathmanathan [5] reported the increased frequency and cost of equipment downtime induced by the negligence of the operator and lack of proper training and knowhow on the part of the equipment supervisor. Arditi *et al.* [6] also consider operation uncertainty (i.e. different environmental conditions) as well as design complexity as the causes of greater risk for equipment downtime. Nepal and Park [7] explored the impact of downtime on construction projects duration and related costs. The analysis highlights how various factors and processes interact with each other to create downtime, and mitigate or exacerbate its impact on project performance.

Thus, effective machine maintenance is an inevitable function in a mining industry. The implementation of this type of vibrating screen will be advantageous in such a way to allow highly flexible and reconfigurable production on a very long term basis. Reconfigurable manufacturing system as designed by Koren *et al.* [8], is referred to as a manufacturing system designed from the outset for rapid change in structure, including software and hardware components, in order to adjust production capacity and functionality quickly, in response to uncertainty in customer requirements. In addition, reconfigurability concepts have been described by different researchers. Wiendahl *et al.* [9], defined reconfigurability as phenomenon which exhibits a switch or change in manufacturing systems or configurations with minimal effort and delay for achieving the desired adaptability to the set of subcomponents. Lee [10] also augmented Wiendahl *et al.* [9] by emphasizing that RMS is completely achieved on a manufacturing system, when it is been produced at optimum costs for its different configurations. In view of this, Setchi and Lagos [11], reported that the aim of reconfigurability is to achieve responsiveness in manufacturing systems with respect to changing market conditions. Daniyan *et al.* [12] stated that the essence of a reconfigurable fixture is to balances operator's safety and comfort with cost effectiveness, accuracy and precision, as well as smart location. Furthermore, Galan *et al.* [13] highlighted that market or customer does not necessarily impact the need for reconfigurability, but its sometimes based on companies own preference or relevance.

Thus, the concept of reconfigurability and its application in manufacturing industries was explored and investigated to develop an RVS which will be beneficial to the mineral processing industries. The RVS is defined by the authors as newly improved beneficiation equipment for use in classifying materials such as bulk granular and particulate materials and wet slurries, through the theory of reconfigurability to increase or decrease its capacity as a result enhancing the productivity of the equipment in response to ever changing customer demands. The characteristics of RMS were meticulously explained by Mehrabi *et al.* [14]. Convertibility: is the ability to easily transform the functionality of existing systems and machines to suit new production requirements. Scalability: is the ability to easily modify production capacity by adding or subtracting manufacturing resources (e.g. machines) and/or changing components of the system. Modularity: is the compartmentalization of operational functions into units that can be manipulated between alternate production schemes for optimal arrangement. Integrability: is the ability to integrate modules rapidly and precisely by a set of mechanical, informational, and control interfaces that facilitate integration and communication. Customization: is the ability to produce a particular product based on the customers' requirements, designs, specifications and configuration in order to ensure customers satisfaction. Diagnosability: is the ability to automatically read the current state of a system to detect and diagnose the root causes of output product defects, and quickly correct operational defects. These characteristics of RMS were utilized in the design and the development of RVS.

For the RVS operation, the crushed granite mineral particles obtained from the jaw crusher are processed using a RVS. The blasted mineral particles from the mine are fed through the hopper of the screen plant by means of Load Haul Dump (LHD) trucks and then transported to the crushing machines through a conveyor belt. The jaw crusher crushes the granite rocks into smaller sizes for further processing. The crushed mineral particles are transported by a conveyor belt to the screening section. The process is continuously repeated in order to compare subsequent processing variations as depicted in Fig. 1. The undersize mineral particles are stored in groups called stockpiles, while the oversized mineral particles are returned back to the crusher. Based on the inferred mineral resource generated throughout the process, industries establish their production targets based on the tonnage, mineral content and the grade of the mineral particles. When companies establish production targets it is mainly on reasonable grounds that are likely to be achieved. Change and uncertainty is a dominant factor affecting mining industries. The demand for processed mineral particles is increasing every day however in some instances they may be decrease in demand such as during the global recession. Regardless of increase or decrease in demand, costumers are still expecting to get mineral particles processed at an optimum cost and at the right time. The change, uncertainty and production targets set by the mining industries have created the need for reconfigurable or adaptable mining machineries that are able to carter for different production variations as shown in Fig. 1.

According to Wills and Napier-Munn [15], the size of the screen length should be double or three times the screen width. In situations where the space is limited an RVS will be an alternative solution. Barabady *et al.* [16], stated that a major part of the mining systems operating costs is due to unplanned system stoppages. Samanta *et al.* [17] further justified that the reduction in the downtime cost due to unnecessary machine breakdown plays a very important role in the profitability of the company. Hence, a RVS can be deployed to address some of these challenges.

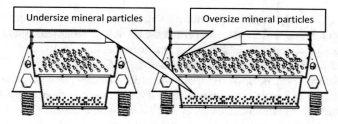

Fig. 1. Variation in processing small and large capacity.

This paper presents the current state of the newly designed beneficiation machine amidst its counterparts according to the University Research Innovation Ratings. The aim of this paper is to investigate the need for reconfigurable systems in meeting fluctuating production demands in small to medium mining industries. The paper first provides an overview of RMS and its application in the industry, then establishes production targets with relation to the challenges and trends. In addition, the paper discusses the theoretical aspects of the RVS, then the discussion continues on how each feature can solve industry problems. The paper then concludes with a technical comparative analysis and market analysis of the RVS compared to the conventional methods.

2 Methodology

Ideally in order to design a machine or technology that can meet production targets numerous number of factors have to be considered, such as its maintainability, reliability, ease of operation and last but not least safety. Figure 2 presents the features of the developed RVS.

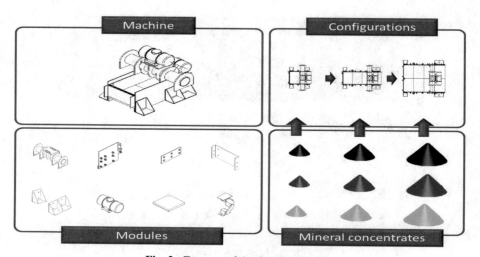

Fig. 2. Features of the developed RVS.

There are currently different technologies that exist that can also address the issue of production targets; mobile screen is currently an obvious alternative that companies consider. The market analysis of this newly designed machine was done by the Tshwane University of Technology (TUT) Research Innovation Committee.

This committee is made up of 10 experts in different disciplines such as law, engineering, management and technology innovation. This committee thoroughly investigates and assess through reconnaissance survey carried out with different experts involved in the production of products similar to the innovative product. The essence is to bench mark for the strength, weakness, opportunities and threats among its counter parts. The information obtained by this committee gives a clear indication to the potential funders of the innovative product of the product feasibility or viability when produced in the market. Proposing the design to be applied to mineral processing industry, a comparative market assessment with the existing method of mineral beneficiation was conducted by the described and aforementioned TUT Research Innovation Committee. The rating was done on a scale of 1–6, 6 being the highest rating. The rating was based on capacity, energy efficiency, reliability, versatility, cost and equipment maintenance as key performance indices. Figure 3 depicts a market analysis carried out by the TUT University Research Innovation Committee. The committee thoroughly investigated and reported that as at this present time, the capacity rating of RVS among its counterparts is 5, which makes it have high competitive strength with Pilot Crushtec and KPI-JCI Mobile Screens of the same rating. The reliability rating among its counterparts is 4, which makes it have high competitive strength with Linear Motion Screens and Exciter Driven Vibrating Screen of the same rating. The versatility rating among its counterparts is 5, which makes it have high competitive strength with Pilot Crushtec and KPI-JCI Mobile Screens of the same rating while the operating cost and equipment maintenance cost rating is 4, indicating that RVS is maintained and operated at low cost compared with its other counterparts.

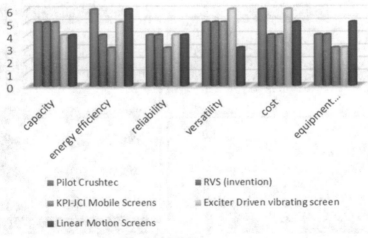

Fig. 3. Market assessment.

3   Results and Discussion

From the comparison of the efficiency and productivity of RVS and conventional vibrating screens it can be affirmed as seen in Table 1, that RVS can achieve variations in mineral volume productivity due to its variable screen structure.

Table 1. RVS model configurations aimed at meeting production targets in mining industries.

	Screen surface	Modules dimension added	Capacity (ton/h) Raw material (30 mm aperture)
1st Configuration	2000 × 1200 mm	-	30
2nd Configuration	2000 × 1600 mm	400 mm	60
3rd Configuration	2000 × 2000 mm	400 mm	80
4th Configuration	2000 × 2500 mm	500 mm	100

The screen surface of 2000 mm by 1600 mm, 2000 mm by 2000 mm and 2000 mm by 2500 mm can produce 60 tons, 80 tons and 100 tons of mineral concentrates per hour respectively, this is achieved through screen extensions modules in its width of 400 mm, 400 mm and 500 mm sequentially. This increases the mineral concentrates productivity in the mineral processing industry for 2nd, 3rd and 4th configurations of RVS achievable by 2 times, 2,7 times and 3.3 times that of the conventional vibrating screens, which is the first configuration of RVS. In view of this variation in configurations, which results in increase in mineral volume productivity, most of down time lost due to mining machine failure and unforeseen contingencies will be recovered when RVS is being utilized in mining industries. Also, the ability to meet variations in terms of decrease or increase in mineral concentrates demands at any time (t) is being achieved at very low production cost. Furthermore, the results from the market assessment indicated that the reconfigurable vibrating screen has an equal advantage over the existing conventional vibrating screens. (Note that the capacity of different modules attached to the standard RVS as different scenarios against the conventional vibrating screen may be compared).

4   Conclusion

The aim of this paper was to investigate the need for reconfigurable systems in meeting fluctuating production demands in small to medium mining industries. This was achieved through a market assessment of the proposed RVS and other existing screening technologies performed by experts. After comparing the market analysis of the existing conventional vibrating screens with the newly developed vibrating screen, the RVS proved that it is capable of adjusting its structure according to industrial requirements, thus, achieving a higher processing capacity as compared to existing conventional vibrating screens. In this regard, the RVS is considered to be a cost-effective approach and it is concluded that it is the technology to meet production targets. The issue of meeting production targets in mining industries can be addressed through the deployment of a reconfigurable beneficiation technology. Future works can test the RVS in other mineral processing industries for more performance evaluation.

Funding. The authors disclosed receipt of the following financial support for the research: Technology Innovation Agency (TIA) South Africa, Gibela Rail Transport Consortium (GRTC), National Research Foundation (NRF grant 123575) and the Tshwane University of Technology (TUT).

References

1. Bi, Z.M., Lang, S.Y.T, Verner, M., Orban, P.: Development of reconfigurable machines. Int. Journal Advance Manufacturing Technology, **39**, 1227–1251 (2008)
2. Vorster, M.C., De La Garza, J.M.: Consequential equipment costs associated with lack of availability and DT. J. Construction Eng. Manage. **116**(4), 656–669 (1990)
3. Edwards, D.J., Holt, G.D., Harris, F.C.: Predicting downtime costs of tracked hydraulic excavators operating in the UK opencast mining industry. Construction Manage. Economics **20**, 581–91 (2002)
4. Vorster, M.C., Sears, A.S.: Model for retiring, replacing, or reassigning construction equipment. J. Constr. Eng. Manag. **113**(1), 125–137 (1987)
5. Pathmanathan, V.: Construction equipment downtime costs. J. Construction Div. **106**(4), 604–607 (1980)
6. Arditi, D., Kale, S., Tangkar, M.: Innovation in construction equipment and its flow into the construction industry. J. Construction Eng. Manage. **123**(4), 371–378 (1997)
7. Nepal, M.P., Park, M.: Downtime model development for construction equipment management. Eng. Construction Architectural Manage. **11**(3), 199–210 (2004)
8. Koren, Y., et al.: Reconfigurable manufacturing systems. Annals of the CIRP **48**(2), 527–540 (1999)
9. Wiendahl, H.P., et al.: Changeable manufacturing - classification, design and operation. Annals of the CIRP **56**(1), 786–809 (2007)
10. Lee, G.H.: Reconfigurability consideration design of components and manufacturing systems. Int. J. Advanced Manufacturing Technol. **13**, 376–386 (1997)
11. Setchi, R.M., Lagos, N.: Reconfigurability and reconfigurable manufacturing systems – state of the art review. In: Proceedings of IEEE Conference on Industrial Informatics, INDIN'04, pp. 529–535, Berlin (2004)
12. Daniyan, I.A., Adeodu, A.O., Oladapo, B.I., Daniyan, O.L., Ajetomobi, O.R.: Development of a reconfigurable fixture for low weight machining operations. Cogent Eng. **6**(1), 1579455 (2019)
13. Galan, R., Racero, J., Eguia, I., Canca, D.: A methodology for facilitating reconfiguration in manufacturing: the move towards reconfigurable manufacturing systems. Int. J. Advanced Technol. **33**, 345–353 (2007)
14. Mehrabi, M.G., Ulsoy, A.G., Koren, Y.: Reconfigurable manufacturing systems: key to future manufacturing. J. Intell. Manuf. **11**, 403–419 (2000)
15. Wills, A.B., Napier-Munn, T.: Mineral processing technology: an introduction to the practical aspects of ore treatment and mineral recovery. 8th Ed., Butterworth Heinemann, UK (2006)
16. Barabady, J., Kumar, U.: Reliability characteristics based maintenance scheduling, a case study of a crushing plant. Int. J. Perform. Eng. **3**(3), 319–328 (2007)

17. Lama, O., Alayo, T., Aparicio, E., Nunura, E.: Improvement of the global efficiency of mining equipment through total productive maintenance – TPM: In: book: Advances in Manufacturing, Production Management and Process Control, Proc. of the AHFE 2021 Virtual Conferences on Human Aspects of Advanced Manufacturing, Advanced Production Management and Process Control, and Additive Manufacturing, Modeling Systems and 3D Prototyping (2021)

Manufacturing Systems

Deep Anomaly Detection for Endoscopic Inspection of Cast Iron Parts

Ole Schmedemann$^{(\boxtimes)}$ (iD), Maximilian Miotke (iD), Falko Kähler (iD),
and Thorsten Schüppstuhl (iD)

Institute of Aircraft Production Technology, Hamburg University of Technology, TUHH,
Denickestr. 17, 21073 Hamburg, Germany
`ole.schmedemann@tuhh.de`

Abstract. Detecting anomalies in image data plays a key role in automated industrial quality control. For this purpose, machine learning methods have proven useful for image processing tasks. However, supervised machine learning methods are highly dependent on the data with which they have been trained. In industrial environments data of defective samples are rare. In addition, the available data are often biased towards specific types, shapes, sizes, and locations of defects. On the contrary, one-class classification (OCC) methods can solely be trained with normal data which are usually easy to obtain in large quantities. In this work we evaluate the applicability of advanced OCC methods for an industrial inspection task. Convolutional Autoencoders and Generative Adversarial Networks are applied and compared with Convolutional Neural Networks. As an industrial use case we investigate the endoscopic inspection of cast iron parts. For the use case a dataset was created. Results show that both GAN and autoencoder-based OCC methods are suitable for detecting defective images in our industrial use case and perform on par with supervised learning methods when few data are available.

Keywords: Anomaly detection · Surface inspection · Endoscopy · Deep learning · Defect detection · Convolutional autoencoder · OCC · GAN · CNN

1 Introduction

A higher degree of automation of endoscopic inspection procedures by machine vision (MV) is often desirable, as cost and time savings are expected. In addition, the quality of inspection can be increased with MV, as manual inspection is often monotonous and therefore prone to fatigue-related errors. When inspecting cavities with visual endoscopy, the miniaturized imaging hardware (sensory, illumination) results in low image quality. Furthermore, endoscopic images are characterized by high variance due to varying relative positioning between the probe and the surface of a part. This makes the setup of classic MV systems challenging because they are based on manually engineered features.

Machine learning approaches promise to reduce the effort needed for setting up a MV system by learning relevant features from provided image data. In addition to reducing complexity in the setup phase, the generalizability of these approaches allows them to respond to high variance in the scene under investigation.

K.-Y. Kim et al. (Eds.): FAIM 2022, LNME, pp. 91–98, 2023.
https://doi.org/10.1007/978-3-031-18326-3_9

The performance of machine learning methods is highly dependent on the available data. In industrial environments, the availability of anomalous data is often limited or of insufficient quality. Collecting a large number of defective samples is costly [1]. In addition, expert knowledge is required to manually annotate the data. Anomalous data are often biased because rare defects are less common in certain positions or shapes. Others have demonstrated the applicability of deep supervised machine learning methods for automatic endoscopic inspection tasks [2, 3]. Martelli et al. [3] address the lack of suitable anomalous training data by manually creating anomalous samples. However, this is a tedious process and not reproducible for every defect type. Normal data from defect-free samples can be easily obtained in large quantities. Therefore, anomaly detection methods based solely on normal training data have proven to be a promising alternative for visual inspection tasks [4]. These methods learn the structure of the normal data. Images with defective surfaces are recognized as deviating from the learned structure.

In this work, the applicability of these methods for findings of endoscopic images in industrial surface inspection is investigated. A dataset is created from a real inspection task showing defective and normal images with surface defects such as crazes and voids. State-of-the-art anomaly detection methods are used and compared with the performance of supervised methods through the use of CNNs.

2   Related Work

2.1   One-Class Classification (OCC) for Visual Anomaly Detection

Anomaly detection refers to finding outliners patterns in data that do not correspond to a defined notion of normal data. When only normal data is used to train a classifier, then these methods are referred to as one-class classification. In this work, we focus on methods that use image data as an input. One can differ between shallow anomaly detection methods, i.e. Semi-Supervised One-Class Support Vector Machines [5], or deep anomaly detection methods. While deep methods are an active field of research, they have already shown the potential to outperform shallow methods [6]. Therefore, we focus on deep methods in this work. Often used methods are Convolutional Autoencoder (CAE) [7] or Generative Adversarial Networks (GANs) [8–10].

2.2   One-Class Classification Anomaly Detection for Industrial Inspection

Other have shown the applicability of CAEs and GANs for several inspection tasks. Liu et al. [11] used CAEs for an automated optical quality inspection task. They constructed a CAE to inspect surface defects on aluminum profiles. Tang et al. [12] investigated the applicability of OCC methods for the inspection of x-ray images of die castings. They successfully adopted a CAE and achieved a high classification accuracy of 97.45%. Kim et al. [13] uses a CAE with skip connections to inspect printed circuit boards. They achieved a high detection rate of 98% while keeping the false pass rate below 2%. GAN-based approaches have also been studied recently for industrial use cases.

However, our endoscopic inspection task differs from the presented tasks. There is more variance in our image datasets. Additionally, due to the miniaturized sensory the image quality is reduced. We intuitively assume that this implies a more challenging anomaly detection. Thus, we want to investigate the performance of state of the art OCC methods on the endoscopic use case.

3 Use Case 'Endoscopic Cavity Inspection of Cast Iron Part'

For this work, real images of a turbocharger housing from the automotive industry have been acquired. The casting must be visually inspected, including the cavities. Today, the part is manually inspected by a human worker. The complex free-form surfaces of the component make automatic visual inspection difficult because of variable distance and angle between sensor, illumination and part. Additionally, the cavity poses challenges to the inspection, such as difficult accessibility, low position accuracy, miniaturized sensory, and insufficient illumination.

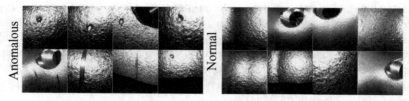

Fig. 1. Examples of the real-world images of the endoscopic images.

For the dataset, the endoscope was inserted by hand through the openings of the component. The dataset[1] consist of 1075 images split in anomalous and normal or defect-free images. Examples can be seen in Fig. 1. Both datasets show endoscopic images of the parts cavities. A 4 mm wide 90°-side view chip-on-tip endoscope is used with a 400x400 pixel resolution and an integrated LED illumination. The images in the dataset were randomly split in training, validation, and testing datasets, see Table 1.

Table 1. Split of endoscopic image surface defect dataset for model training.

	Training	Validation	Test	Sum
Anomalous	/	107	108	215
Normal	645	107	108	860

4 Proposed Approach

4.1 Experimental Setup

Multiple GAN-based approaches in the field of deep OCC anomaly detection have been published, i.e. AnoGAN [8], GANomaly [9], Skip-GANomaly [10], or DAGAN [14].

---

[1] The datasets may be requested from the corresponding author.

Others have investigated performance differences between the different architectures by investigating the performance on different datasets [4, 9, 10, 14]. But a general statement about the best performing architecture could not be derived. For this work, GANomaly and Skip-GANomaly are used. Firstly, they have achieved some of the best results in several comparative studies and secondly, the program code was provided by the authors via a source code host for reproducibility.

GANomaly. Akcay et al. [9] developed GANomaly. An Autoencoder is used as the generator. The encoder uses Leaky ReLU, Convolutional layers, and batch normalization and the decoder uses ReLU, Transposed-Convolutional layers to reconstruct the original image. Then, both the generated and the original image are mapped again into the latent representation using an encoder network. For training, the Adversarial Loss and the Contextual Loss are formed. The Adversarial Loss is used to improve the reconstruction abilities during the training. To explicitly learn this contextual information and thus capture the underlying data structure of the normal data instances, the $L1$-norm is applied to the input and the reconstructed output. This normalization ensures that the model is able to produce contextually similar images to normal data instances. The third and last part of the training function is the encoder loss. The encoder loss aims to minimize the distance between the feature representations from the input image and the generated image. The higher-level training function is ultimately composed of a weighted sum of the three distinct sub-training functions.

To identify anomalous data instances with the model in the test phase, an anomaly score is calculated. It is based on the reconstruction error, which measures the contextual similarity between the real and the generated image, as well as the similarity of the latent representation of the real and the generated image. Depending on the threshold chosen, sample data instances are classified as normal or anomalous.

Skip-GANomaly. Skip-GANomaly [10] is an improved version of GANomaly. Main modifications are added skip connections between the encoder and decoder network. Due to the direct information transfer between the layers, both local and global information is preserved and thus an overall better reconstruction of the input data is possible.

CAE. The CAE used in this work consist of the decoder core network of the Skip-GANomaly architecture. Therefore, our CAE uses likewise skip connections between each down-sampling and up-sampling layer. In total the architecture consists of five blocks with each having a Convolutional and batch-normalization layers as well as Leaky ReLU activation function. With a symmetrical setup the outputted latent representation is up sampled back to the original dimension and the image is reconstructed. For calculating the reconstruction error an $L1$ loss between the input and the reconstructed output is used.

CNN. We used a Convolutional Neural Network to create a benchmark and compare the OCC methods with supervised methods. We used an 18-layer ResNet [15] architecture as a binary classifier with one class being the anomalous and one class being the normal images.

4.2 Experimental Results

Firstly, we investigate the performance of the three aforementioned anomaly detection methods on the dataset. Anomaly scores are used to classifier a sample depending on a chosen threshold, see Fig. 2. We use the receiver operating characteristic curve (ROC-curve) to evaluate the performance of a classifier independent from the threshold. In the ROC-curve, the True Positive Rate (TPR) or recall is mapped above the False Negative Rate (FNR) by forming these metrics for different thresholds (yellow values in the right figure). The area under the ROC-curve (AUC) is used as a measure of the performance of the classification model.

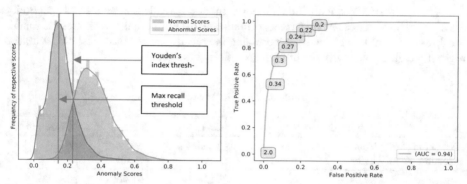

Fig. 2. Left: Histogram of normalized anomaly scores for normal and anomalous samples from the validation dataset. The two methods for threshold selection are plotted. Right: Corresponding receiver operating characteristic curve (ROC-curve) used to calculate the area under the curve (AUC). Yellow: Thresholds.

We conducted a parameter search and evaluated combinations of learning rates (2e-2, 2e-3, 2e-4), batch sizes (32, 64, 128), and input image sizes (32^2, 64^2, 128^2). The models are trained with the training data set and evaluated with the validation data set respectively, see Table 1. The highest AUC values for each method and dataset are listed in Table 2.

Table 2. Area under the ROC curve (AUC) for three anomaly detection methods.

Method	GANomaly	Skip-GANomaly	CAE
AUC	0,771	0,966	**0,973**

High AUC results indicate that the anomaly scores for the two classes differ significantly. Best results are achieved with the CAE. The GANomaly method performs worst and is not considered further in this work.

4.3   Threshold Selection and Comparison with Supervised Learning

While the AUC is a measure for the overall efficiency of a binary classification model, the choice of threshold is decisive for the use of the model in an application. We use two approaches to determine the threshold value, (a) Youden's index and (b) maximization of recall. Both approaches are plotted qualitatively in Fig. 2. The Youden's index, see Eq. 1, is defined for all points of the ROC curve, and the maximum value is used as a criterion for selecting the threshold.

$$J = \max(\text{recall} + \text{specificity} - 1) \tag{1}$$

When using the Youden's index, the assumption is made that recall and specificity are of equal importance. When it is more important that all anomalous samples are found, the threshold can be selected by maximizing the recall. Among the thresholds with the highest recall, the one with the highest specificity is selected.

Table 3. Threshold selection.

Method	Skip-GANomaly		CAE		CNN
Threshold selection	Youden's index	Max recall	Youden's index	Max recall	/
Accuracy	87.04%	**88.89%**	**92.59%**	87.96%	93.98%
Recall	80.56%	100%	91.67%	100%	88.89%
Specificity	93.52%	77.78%	93.52%	75.93%	99.07%

For the best performing model according to Table 2, both thresholds are set on the validation dataset. Subsequently, the anomaly scores for the test dataset, see Table 1, are determined with the model. Based on the selected thresholds, the quantitative measures for the assessment of a classifier are determined: accuracy, recall, and specificity, see Table 3. In this way, it is ensured that the threshold is tested on data that did not play a role in the determination of the threshold.

In order to benchmark the anomaly detection methods to a supervised method we trained a CNN. Therefore the 214 images in the validation dataset are split in an 80/20 ratio in a train and a validation dataset. We conducted a small hyperparameter study and evaluated combinations of three learning rates (1e-2, 1e-3, and 1e-4) and three batch sizes (8, 16, and 32). We trained for 25 epochs with an SGD optimizer. The best performing network was then tested on the test dataset with the results being presented in Table 3.

4.4   Discussion

The CAE performs on par with the CNN when Youden's index is used for threshold selection. This is astonishing considering that the CAE has not seen anomalous samples during training. Therefore we believe that CAEs are reasonable alternative to CNNs when few anomalous samples are available. Threshold selection is key for the application of anomaly detection methods. With both Skip-GANomaly and CAE methods we were able to train a classifier with a 100% recall or true positive rate while the specificity decreased to 77.8% and 75.9% respectively. These models could be used to presort acquired images reducing the overall scope of images that need to be examined.

Using GANs for anomaly detection has not achieved any added value for our use case. On the one hand, the trained GAN classifier performed worse or almost on the same level as the CAE. On the other hand the training effort in order to train GANs was significantly higher, because GANs are more difficult to converge, making the training process more challenging.

5 Conclusion and Outlook

In this work deep anomaly detection methods were applied to detect defects on the surface in the cavity of casting part using endoscopes. To counteract the challenge of insufficient anomalous training samples, one-class classification methods were investigated that rely solely on normal images for training. Two GAN based methods and one Convolutional Autoencoder were trained on our endoscopic dataset.

Results show high accuracy of 92.6% for the Autoencoder which outperformed the GAN-based approaches. When compared to supervised trained models, we could show that the Autoencoder performs on the same level as the trained CNN considering the small dataset used.

Our results are promising but do not indicate that OCC methods can be used without human support for the presented use case. We can identify two application potentials for the investigated anomaly detection methods. On the one hand, the models trained in this way can support a human worker during endoscopic inspection as an assistance system due to their already high correct classification rate. It should be emphasized that only a few defect-free components are required for data acquisition and that the models can therefore be trained and adapted to a task quickly and easily. A second application scenario is the partial automation of the inspection process. The demonstrated ability to pre-sort captured images can significantly reduce the number of images to be inspected by a human in the real process.

Acknowledgements. This research was funded by the German Federal Ministry for Economic Affairs and Climate Action under grant number ZF4736301LP9.

References

1. Peres, R.S., Guedes, M., Miranda, F., et al.: Simulation-based data augmentation for the quality inspection of structural adhesive with deep learning. IEEE Access **9**, 76532–76541 (2021). https://doi.org/10.1109/ACCESS.2021.3082690
2. Aust, J., Shankland, S., Pons, D., et al.: Automated defect detection and decision-support in gas turbine blade inspection. Aerospace **8**, 30 (2021). https://doi.org/10.3390/aerospace8020030

3. Martelli, S., Mazzei, L., Canali, C., et al.: Deep endoscope: intelligent duct inspection for the avionic industry. IEEE Trans. Ind. Inf. **14**, 1701–1711 (2018). https://doi.org/10.1109/TII.2018.2807797
4. Bergmann, P., Batzner, K., Fauser, M., Sattlegger, D., Steger, C.: The MVTec anomaly detection dataset: a comprehensive real-world dataset for unsupervised anomaly detection. Int. J. Comput. Vision **129**(4), 1038–1059 (2021). https://doi.org/10.1007/s11263-020-01400-4
5. Mũnoz-Marí, J., Bovolo, F., Gómez-Chova, L., et al.: Semisupervised one-class support vector machines for classification of remote sensing data. IEEE Trans. Geosci Remote Sensing **48**, 3188–3197 (2010). https://doi.org/10.1109/TGRS.2010.2045764
6. Chalapathy, R., Chawla, S.: Deep Learning for Anomaly Detection: A Survey (2019)
7. Zhou,C,. Paffenroth, R.C.: Anomaly detection with robust deep autoencoders. In: Matwin, S., Yu, S., Farooq, F. (eds): Proceedings of the 23rd ACM SIGKDD International Conference on Knowledge Discovery and Data Mining. ACM, New York, NY, USA, pp . 665–674 (2017)
8. Schlegl, T., Seeböck, P., Waldstein, S.M., et al.: Unsupervised Anomaly Detection with Generative Adversarial Networks to Guide Marker Discovery (2017)
9. Akcay, S., Atapour-Abarghouei, A., Breckon, T.P.: GANomaly: Semi-Supervised Anomaly Detection via Adversarial Training (2018)
10. Akcay, S., Atapour-Abarghouei, A., Breckon, T.P.: Skip-GANomaly: skip connected and adversarially trained encoder-decoder anomaly detection. In: 2019 International Joint Conference on Neural Networks (IJCNN). IEEE, pp. 1–8 (2019)
11. Liu, J., Song, K., Feng, M., et al.: Semi-supervised anomaly detection with dual prototypes autoencoder for industrial surface inspection. Optics and Lasers in Eng. **136**,106324 (2021). https://doi.org/10.1016/j.optlaseng.2020.106324
12. Tang, W., Vian, C.M., Tang, Z., Yang, B.: Anomaly detection of core failures in die casting X-ray inspection images using a convolutional autoencoder. Mach. Vis. Appl. **32**(4), 1–17 (2021). https://doi.org/10.1007/s00138-021-01226-1
13. Kim, J., Ko, J., Choi, H., et al.: Printed circuit board defect detection using deep learning via a skip-connected convolutional autoencoder. Sensors (Basel) 21 (2021). https://doi.org/10.3390/s21154968
14. Tang, T.-W., Kuo, W.-H., Lan, J.-H., et al.: Anomaly detection neural network with dual auto-encoders gan and its industrial inspection applications. Sensors (Basel) 20 (2020). https://doi.org/10.3390/s20123336
15. He, K., Zhang, X., Ren, S., et al.: Deep Residual Learning for Image Recognition

Classification and Detection of Malicious Attacks in Industrial IoT Devices via Machine Learning

Mohammad Shahin[✉], F Chen, Hamed Bouzary, Ali Hosseinzadeh,
and Rasoul Rashidifar

The University of Texas at San Antonio, San Antonio, TX 78249, USA
mshahin86@ymail.com, FF.Chen@utsa.edu

Abstract. The term "the Industrial Internet of Things" has become increasingly more pervasive in the context of manufacturing as digitization has become a business priority for many manufacturers. IIoT refers to a network of interconnected industrial devices, resulting in systems that can monitor, collect, exchange, analyze, and deliver valuable data and new insights. These insights can then help drive smarter, and faster business decisions for manufacturers. However, these benefits have come at the cost of creating a new attack vector for the malicious agents that aim at stealing manufacturing trade secrets, blueprints, or designs. As a result, cybersecurity concerns have become more relevant across the field of manufacturing. One of the main tracks of research in this field deals with developing effective cyber-security mechanisms and frameworks that can identify, classify, and detect malicious attacks in industrial IoT devices. In this paper, we have developed and implemented a classification and detection framework for addressing cyber-security concerns in industrial IoT which takes advantage of various machine learning algorithms. The results prove the satisfactory performance and robustness of the approach in classifying and detecting the attacks.

Keywords: Malicious attacks · Industrial IoT · Machine learning · Classification and detection

1 Introduction

Cyber-Physical Systems (CPS) are defined as systems in which a tight integration between the real-world and cyberspace exists [1]. Cyberspace is the virtual medium responsible for facilitating interconnections between users through telecommunications and computers to store, modify, or exchange data [2]. Once a CPS device is connected to the internet, it is referred to as the Internet of Things (IoT) [3]. IoT allows the interaction and cooperation of inter-networked physical objects to collect and exchange data over the Internet [4]. Advancements in IoT devices are urging traditional manufacturing systems to be integrated into cyberspace to take advantage of this emerging interaction and cooperation [5]. These systems are then can be replaced by a geographically dispersed network of services that are connected to the shop floor through the power of IoT. This spread or decentralization in manufacturing systems can help with providing

© The Author(s) 2023
K.-Y. Kim et al. (Eds.): FAIM 2022, LNME, pp. 99–106, 2023.
https://doi.org/10.1007/978-3-031-18326-3_10

more flexibility, agility, and adaptivity through a faster responsivity in processing shop floor data and thus can effectively overcome the challenges corresponding with traditional manufacturing systems. However, this higher connectivity can come at the cost of an increase in the number of cyber-attacks [6–8]. These attacks showed that given enough resources, all systems can be breached, with manufacturing systems being no exception with one in every three cyber-physical attacks happening in the manufacturing sector according to the Industrial Control Systems Monito Newsletter issued by the U.S. Department of Homeland Security [9, 10]. The rapid occurrence of such attacks on manufacturing and business operations and their information systems and the resulting damages and costs associated with them have urged scholars to consider new ways of detecting such attacks [11]. As the continuation of such efforts, we intend to show how appropriate machine learning approaches can be utilized to enhance the deterrence level of malicious attacks in industrial IoT devices in manufacturing. To this end, we have implemented a set of preprocessing and data analytics techniques on a new dataset in which various cyber-security attacks have been successfully detected via classification algorithms.

2   Background

Machine learning methods have been applied in many aspects of today's manufacturing enterprises. Many scholars are now focusing on the use of these techniques to improve cybersecurity by monitoring and conducting surveillance of real-time network streams and real-time detection of threat patterns [12]. These methods can learn from historical data and train a model to correlate events, identify patterns, and detect anomalous behavior. Apart from the algorithm implementation and development, various efforts have been put forward by researchers in this field to simulate breach scenarios and record the subsequent data. These studies have resulted in a variety of data sets existing in the field within each different pre-processing technique have been coupled. As a result, a detailed literature review is needed to summarize the state-of-the-art of the field and identify the potential areas of improvement. The following paragraphs summarize the most notable research works done in this field to date.

Terzi, Terzi & Sagiroglu [13] have used an unsupervised anomaly detection approach and Principal Component Analysis (PCA) to identify anomalies in public big network data to understand network behavior to distinguish cyber-attacks and to provide better detection in the future. Autoencoder has been used with dimension reduction to detect cyber-attack anomalies [14]. In another study, Wan et al. [15] showed that using Wavelet Neural Network (WNN) to detect anomalies in industrial control communication systems can lead to better accuracy compared to using Back Propagation Neural Network (BPNN) in addition to being more adequate in real-time analysis.

The denial of service category (DoS) in KDD CUP 1999 (KDD) and CSE-CIC-IDS2018 data sets have been used by Kim et al. [15] to develop Convolutional Neural Network (CNN) models to detect DoS intrusion attacks resulting in a high accuracy detection that ranged between 89%–99%. Wang et al. [16], McLaughlin et al.[17], and Gibert [18] have also used a CNN approach to detect malware. The latter evaluated their technique using the MalImg dataset and the Microsoft Malware Classification Challenge dataset and managed to outperform other methods in terms of accuracy and classification time.

Deep Neural Network (DNN) has been deployed to detect malware [19] on large scales data sets such as the Internal Microsoft dataset with over 2.6 million labeled samples with results for a two-class error rate of 0.49% for a single neural network and 0.42% for an ensemble of neural networks [20]. Xu et al. [21] combined DNN with Multiple Kernel Learning (MKL) to detect malware in applications run by users of Android devices. Aside from the aforementioned studies, there exist other studies that attempt to address the problem from aspects other than algorithm development. For instance, Elhabashy et al. [9] have proposed an attack taxonomy to better understand the relationships between quality control systems, manufacturing systems, and cyber-physical attacks. In another study, Wu et al. [22] have utilized anomaly detection and Random Forest algorithm to detect 3D printing and CNC milling machine malicious attacks.

3   Dataset and Methodology

In this paper, we used a dataset called "N-BaIoT" that was initially generated by Meidan et al. from network traffic patterns [23]. The initial data was gathered from nine commercial IoT devices infected by two different botnets. They have deployed two of the most common IoT botnet families namely, Gafgyt and Mirai, and collected traffic data before and after the infection. Gafgyt (also known as BASHLITE, Q-Bot, Torlus, Lizard-Stresser, and Lizkebab) is one of the most infamous types of IoT botnets. To launch an attack, the botnet infects Linux-based IoT devices by brute-forcing default credentials of devices with open Telnet ports. Mirai is the second botnet that has been deployed in this isolated network. The experimental setup included a C&C server and a server with a scanner and loader. The scanner and loader components were responsible for scanning and identifying vulnerable IoT devices, and loading the malware to the vulnerable IoT devices detected. Once a device was infected, it automatically started scanning the network for new victims while waiting for instructions from the C&C server [23]. In our analysis, we only use seven of the devices out of the nine that exist in this data set. We have implemented and chosen the most effective classifiers for this specific data set which turned out to be KNN, DT, and RF. A brief description of these algorithms is described below:

1. K-Nearest Neighbors (KNN): KNN is a supervised machine learning algorithm that can be used to solve both classification and regression problems. KNN assumes that similar data points exist nearby. In other words, similar data points are near to each other. KNN searches the entire data set for the k number of most neighbors and calculates distances for proximities before sorting the calculated distances in ascending order from smallest to largest and picking the first K with its feature that is associated with the smallest distance. KNN uses a large amount of training data, where data points are plotted in a high-dimensional space, where each axis in the space corresponds to an individual variable that characterizes that data point [24]. KNN has been used in intelligent mechanical systems to detect online fraud [25] and has been successfully implemented in a large number of business problems [26, 27].

2. Decision Tree (DT): DT is a set of rules for dividing a large heterogeneous popula-
    tion into smaller, more homogeneous groups concerning a particular output feature.
    DT is one of the most common Data Mining (DM) techniques that is widely being
    used for both classification and regression analysis. DT comes in many types of
    decision algorithms, some of which are binary trees that always produce two cat-
    egories (binary-split) at any level of the tree-like CART and QUEST. Others like
    CHAID and C5.0 are non-binary trees that often produce more than two categories
    at any level in the tree. Other minor differences exist between these four main DT
    algorithms such as, how to deal with missing value, variable selection, capacity to
    handle a huge number of classes in variables, and pruning methods [28–30]. DT has
    been used in phishing detection [31] and Adversarial detection [32].
3. Random Forest (RF): RF is a type of ensemble learning method that have been
    widely used in many fields, such as computer vision and data mining. MRF performs
    very well with a large data set in a short time compared with other techniques.
    MRF is easy to interpret and understand, can handle both numerical and categorical
    data. MRF consists of a large number of individual decision trees that operate as a
    group producing a single effect (ensemble). Each decision tree is built by randomly
    selecting observations and specific features and averaging the results at the end. Thus,
    allowing it to limit overfitting without a substantial increase in the generalization
    error [33, 34]. RF has been used to detect ransomware and achieved a high accuracy
    level of 97.74% in detecting ransomware [35]. At the same time, RF was used as a
    feature selection tool when building an Auto-Encoder Intrusion Detection System
    (AE-IDS). The results showed that using RF helped in reducing the detection time
    and effectively improved the prediction accuracy [36].

4   Results and Discussion

A 90/10 split has been used to form the training and test data sets considering the large
scale of the data set. Also, in all of the experiments, a 5-fold cross-validation has been
used for model validation. The accuracy results for each of these classifiers can be
found in Fig. 1. As one can see from Fig. 1, the algorithms have been implemented on
three different IoT devices (Ecobee Thermostat, Philips B120N10 Baby Monitor, and
Provision PT737E Security Camera) compromised by two different bots (Mirai, and
Gafgyt). The results indicate that the determining factor in the final accuracy of attack
classification is the type of bot rather than the device type. In other words, the accuracy
results show a similar pattern among three different devices compromised by a similar
bot. According to the results, for devices attacked by Mirai bot, RF algorithm delivers
the highest accuracy followed by the DT, and KNN. In particular, the accuracy achieved
by the KNN algorithm dealing with the Thermostat compromised by the Mirai bot is
the lowest among any other scenarios as this algorithm is only capable of accurately
classifying the data in 0.755426 of the test data instances. This translates to a significant
number of misclassification instances (12846 out of the 52525 instances in the test
dataset) which underlines the poor performance of this algorithm in this specific scenario.
On the other hand, for the Gafgyt bot, RF outperforms the other two algorithms while DT
performs worst among them. As opposed to the left-hand side scenarios corresponding

with the Mirai bot, even the worst-performing algorithm dealing with the Gafgyt bot (DT) is capable of accurately classifying the attacks in more than 0.99 of the test data instances.

It is important to note that even though the accuracy values for different algorithms look reasonably close, they translate to a significantly different number of misclassifications due to the large size of the dataset. This can be very critical in real-world scenarios as even a single cyber-security breach can result in a significant amount of loss from security and/ or economic points of view. The corresponding misclassification values can be found in Table 1.

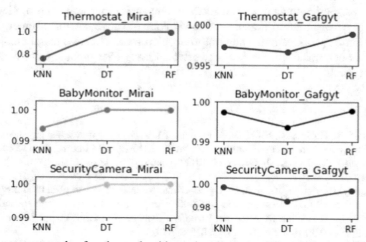

Fig. 1. Accuracy results for three algorithms detecting six different device and bot type combinations.

Table 1. Misclassification results.

Device/Attack type	KNN	DT	RF
EcobeeThermostat_Mirai	12846/52525	5/52525	0
EcobeeThermostat_Gafgyt	90/32374	109/32374	36/32374
BabyMonitor_Mirai	474/78595	4/78595	0
BabyMonitor_ Gafgyt	118/43786	321/43786	105/43786
SecurityCamera_Mirai	230/49816	8/49816	1/49816
SecurityCamera_ Gafgyt	106/39225	579/39225	221/39225

5 Conclusion

We proposed a machine learning-based framework for attack classification and detection in IIoT devices. The experiments have shown the successful adoption of artificial intelligence to cybersecurity, which has led to an effective and robust approach for identifying, classifying, and detecting two different types of botnet attacks compromising three different IIoT devices. The evaluation process has employed accuracy as a performance metric to show the effectiveness of this approach. The experiments have demonstrated that a combination of various machine learning algorithms is capable of accurately detecting and classifying the attacks in more than 99.9% of the instances in the test data set employed. Future endeavors can focus on enhancing our approach by developing deep neural network-based models and also taking advantage of other emerging IIoT data sets. Future work can also attempt to develop more effective feature engineering methods that can transform the raw network data into richer input sources for building learning methods.

References

1. Chhetri, S.R., Rashid, N., Faezi, S., Al Faruque, M.A.: Security trends and advances in manufacturing systems in the era of industry 4.0. In: 2017 IEEE/ACM International Conference on Computer-Aided Design (ICCAD), pp. 1039–1046 (2017). https://doi.org/10.1109/ICCAD.2017.8203896
2. Koppisetty, H., Potdar, K., Jain, S.: Cyber-crime, forensics and use of data mining in cyber space: a survey. In: 2019 International Conference on Smart Systems and Inventive Technology (ICSSIT), Smart Systems and Inventive Technology (ICSSIT), pp. 722–727 (2019). https://doi.org/10.1109/ICSSIT46314.2019.8987921
3. Jazdi, N.:Cyber physical systems in the context of industry 4.0. In: 2014 IEEE International Conference on Automation, Quality and Testing, Robotics, pp. 1–4, May 2014. https://doi.org/10.1109/AQTR.2014.6857843
4. Atzori, L., Iera, A., Morabito, G.: The Internet of Things: a survey. Comput. Netw. **54**(15), 2787–2805 (2010). https://doi.org/10.1016/j.comnet.2010.05.010
5. Shahin, M., Chen, F.F., Bouzary, H., Krishnaiyer, K.: Integration of lean practices and Industry 4.0 technologies: smart manufacturing for next-generation enterprises. Int. J. Adv. Manufact. Technol. **107**(5–6), 2927–2936 (2020). https://doi.org/10.1007/s00170-020-05124-0
6. Rauch, E., Dallasega, P., Matt, D.T.: Distributed manufacturing network models of smart and agile mini-factories. Int. J. Agile Syst. Manage. **10**(3–4), 185–205 (2017)
7. Elhabashy, A.E., Wells, L.J., Camelio, J.A.: Cyber-physical security research efforts in manufacturing - a literature review. Procedia Manufact. **34**, 921–931 (2019). https://doi.org/10.1016/j.promfg.2019.06.115
8. Shahin, M., Chen, F.F., Bouzary, H., Zarreh, A.: Frameworks proposed to address the threat of cyber-physical attacks to lean 4.0 systems. Procedia Manufact. **51**, 1184–1191 (2020). https://doi.org/10.1016/j.promfg.2020.10.166
9. Elhabashy, A.E., Wells, L.J., Camelio, J.A., Woodall, W.H.: A cyber-physical attack taxonomy for production systems: a quality control perspective. J. Intell. Manuf. **30**(6), 2489–2504 (2018). https://doi.org/10.1007/s10845-018-1408-9
10. ICS Monitor Newsletters | CISA. https://www.us-cert.gov/ics/monitors. Accessed 20 Oct. 20
11. Culot, G., Fattori, F., Podrecca, M., Sartor, M.: Addressing industry 4.0 cybersecurity challenges. IEEE Eng. Manage. Rev. **47**(3), 79–86, thirdquarter (2019). https://doi.org/10.1109/EMR.2019.2927559

12. Mahmood, T., Afzal, U.: Security analytics: big data analytics for cybersecurity: a review of trends, techniques and tools. In: 2013 2nd National Conference on Information Assurance (NCIA), pp. 129–134 (2013). https://doi.org/10.1109/NCIA.2013.6725337

13. Terzi, D.S., Terzi, R., Sagiroglu, S.: Big data analytics for network anomaly detection from net-flow data. In: 2017 International Conference on Computer Science and Engineering (UBMK), pp. 592–597 (2017). https://doi.org/10.1109/UBMK.2017.8093473

14. Gaggero, G.B., Rossi, M., Girdinio, P., Marchese, M.: Neural network architecture to detect system faults/cyberattacks anomalies within a photovoltaic system connected to the grid. In: 2019 International Symposium on Advanced Electrical and Communication Technologies (ISAECT), pp. 1–4 (2019). https://doi.org/10.1109/ISAECT47714.2019.9069683

15. Wan, M., Song, Y., Jing, Y., Wang, J.: Function-aware anomaly detection based on wavelet neural network for industrial control communication. Secur. Commun. Netw. (2018). https://doi.org/10.1155/2018/5103270

16. Wang, W., Zhu, M., Zeng, X., Ye, X., Sheng, Y.: Malware traffic classification using convolutional neural network for representation learning. In: 2017 International Conference on Information Networking (ICOIN), pp. 712–717. IEEE (2017). https://doi.org/10.1109/ICOIN.2017.7899588

17. McLaughlin, N., et al.: Deep Android Malware Detection, pp. 301–308 (2017). https://doi.org/10.1145/3029806.3029823

18. Gibert, D., Mateu, C., Planes, J., Vicens, R.: Using convolutional neural networks for classification of malware represented as images. J. Comput. Virol. Hack. Tech. **15**(1), 15–28 (2018). https://doi.org/10.1007/s11416-018-0323-0

19. Grosse, K., Papernot, N., Manoharan, P., Backes, M., McDaniel, P.: Adversarial perturbations against deep neural networks for malware classification (2016). arXiv:1606.04435 [cs], http://arxiv.org/abs/1606.04435. Accessed 18 Jun 2020

20. Dahl, G.E., Stokes, J.W., Deng, L., Yu, D.: Large-scale malware classification using random projections and neural networks. In: 2013 IEEE International Conference on Acoustics, Speech and Signal Processing, pp. 3422–3426, May 2013. https://doi.org/10.1109/ICASSP.2013.6638293

21. Xu, L., Zhang, D., Jayasena, N., Cavazos, J.: HADM: hybrid analysis for detection of malware. In: Bi, Y., Kapoor, S., Bhatia, R. (eds.) IntelliSys 2016. LNNS, vol. 16, pp. 702–724. Springer, Cham (2018). https://doi.org/10.1007/978-3-319-56991-8_51

22. Wu, M., Song, Z., Moon, Y.B.: Detecting cyber-physical attacks in CyberManufacturing systems with machine learning methods. J. Intell. Manuf. **30**(3), 1111–1123 (2017). https://doi.org/10.1007/s10845-017-1315-5

23. Meidan, Y., et al.: N-BaIoT: network-based detection of IoT botnet attacks using deep autoencoders. IEEE Pervasive Comput. **17**(3), 12–22 (2018). https://doi.org/10.1109/MPRV.2018.03367731

24. Samui, P., Sekhar, S., Balas, V.E.: Handbook of Neural Computation. Elsevier (2017). https://doi.org/10.1016/C2016-0-01217-2

25. Kannagi, A., Mohammed, J.G., Murugan, S.S.G., Varsha, M.: Intelligent mechanical systems and its applications on online fraud detection analysis using pattern recognition K-nearest neighbor algorithm for cloud security applications. Mater. Today: Proc. (2021). https://doi.org/10.1016/j.matpr.2021.04.228

26. Greenwell, B.B.B.: Hands-On Machine Learning with R. 2020. https://bradleyboehmke.github.io/HOML/knn.html. Accessed 17 Jun 2020

27. Cahyani, D.E., Nuzry, K.A.P.: Trending topic classification for single-label using multinomial naive bayes (MNB) and multi-label using k-nearest neighbors (KNN). In: 2019 4th International Conference on Information Technology, Information Systems and Electrical

Engineering (ICITISEE), Information Technology, Information Systems and Electrical Engineering (ICITISEE), 2019 4th International Conference on, pp. 547–552 (2019). https://doi.org/10.1109/ICITISEE48480.2019.9003944

28. Kass, G.V.: An exploratory technique for investigating large quantities of categorical data. J. R. Stat. Soc. Ser. C (Appl. Stat.) **29**(2), 119–127 (1980). https://doi.org/10.2307/2986296
29. Salzberg, S.L.: C4.5: programs for machine learning by J. Ross Quinlan. Morgan Kaufmann Publishers Inc, 1993. Mach Learn **16**(3), 235–240 (1994). https://doi.org/10.1007/BF00993309
30. Loh, W.-Y., Shih, Y.-S., Loh, W.-Y., Shih, Y.-S.: Split selection methods for classification trees. **7**(4) (1997). http://www3.stat.sinica.edu.tw/statistica/j7n4/j7n41/j7n41.htm . Accessed 17 Jun 2020
31. Zhu, E., Ju, Y., Chen, Z., Liu, F., Fang, X.: DTOF-ANN: an artificial neural network phishing detection model based on decision tree and optimal features. Appl. Soft Comput. **95**, 106505, (2020). https://doi.org/10.1016/j.asoc.2020.106505
32. Appiah, B., Qin, Z., Abra, A.M., Kanpogninge, A.J.A.: Decision tree pairwise metric learning against adversarial attacks. Comput. Secur. **106**, 102268, (2021). https://doi.org/10.1016/j.cose.2021.102268
33. Bai, J., Li, Y., Li, J., Yang, X., Jiang, Y., Xia, S.-T.: Multinomial random forest. Pattern Recogn. **122**, 108331, (2022). https://doi.org/10.1016/j.patcog.2021.108331
34. Breiman, L.: Random forests. Mach. Learn. **45**(1), 5–32 (2001). https://doi.org/10.1023/A:1010933404324
35. Khammas, B.M.: Ransomware detection using random forest technique. ICT Express **6**(4), 325–331 (2020). https://doi.org/10.1016/j.icte.2020.11.001
36. Li, X., Chen, W., Zhang, Q., Wu, L.: Building auto-encoder intrusion detection system based on random forest feature selection. Comput. Secur. **95**, 10185110 (2020). https://doi.org/10.1016/j.cose.2020.101851

Implementation of a Novel Fully Convolutional Network Approach to Detect and Classify Cyber-Attacks on IoT Devices in Smart Manufacturing Systems

Mohammad Shahin, FFrank Chen$^{(\boxtimes)}$, Hamed Bouzary, Ali Hosseinzadeh, and Rasoul Rashidifar

The University of Texas at San Antonio, San Antonio, TX 78249, USA
FF.Chen@utsa.edu

Abstract. In recent years, Internet of things (IoT) devices have been widely implemented and industrially improved in manufacturing settings to monitor, collect, analyze, and deliver data. Nevertheless, this evolution has increased the risk of cyberattacks, significantly. Consequently, developing effective intrusion detection systems based on deep learning algorithms has proven to become a dependable intelligence tool to protect Industrial IoT devices against cyber-attacks. In the current study, for the first time, two different classifications and detection long short-term memory (LSTM) architectures were fine-tuned and implemented to investigate cyber-security enhancement on a benchmark Industrial IoT dataset (BoT-IoT) which takes advantage of several deep learning algorithms. Furthermore, the combinations of LSTM with FCN and CNN demonstrated how these two models can be used to accurately detect cyber security threats. A detailed analysis of the performance of the proposed models is provided. Augmenting the LSTM with FCN achieves state-of-the-art performance in detecting cybersecurity threats.

Keywords: Smart manufacturing · Industrial IoT · Machine learning · Cybersecurity

1 Introduction

In the last two decades, there has been growing interest in smart Internet of things (IoT) devices in many applications of Industry 4.0 [1] such as smart manufacturing due to increasing the integration of cyber-physical systems (CPS) into the internet [2]. Generally, large-scale CPS networks made smart manufacturing systems that are safety-critical and rely on networked and distributed control architectures [3]. Recently, with decreasing cost of sensors and superior access to high bandwidth wireless networks, the usage of IoT devices in manufacturing systems has increased significantly [4]. Nevertheless, the implementation of IoT devices into manufacturing systems increases the risk of cyber-attacks. Therefore, the security of IoT systems has become a vital concern to businesses.

© The Author(s) 2023
K.-Y. Kim et al. (Eds.): FAIM 2022, LNME, pp. 107–114, 2023.
https://doi.org/10.1007/978-3-031-18326-3_11

According to the report of Industrial Control Systems Monito Newsletter, approximately one-third of the cyber-attacks target the manufacturing sector [5]. Furthermore, based on the National Institute of Standards and Technology (NIST), these attacks via cyberspace, target an enterprise's use of cyberspace to destroy, or maliciously control a computing infrastructure [6].

Realistic security and investigation countermeasures, such as network intrusion detection and network forensic systems, must be designed effectively to face the rising threats and challenges of cyber-attacks [7]. Today, data analytics is at the forefront of the war against cyber-attacks. Cybersecurity experts have been employing data analytics not only to improve the cybersecurity monitoring levels over their network streams but also to increase real-time detection of threat patterns [8, 9].

Neural Networks (NN) were inspired by the way the human brain works. NN algorithms are well-suited for usage in a variety of Artificial Intelligence (AI) and (Machine Learning) ML applications because they are made up of several data layers. Recurrent Neural Networks (RNNs) transmit data back and forth from later processing stages to earlier stages (networks with cyclic data flows that may be employed in natural language processing and speech recognition) [10]. RNN was used to achieve a true positive rate of 98.3% at a false positive rate of 0.1% in detecting malware [11]. In another paper, Shibahara et al. [12] utilized RNN to detect malware based on network behavior with high precision. Also, despite many advantages, one problem with RNN is that it can only memorize part of the time series which results in lower accuracy when dealing with long sequences (vanishing information problem). To solve this problem, the RNN architecture is combined with Long Short-Term Memory (LSTM) [13]. An RNN-LSTM approach has been used in intrusion detection systems to detect botnet activity within consumer IoT devices and networks [14].

LSTM [13] refers to neural networks capable of learning order dependency in sequence prediction and remembering a large amount of prior information via Back Propagation (BP) or previous neuron outputs and incorporating them into present processing. LSTM can be leveraged with various other architectures of NN. The most notable application for such network builds is seen in text prediction, machine translation, speech recognition, and more [10]. By replacing the hidden layer nodes that act on memory cells through the Sigmoid function, LSTM proposes an enhancement to the RNN model. These memory cells are in charge of exchanging information by storing, recording, and updating previous information [15].

Convolutional Neural Network (CNN) uses a feed-forward topology to propagate signals, CNN is more often used in classification and computer vision recognition tasks [10]. In a unique study, Yu et al. [16] suggested a neural network architecture that combines CNN with autoencoders to evaluate network intrusion, detection models. Also, Kolosnjaji et al. [17] proposed neural network architecture that consisted of CNN combined with RNN to better detect malware from a VirusShare dataset showing that this newly developed architecture was able to achieve an average precision of 85.6%. In conclusion, CNN has a Deep Learning (DL) network architecture that learns directly from data without the necessity of manual feature extraction.

Fully Convolutional Neural Network (FCN) is a CNN without fully connected layers [18]. A major advantage of using FCN models is that it does not require heavy preprocessing or feature engineering since the FCN neuron layers are not dense (fully connected) [19]. FCN has been used [20] to detect fake fingerprints and it was shown that FCN provides high detection accuracy in addition to less processing times and fewer memory requirements compared to other NN.

Although progress has been made to solve and decrease the risk of cyber-attacks with different machine learning models and algorithms, it is necessary to implement novel and efficient methods to keep protections updated. In this study, for the first time, we propose and compare the use of two novel models, reliable, and effective data analytics algorithms for time-series classification on a Bot-IoT dataset. The first approach is Long Short-Term Memory Fully Convolutional Network (LSTM-FCN) and the second approach is Convolutional Neural Network with Long Short Term Memory (CNN-LSTM). The results of the current study show how such approaches can be utilized to enhance the deterrence level of malicious attacks in industrial IoT devices. This paper shows how DL algorithms can be vital in detecting cybersecurity threats by proposing novel algorithms and evaluating their efficiency and fidelity on a new dataset. The next three sections discuss the preprocessing methodology of the dataset, the results and analysis of this paper, and the conclusion.

2 Preprocessing of Datasets

Network Intrusion Detection Systems (NIDS) based on DL algorithms have proven to be a reliable network protection tool against cyber-attacks [14]. In this paper, we applied state-of-the-art DL algorithms on a benchmark NIDS dataset known as BoT-IoT [11]. This dataset was released by The Cyber Range Lab of the Australian Centre for Cyber Security (ACCS) in 2018.

The Bot-IoT dataset [11] contains roughly 73 million records (instances). The BoT-IoT dataset was created by the Cyber Range Lab of UNSW Canberra. The process involved designing a realistic network environment that incorporated a combination of normal and botnet traffic. For better handling of the dataset, only 5% of the original set was randomly extracted using MySQL queries. The extracted 5%, is comprised of 4 files of approximately 1.07 GB total size, and about 3 million records, [21]. The dataset includes a range of attack categories such as Denial-of-Service (DoS), Distributed Denial-of-Service (DDoS), Operating System Scan (OS Scan) also known as Reconnaissance or Prope, Keylogging (Theft), Data Exfiltration (Theft), Benign (No attack).

This dataset contains 45 explanatory features and one binary response feature (attack or benign), only 16 of the 45 features were used as input to our models. In all conducted deep learning models, feature selection was employed when the algorithm itself extracts the important features. Furthermore, an upsampling technique [46] was used to overcome the heavily imbalanced binary response feature. The feature contained only 13859 minority counts of benign compared to a whapping 586241 majority counts of attack. Upsampling procedure prevents the model from being biased toward the majority label. The existing data points corresponding to the outvoted labels were randomly selected and duplicated into the training dataset.

Since input numerical features have different units which means that they have different scales, the SKlearn Standard Scaler was utilized to standardize numerical features by subtracting the mean and then scaling to unit variance by dividing all the values by the standard deviation [22]. DL models require all features to be numeric. For categorical features where no ordinal relationship is in existence, the integer encoding (assigning an integer to each category) can be misleading to the model and results in poor performance or unexpected results (predictions halfway between categories) as it allows the model to assume a natural ordering between categories. In this case, a one-hot encoding can be applied to the categorical representation [23].

3   Results and Analysis

To create our four main models on the dataset, two basic architectures were proposed: CNN-LSTM. The suggested CNN-LSTM architecture employs a one-dimensional convolutional hidden layer with three filters (collection of kernels used to store values learned during the training phase) and a kernel size of 32 that operates over a 1D sequence. Batch normalization is used in conjunction with the convolutional hidden layer to normalize its input by implementing a transformation that keeps the mean output near 0 and the output standard deviation close to 1. The hidden layer is used to extract features. In the hidden layers of a neural network, an activation function is employed to allow the model to learn increasingly complicated functions. Rectified Linear Activation (ReLU) was utilized in our design to improve the training performance. The ReLU is then followed by a MaxPooling1D layer, to minimize the learning time by filtering the input (prior layer's output) to the most important new output. A dropout layer was included to prevent overfitting, which is a typical problem with LSTM models. The added dropout layer has a probability of 0.2, which means that the layer's outputs are dropped out. The dropout layer's output is subsequently sent into the LSTM block. A single hidden layer made up of 8 LSTM units and an output layer are used to create the LSTM block. After the LSTM block, a Dense layer (which gets input from all neurons in the preceding LSTM output layer) produces one output value for the sigmoid activation function. The sigmoid function's input values are all real integers, and its output values are in the range of (0, 1), a binary result that reflects (benign, attack). As part of the optimization of the algorithm, a Binary Cross-Entropy loss function was used to estimate the loss of the proposed architecture on each iteration so that the weights can be updated to reduce the loss on the next iteration [24, 25].

LSTM-FCN combines the exact classification of LSTM Neural Networks with the quick classification performance of temporal convolutional layers [26]. For time series classification tasks, temporal convolutions have proven to be an effective learning model [19]. The proposed LSTM-FCN has a similar architecture to the proposed CNN-LSTM architecture but instead, it utilizes a GlobalAveragePooling1D layer to retain much information about the "less important" outputs [27]. The layers are then concatenated into a single Dense final layer with Sigmoid activation.

Both models have utilized Adam Optimization Algorithm [28] with a steady learning rate of 0.03 (the proportion that weights are updated throughout the 3 epochs of the proposed architecture). The 0.03 is a mid-range value that allows for steady learning. There was no need to optimize the hyperparameters (finding the optimal number of LSTM cells) due to the almost 0% misclassification rate of the proposed models. The default weight initializer that was used in the proposed architecture is Xavier Uniform. Since k-fold cross-validation (CV) is not commonly used in DL, here it is introduced on each model to investigate if it produces different results by preventing overfitting. Moreover, the k value is chosen as 5 which is very common in the field of ML [29, 30]. The models have utilized the StratifiedKFold [31] to ensure that each fold of the dataset has the same proportion of observations (balanced) with the response feature. In the case where k-fold CV was not introduced, the train_test_split function from Scikit-learn [32] was utilized to split data into 80% for training and 20% for testing. A summary of the accuracy and loss results for the applied models is listed in Table 1.

Accuracy describes just what percentage of test data are classified correctly. In any of these models, there is a binary classification of Attack or Benign. When accuracy is 99.99%, it means that out of 10000 rows of data, the model can correctly classify 9999 rows. Table 2 shows that very high accuracy levels (-99.99%) were achieved for the BoT-IoT datasets. The proposed LSTM-FCN models have shown slightly better performance than the proposed CNN-LSTM models in detecting attacks using the BoT-IoT dataset (100% vs 99.99%).

Table 1. Accuracy and Loss values for different methods.

Methods	Accuracy	Loss
CNN-LSTM	99.99%	0.0016
LSTM-FCN	100%	0.0068
CNN-LSTM 5-folds CV	99.99%	0.0020
LSTM-FCN 5-folds CV	100%	0.0015

The models use probabilities to predict binary class Attacks or Benign between 1 and 0. So if the probability of Attack is 0.6, then the probability of Benign is 0.4. In this case, the outcome is classified as an Attack. The loss will be the sum of the difference between the predicted probability of the real class of the test outcome and 1. Table 2 shows that very low loss values were achieved for the BoT-IoT dataset. At the same time, using 5-folds CV reduced the loss values for the FCN-LSTM from 0.0068 to 0.0015.

The Area Under the Receiver Operating Characteristics (AUROC) is a performance measurement for classification models. The AUROC reveals the model probability of separating between various classes, Attack or Benign in this case. The AUROC is a probability that measures the performance of a binary classifier averaged across all possible decision thresholds. When AUROC value is 1, it indicates that the model has an ideal capacity to distinguish between Attack or Benign. When the AUROC value is 0, it indicates that the model is reciprocating the classes. Table 2 shows a summary of AUROC values for all proposed models. The CNN-LSTM and LSTM-FCN models showed high capacity (AUROC = 1.00) of predicting Attack or Benign classes.

Table 2. Summary of AUROC values from different models.

CNN-LSTM	LSTM-FCN	CNN-LSTM 5-folds CV					LSTM-FCN 5-folds CV				
		1	2	3	4	5	1	2	3	4	5
1.00	1.00	0.500	0.500	0.500	0.500	0.500	0.998	0.976	0.987	0.993	0.998

4   Conclusions

In this paper, novel deep learning models for attack classification and detection were proposed utilizing the Industrial IoT dataset (BoT-IoT). The results revealed cutting-edge performance in terms of detecting, classifying, and identifying cybersecurity threats. The evaluation process has utilized accuracy and AUROC values as performance metrics to show the effectiveness of the proposed models on the three benchmark datasets. Deep learning algorithms were shown to be capable of successfully identifying and categorizing assaults in more than 99.9% of cases in two of the three datasets used. With the Attention LSTM block, future researchers may investigate the use of attention processes to enhance time series classification. Future research might look at whether having a similar or distinct collection of characteristics across different datasets affects the NIDS' performance using DL methods.

References

1. Shahin, M., Chen, F.F., Bouzary, H., Krishnaiyer, K.: Integration of lean practices and Industry 4.0 technologies: smart manufacturing for next-generation enterprises. Int. J. Adv. Manufact. Technol. **107**(5–6), 2927–2936 (2020). https://doi.org/10.1007/s00170-020-05124-0
2. Zheng, Y., Pal, A., Abuadbba, S., Pokhrel, S.R., Nepal, S., Janicke, H.: Towards IoT security automation and orchestration. In: 2020 Second IEEE International Conference on Trust, Privacy and Security in Intelligent Systems and Applications (TPS-ISA), Trust, Privacy and Security in Intelligent Systems and Applications (TPS-ISA), 2020 Second IEEE International Conference on, TPS-ISA, pp. 55–63 (2020). https://doi.org/10.1109/TPS-ISA50397.2020.00018
3. Baumann, D., Mager, F., Wetzker, U., Thiele, L., Zimmerling, M., Trimpe, S.: Wire-less control for smart manufacturing: recent approaches and open challenges. Proc. IEEE **109**(4), 441–467 (2021). https://doi.org/10.1109/JPROC.2020.3032633
4. Donnal, J., McDowell, R., Kutzer, M.: Decentralized IoT with wattsworth. In: 2020 IEEE 6th World Forum on Internet of Things (WF-IoT), Internet of Things (WF-IoT), 2020 IEEE 6th World Forum on, pp. 1–6, (2020). https://doi.org/10.1109/WF-IoT48130.2020.9221350
5. Elhabashy, A.E., Wells, L.J., Camelio, J.A., Woodall, W.H.: A cyber-physical attack taxonomy for production systems: a quality control perspective. J. Intell. Manuf. **30**(6), 2489–2504 (2018). https://doi.org/10.1007/s10845-018-1408-9
6. O'Reilly, P., Rigopoulos, K., Feldman, L., Witte, G.: 2020 Cybersecurity and privacy annual report. Natl. Inst. Stand. Technol. (2021). https://doi.org/10.6028/NIST.SP.800-214
7. Shahin, M., Chen, F.F., Bouzary, H., Zarreh, A.: Frameworks proposed to address the threat of cyber-physical attacks to lean 4.0 systems. Procedia Manufact. **51**, 1184–1191 (2020). https://doi.org/10.1016/j.promfg.2020.10.166

8. Mahmood, T., Afzal, U.: Security analytics: big data analytics for cybersecurity: a review of trends, techniques and tools. In: 2013 2nd National Conference on Infor-mation Assurance (NCIA), pp. 129–134 (2013). https://doi.org/10.1109/NCIA.2013.6725337

9. Gaggero, G.B., Rossi, M., Girdinio, P., Marchese, M.: Neural network architecture to detect system faults/cyberattacks anomalies within a photovoltaic system connected to the grid. In: 2019 International Symposium on Advanced Electrical and Communication Technologies (ISAECT), pp. 1–4 (2019). https://doi.org/10.1109/ISAECT47714.2019.9069683

10. Ciaburro, G.: Neural Networks with R. Packt Publishing (2017). https://libproxy.txs tate.edu/login?, https://search.ebsco-host.com/login.aspx?direct=true&db=cat00022a&AN= txi.b5582708&site=eds-live&scope=site. Accessed 18 Oct 2021

11. Pascanu, R., Stokes, J.W., Sanossian, H., Marinescu, M., Thomas, A.: Malware classification with recurrent networks. In: 2015 IEEE International Conference on Acoustics, Speech and Signal Processing (ICASSP), pp. 1916–1920 (2015). https://doi.org/10.1109/ICASSP.2015. 7178304

12. Shibahara, T., Yagi, T., Akiyama, M., Chiba, D., Yada, T.: Efficient dynamic mal-ware analysis based on network behavior using deep learning. In: 2016 IEEE Global Communications Conference (GLOBECOM), pp. 1–7 (2016). https://doi.org/10.1109/GLOCOM.2016. 7841778

13. Hochreiter, S., Schmidhuber, J.: Long short-term memory. Neural Comput. **9**(8), 1735–1780 (1997). https://doi.org/10.1162/neco.1997.9.8.1735

14. Kim, J., Kim, J., Thu, H.L.T., Kim, H.: Long short term memory recurrent neural network classifier for intrusion detection. In: 2016 International Conference on Plat-form Technology and Service (PlatCon), pp. 1–5 (2016). https://doi.org/10.1109/PlatCon.2016.7456805

15. Zhao, Q., Zhu, Y., Wan, D., Yu, Y., Cheng, X.: Research on the data-driven quality control method of hydrological time series data. Water (Switzer-land), **10**(12), 23 (2018). https://doi. org/10.3390/w10121712

16. Yu, Y., Long, J., Cai, Z.: Network intrusion detection through stacking dilated con-volutional autoencoders. Secur. Commun. Netw. **16** (2017). https://www.hindawi.com/journals/scn/ 2017/4184196/. Accessed 20 Jun 2020

17. Kolosnjaji, B., Zarras, A., Webster, G., Eckert, C.: Deep learning for classification of malware system call sequences. In: Kang, B.H., Bai, Q. (eds.) AI 2016. LNCS (LNAI), vol. 9992, pp. 137–149. Springer, Cham (2016). https://doi.org/10.1007/978-3-319-50127-7_11

18. Karim, F., Majumdar, S., Darabi, H.: Insights into LSTM fully convolutional networks for time series classification. IEEE Access **7**, 67718–67725 (2019). https://doi.org/10.1109/ACC ESS.2019.2916828

19. Zhiguang, W., Yan, W., Oates, T.: Time series classification from scratch with deep neural networks: a strong baseline. In: 2017 International Joint Conference on Neural Networks (IJCNN), Neural Networks (IJCNN), pp. 1578–1585 (2017). https://doi.org/10.1109/IJCNN. 2017.7966039

20. Park, E., Cui, X., Nguyen, T.H.B., Kim, H.: Presentation attack detection using a tiny fully convolutional network. IEEE Trans. Inform. Forensic Secur. **14**(11), 3016–3025 (2019). https:// doi.org/10.1109/TIFS.2019.2907184

21. Peterson, J.M., Leevy, J.L., Khoshgoftaar, T.M.: A review and analysis of the Bot-IoT dataset. In: 2021 IEEE International Conference on Service-Oriented System En-gineering (SOSE), Service-Oriented System Engineering (SOSE), pp. 20–27 (2021). https://doi.org/10.1109/ SOSE52839.2021.00007

22. Bishop, C.M.: Neural networks for pattern recognition. Oxford University Press (1995). https://lib-proxy.txstate.edu/login?, https://lib-proxy.txstate.edu/login?. Accessed 11 Dec 2021

23. Zheng, A., Casari, A.: Feature engineering for machine learning : principles and techniques for data scientists, First edition. O'Reilly Media (2018). https://libproxy.txstate.edu/login?, https://search.ebsco-host.com/login.aspx?direct=true&db=cat00022a&AN=txi.b5167004&site=eds-live&scope=site. Accessed 11 Dec 2021

24. Livieris, I.E., Pintelas, E., Pintelas, P.: A CNN–LSTM model for gold price time-series forecasting. Neural Comput. Appl. **32**(23), 17351–17360 (2020). https://doi.org/10.1007/s00521-020-04867-x

25. Srivastava, N., Hinton, G., Krizhevsky, A., Sutskever, I., Salakhutdinov, R.: Dropout: a simple way to prevent neural networks from overfitting. J. Mach. Learn. Res. **15**, 1929–1958 (2014)

26. Karim, F., Majumdar, S., Darabi, H., Chen, S.: LSTM fully convolutional networks for time series classification. IEEE Access **6**, 1662–1669 (2018). https://doi.org/10.1109/ACCESS.2017.2779939

27. Chollet, F.: Deep learning with Python. Manning Publications (2018). https://libproxy.txstate.edu/login?, https://search.ebsco-host.com/login.aspx?direct=true&db=cat00022a&AN=txi.b5162307&site=eds-live&scope=site. Accessed 12 Dec 2021

28. Kingma, D.P., Ba, J.: Adam: A Method for Stochastic Optimization (2017). arXiv:1412.6980 [cs]. http://arxiv.org/abs/1412.6980. Accessed 13 Dec 2021

29. Kuhn, M., Johnson, K.: Applied predictive modeling. Springer (2013). https://libproxy.txstate.edu/login?, https://search.eb-scohost.com/login.aspx?direct=true&db=cat00022a&AN=txi.b2605857&site=eds-live&scope=site. Accessed 13 Dec 2021

30. Alpaydin, E.: Introduction to Machine Learning, vol. Third edition. Cambridge, MA: The MIT Press (2014). https://lib-proxy.txstate.edu/login?, https://search.ebscohost.com/login.aspx?di-rect=true&db=nlebk&AN=836612&site=eds-live&scope=site. Accessed 13 Dec 2021

31. Adagbasa, E.G., Adelabu, S.A., Okello, T.W.: Application of deep learning with stratified K-fold for vegetation species discrimination in a protected mountainous region using Sentinel-2 image. Geocarto Int. **37**(01), 142-162 (2019). https://doi.org/10.1080/10106049.2019.1704070

32. scikit-learn: machine learning in Python — scikit-learn 1.0.2 documentation. https://scikit-learn.org/stable/index.html. Accessed 08 Jan 2022

Application of ARIMA-LSTM for Manufacturing Decarbonization Using 4IR Concepts

Olukorede Tijani Adenuga[(✉)] [iD], Khumbulani Mpofu[iD], and Ragosebo Kgaugelo Modise[iD]

Tshwane University of Technology, Staartsartillirie Road, Pretoria West, Pretoria 0183, South Africa
olukorede.adenuga@gmail.com, mpofuk@tut.ac.za

Abstract. Increasing climate change concerns call for the manufacturing sector to decarbonize its process by introducing a mitigation strategy. Energy efficiency concepts within the manufacturing process value chain are proportional to the emission reductions, prompting decision makers to require predictive tools to execute decarbonization solutions. Accurate forecasting requires techniques with a strong capability for predicting automotive component manufacturing energy consumption and carbon emission data. In this paper we introduce a hybrid autoregressive moving average (ARIMA)-long short-term memory network (LSTM) model for energy consumption forecasting and prediction of carbon emission within the manufacturing facility using the 4IR concept. The method could capture linear features (ARIMA) and LSTM captures the long dependencies in the data from the nonlinear time series data patterns, Root means square error (RMSE) is used for data analysis comparing the performance of ARIMA which is 448.89 as a single model with ARIMA-LSTM hybrid model as actual (trained) and predicted (test) 59.52 and 58.41 respectively. The results depicted RMSE values of ARIMA-LSTM being extremely smaller than ARIMA, which proves that hybrid ARIMA-LSTM is more suitable for prediction than ARIMA.

Keywords: Manufacturing decarbonization · Energy efficiency · Prediction · ARIMA · LSTM

1 Introduction

Automotive manufacturing industries are faced with new challenges in technology adoption, environmental degradation from a significant proportion of carbon emission, supply change for the digital economy, and multi-faceted sustainability drives [1]. Energy efficient system within complex automotive manufacturing has great potential for energy consumption reduction, relative to size, season, and types of manufactured components. Decision makers require fourth industrial revolution (4IR) application techniques to ensure that the industry's operational energy use is efficiently managed and decarbonized.

© The Author(s) 2023
K.-Y. Kim et al. (Eds.): FAIM 2022, LNME, pp. 115–123, 2023.
https://doi.org/10.1007/978-3-031-18326-3_12

Understanding the meaning of 'decarbonization' is important as it is a term that is used for 'reduction or total removal of carbon dioxide (CO_2) emissions [2].

Accurate forecasting is a challenging process and requires statistical or machine learning techniques with a strong capability for predicting energy consumption and carbon emission, which is part of decarbonization process planning. Time series forecasting is known for a collection of past observation data of the same variables, which can be analyzed to develop a model for future prediction [3]. The widely applied prediction model for time series (TS) data stream is a statistical technique; autoregressive integrated moving average (ARIMA), Box Jenkins methodology proposed in 1970 [4, 5]. ARIMA has been applied to any form of process challenges within the different industries to assist decision makers to plan future-based predictions on trusted applications [6–9]. ARIMA ais quite flexible in that they can models several different types of time series data, i.e. pure autoregressive (AR), pure moving average (MA) and combined AR and MA (ARMA) series, with major limitation as pre-assumed linear form of the model [3].

Machine Learning (ML) as a subset of artificial intelligence (AI) is considered for the non-linear application data pattern. ML algorithms are used for data-driven fault prediction technology [10] and consist of the process of building an inductive model that learns from a limited amount of data without specialist intervention [11]. There are various ML techniques such as convolution neural network (CNN), gated recurrent network (GCN), artificial neural network (ANN), recurrent neural network (RNN), and long short-term memory (LSTM). Recently, LSTM has transformed from a modified RNN architecture introduced in 1997 [12] and has attracted attention for its capability to capture non-linear trends and dependencies [5]. LSTM-based deep learning methods have achieved great success in artificial intelligence fields involving large datasets [13, 14].

In this paper, we propose a hybrid time series forecasting approach using ARIMA and LSTM models for an automotive component manufacturing company data using the 4IR concept, based on the following motivation: in practice, it is challenging to determine whether the study is generated from a linear or non-linear process and time series are rarely linear or nonlinear in the data patterns. Comparison of the result of the ARIMA model and ARIMA-LSTM model using predictive evaluation indicators (PMI); mean absolute error (MAE), root mean square error (RMSE), and mean percentage error (MPE) are evaluated to calculate prediction accuracy.

2   Related Work on ARIMA-LSTM

The hybrid models are becoming popular in decision-making. This is due to the combination of linear and nonlinear aspects of the data pattern, which further increases the accuracy of the predictions in an application. In recent years, it has become evident that hybrid methods yielded better results compared to a single method. A summary of studies that applied hybrid model for accurate prediction: Soy Temür et al. [15] proposed a hybrid model which consists of a combination of the linear model (ARIMA), nonlinear model (LSTM), and hybrid (LSTM and ARIMA) model to improve system performance compared to a single model. A prediction method (GA-CNN-LSTM) combines a convolutional neural network (CNN) and a long-short-term memory network (LSTM) and is optimized by a genetic algorithm (GA) [16]. Authors [17] proposed a new hybrid model using long short-term memory (LSTM), a recurrent neural network (RNN) technique, and autoregressive integrated moving average (ARIMA), as time series forecasting technique to capture live stock market data, the method performed very well compared to clairvoyant forecasting library. The work of [18] provided Indonesian governments with an accurate prediction of future exports, a hybrid model that integrates ARIMA and LSTM models based on their specialties, where LSTM was applied to the non-linear component of the data, and ARIMA was applied to the linear component of the data. The results showed that the hybrid (LSTM-ARIMA) model achieved the lowest error metrics among all the tested models. ARIMA and LSTM techniques were used to establish rolling forecast models, which greatly improve the accuracy and efficiency of demand, and inventory forecasting, while the authors proposed ARIMA and LSTM as superior to the manufacturer's empirical model prediction results [19].

The proposed hybrid-based model on deep learning methods integrates ARIMA and LSTM model to improve the accuracy of short-term drought prediction [20] and the results state that the ARIMA-LSTM model has the highest prediction accuracy [21]. In the paper titled, we present a novel hybrid ARIMA-LSTM model for automotive component manufacturing company production data forecasting considering manual operations to establish the advantages of linearity and nonlinearity, which exhibited better results than the individual models.

3 Methodology

3.1 Data Set and Processing

The study uses tier 2 automotive company-generated electricity data retrieved from National Cleaner Production Centre (NCPC) as secondary data obtained through its energy management training program. The carbon emission equivalent was derived using compared country-specific energy guidelines according to Intergovernmental Panel on Climate Change (IPCC) 2019 emission factors for coal mining [22, 23].

3.2 Auto-regressive Integrated Moving Average (ARIMA)

ARIMA model are a transformational statistical method that supports seasonality in data prediction [24], which have gained popularity among researchers due to their vast applications in manufacturing [25]. The models are known for their notable forecasting OF time series data accuracy and flexibility in different applications [26]. ARIMA models use regression equation to determine how variables respond to stochastic dissimilarity [24]. The independent variables are dependent on the lagged value of the previous values of the forecast as proposed by Box and Jenkins in the 1970s [4]. The equation is given as:

$$y'(t) = c + \varphi_i * y'(t-1) + \ldots + \varphi_i p * y'(t-p) + \theta_i * \varepsilon(t-1) + \ldots + \theta_p * \varepsilon(t-q) + \varepsilon t \qquad (1)$$

$y'(t)$ = differenced series, (p = order of lag depicted as autoregression, q is the order of error lag (moving average) and $\varepsilon(t-1)$ is residuals of past observation, φ_i is the coefficient of the first AR term, θ_i is the coefficient of the first MA term.

3.3 Long Short-Term Memory (LSTM)

Recurrent neural networks are difficult to train, as they often suffer from the exploding/vanishing gradient problem [27]. To overcome this shortcoming when learning long-term dependencies, the LSTM architecture [12] was introduced. The LSTM architecture consists of a set of recurrently connected sub-networks, known as memory blocks, the idea behind the memory block is to maintain its state over time and regulate the information flow through non-linear gating units [28]. Figure 1 is the N architecture of LSTM adopted from the work of [15]. Figure 2 depicted LSTM Structure.

3.4 Hybrid ARIMA-LSTM

ARIMA filters linear trends in the TS data and the residual values are passed to the LSTM model for training and residuals predict for the upcoming year. The LSTM has longer memory and works well for the non-stationary section of the data. LSTM works well for the non-stationary portion of the data with a relatively longer memory. LSTM's capability of capturing nonlinear patterns in TS data is one of the method's main advantages, as an attempt to overcome the challenges of obtaining an accurate forecasting model considering the intrinsic characteristics of the demand time series (being nonlinear and non-stationary).

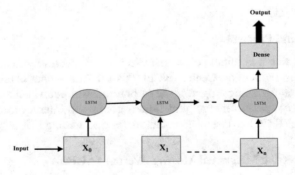

Fig. 1. N architecture of LSTM

X_t represents the input data at t time step and the output of the previous unit, h_t is the hidden units' output, while h_{t-1} is their previous output. The new memory in Eq. 3 is the LSTM unit calculated from Eq. 2

$$f_t^j = \tanh(W_{xc}x_t + W_{ht}h_{h-1} + b_c)^j \tag{2}$$

$$h_t^j = \sigma_t^j tanh\left(c_t^j\right) \tag{3}$$

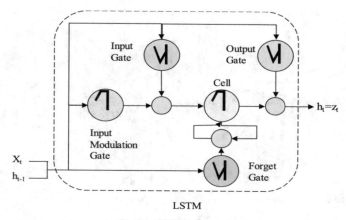

LSTM

Fig. 2. LSTM structure

3.5 Predictive Evaluation Indicators Validation

The time series prediction performance assessment was validated to evaluate the accuracy of trained models and identified efficient models between ARIMA and ARIMA − LSTM hybrid models. The root mean square error (RMSE) is a means of measuring the error in predicting quantitative data, it is a normalized distance between the observed values and the predicted values. The predictive evaluation indicators are engaged heuristically to decrease the non-absolute size of the error iteration from one step to the next.

$$\text{RMSE} = \sqrt{\frac{1}{n} \sum_{i=1}^{n} (x_1(t) - y_t(t))} \tag{4}$$

where yt is the actual value, ft is the predicted value; k is the sample size square error. In this paper, the proposed hybrid formulation is the residuals from the rolling LSTM model analysis for further optimization, given its non-linear form.

4 Results

ARIMA and ARIMA-LSTM models were used to forecast the electricity generated from coal and carbon dioxide emissions. Figure 3 is the initial visual observation of tier 2 automotive component manufacturing electricity generation. RMSE is used as an indicator for analysis comparing ARIMA as a single model and the performance of the hybrid model ARIMA-LSTM. The RMSE for ARIMA is 448.89 from the initial visual observed data in Fig. 4 and the ARIMA-LSTM actual and predicted are 59.52 and 58.41 as presented in Fig. 5. The results of ARIMA and ARIMA-LSTM predictions are depicted in Fig. 6. The RMSE values of ARIMA-LSTM are smaller than ARIMA, which proves that the hybrid ARIMA-LSTM is more suitable for prediction than the single model that is ARIMA.

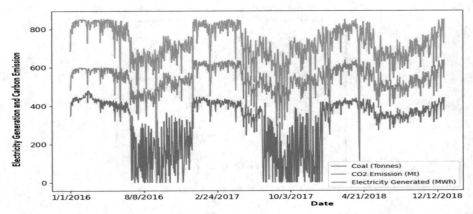

Fig. 3. Initial visual observation of tier 2 automotive component manufacturing electricity generation

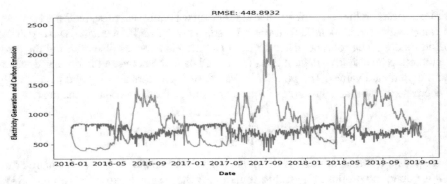

Fig. 4. ARIMA RMSE results

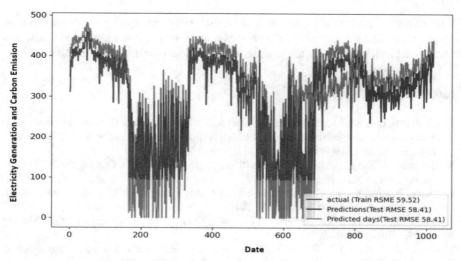

Fig. 5. RMSE for actual train data and predicted test data

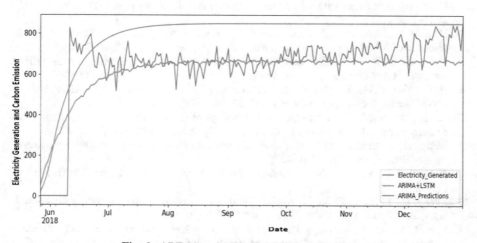

Fig. 6. ARIMA and ARIMA-LSTM predictions

5   Conclusion

The long short-term memory network (LSTM) model was used to forecast energy consumption and carbon emission within the manufacturing facility using the 4IR concept. The method captures linear and nonlinear patterns in time series data, ARIMA captures the linear features and LSTM captures the long dependencies in the data. The obtained result with hybrid models were individually compared, it was observed that they could reduce the general variance or error, even if they are unrelated. Due to this reason, hybrid models are recognized as the most successful models for forecasting tasks [15]. This information will support the decision makers for energy management and decarbonization planning. The objective to construct two models was to test which model will best fit for prediction, using the configuration that gives the lowest root mean square error.

References

1. Katchasuwanmanee, K.: Bateman, R., Cheng, K.: An Integrated approach to energy efficiency in automotive manufacturing systems: quantitative analysis and optimisation. Product. Manufact. Res. **5**(1), 90–98 (2017)
2. Romano, A., Yang, Z.: Decarbonisation of shipping: a state of the art survey for 2000–2020. Ocean Coast. Manag. **214**, 105936 (2021)
3. Zhang, P., Zhang, G.P.: Time series forecasting using a hybrid ARIMA and neural network model. Neurocomputing **50**, 159–175 (2003)
4. Box, G.E.P., Jenkins, G.M., Reinsel, G.C., Ljung, G.M.: Time series analysis forecasting and control.– Fifth edition (2016)
5. Fan, D., Sun, H., Jun, Y., Zhang, K., Yan, X., Sun, Z.: Well production forecasting based on ARIMA-LSTM model considering manual operations. Energy, **220**, 119708 (2021)
6. Sen, P., Roy, M., Pal, P.: Application of ARIMA for forecasting energy consumption and GHG emission: a case study of an Indian pig iron manufacturing organization. Energy, **116**, 1031–1038 (2016)
7. Haneen, A., Mohammed, N., Rawajfih, A.Y., Bareeq, A.A., Abeer, A.A., Fawaz, S.: On the accuracy of ARIMA based prediction of COVID-19 spread. Results Phys. **27**, 104509 (2021)
8. Phan, T.T.H., Nguyen, X.H.: Combining statistical machine learning models with ARIMA for water level forecasting: the case of the red river. Adv. Water Resour. **142** (2020)
9. Jia, L., Zou, Y., He, K., Zhu, B.: Carbon futures price forecasting based with ARIMA-CNN-LSTM model. Procedia Comput. Sci. **162**, 33–38 (2019)
10. Xu, H., Ruizhe, M., Li, Y., Zongmin, M.: Two-stage prediction of machinery fault trend based on deep learning for time series analysis. Digit. Sig. Process. **117**, 103150 (2021)
11. Stetco, A., et al.: Machine learning methods for wind turbine condition monitoring: a review. Renew. Energy, **133**, 620–635 (2019)
12. Hochreiter, S., Schmidhuber, J.: Long short-term memory. Neural Comput. **9**(8), 1735–80 (1997)
13. Zhang, N., Shen, S., Zhou, Y.: Application of LSTM approach for modelling stress–strain behaviour of soil. Appl. Soft Comput. **100**, 106959 (2021)
14. Sak, H., Senior, A., Beaufays, F.: Long short-term memory recurrent neural network architectures for large scale acoustic modeling. In: Proceedings of the Annual Conference of the International Speech Communication Association, INTERSPEECH, pp. 338–342 (2014)
15. Soy Temür, A., Yıldız, Ş.: Comparison of forecasting performance of ARIMA LSTM and HYBRID models for the sales volume budget of a manufacturing enterprise. Istanbul Bus. Res. **50**(1), 15–46 (2021)
16. Lu, W., Rui, H., Liang, C., Jiang, L., Zhao, S., Li, K.: A method based on GA-CNN-LSTM for daily tourist flow prediction at scenic spots. Entropy, **22**, 261 (2020)
17. Sakshi, K.V.A.: An ARIMA- LSTM hybrid model for stock market prediction using live data. J. Eng. Sci. Technol. Rev. **13**(4), 117–123 (2020)
18. Dave, E., Leonardo, A., Jeanive, M., Hanafiah, N.: Forecasting Indonesia exports using a hybrid model ARIMA-LSTM. Procedia Comput. Sci. **179**, 480–487 (2021)
19. Wang, C.C., Chien, C.H., Trappey, A.J.C.: On the application of ARIMA and LSTM to predict order demand based on short lead time and on-time delivery requirements. Processes, **9**(7), 1157 (2021)

20. IPCC, Global warming of 1.5°C, I.P.o.C. Change., Editor, p. 616 (2019)
21. Dea, GHG National Inventory Report South Africa, D.o.E. Affairs, Editor, Pretoria (2015)
22. Chou, J.S., Tran, D.S.: Forecasting energy consumption time series using machine learning techniques based on usage patterns of residential householders. Energy **165**, 709–726 (2018)
23. Hakeem, U.R., Raza, R., Mohsin, N.J., Muhammad, A.Z.C.: Forecasting CO2 emissions from energy, manufacturing and transport sectors in Pakistan: statistical vs. Mach. Learn. Meth. (2018)
24. Zhang, P.: Time series forecasting using a hybrid ARIMA and neural network model. Neurocomputing **50**, 159–175 (2003)
25. Hochreiter, S., Bengio, Y., Frasconi, P., Schmidhuber, J.: Gradient flow in recurrent nets: the difficulty of learning long-term dependencies. a field guide to dynamical recurrent neural networks. IEEE IEEE Trans. Neural Netw. **7**(6), 1329–1338 (2001)
26. Temür, A.S., Akgün, M., Temür, G.: Predicting housing sales in Turkey using arima, LSTM and hybrid models. J. Bus. Econ. Manage. **20**(5), 920–938 (2019)
27. Khashei, M., Reza, H.S., Bijari, M.: A new hybrid artificial neural networks and fuzzy regression model for time series forecasting. Fuzzy Sets Syst. **159**(7), 769–786 (2008)

Online Path Planning in a Multi-agent-Controlled Manufacturing System

Sudha Ramasamy[1]([⊠]) [iD], Mattias Bennulf[1] [iD], Xiaoxiao Zhang[1] [iD],
Samuel Hammar[2] [iD], and Fredrik Danielsson[1] [iD]

[1] Högskolan Väst, 46 132 Trollhättan, Sweden
sudha.ramasamy@hv.se
[2] Graniten, Uddevalla, Sweden

Abstract. In recent years the manufacturing sectors are migrating from mass production to mass customization. To be able to achieve mass customization, manufacturing systems are expected to be more flexible to accommodate the different customizations. The industries which are using the traditional and dedicated manufacturing systems are expensive to realize this transition. One promising approach to achieve flexibility in their production is called Plug & Produce concept which can be realized using multi-agent-based controllers. In multi-agent systems, parts and resources are usually distributed logically, and they communicate with each other and act as autonomous agents to achieve the manufacturing goals. During the manufacturing process, an agent representing a robot can request a path for transportation from one location to another location. To address this transportation facility, this paper presents the result of a futuristic approach for an online path planning algorithm directly implemented as an agent in a multi-agent system. Here, the agent systems can generate collision-free paths automatically and autonomously. The parts and resources can be configured with a multi-agent system in the manufacturing process with minimal human intervention and production downtime, thereby achieving the customization and flexibility in the production process needed.

Keywords: Online path planning · Plug & Produce · Multi-agent systems · Path planner service

1 Introduction

The manufacturing industries are in need of change in the production from producing high volume and low variety to high volume and large variety of customization. This demand increases to adopt the flexibilities for customizing their products, which leads to flexibility in the manufacturing systems [1]. There are several concepts to achieve flexible manufacturing in production. The manufacturing systems have encountered a variety of evolutions. Dedicated Manufacturing Systems (DMS) are expensive to adopt since it requires a lot of time for any change in the manufacturing product specification.

© The Author(s) 2023
K.-Y. Kim et al. (Eds.): FAIM 2022, LNME, pp. 124–134, 2023.
https://doi.org/10.1007/978-3-031-18326-3_13

Because they are manufactured to produce only a specific type of product for the whole lifetime. This enables the manufacturer to manufacture a low-cost product with a huge volume of quantity or mass production. These types of solutions are well established in an optimized way to produce high throughput and energy-efficient production rates for their lifetime. If the manufacturer wanted to change the specification of a product during the production, DMS cannot accommodate any change in the production as these systems are static [2].

Flexible Manufacturing Systems (FMS) are built to produce a range of similar products which can be realized through programmable devices, e.g., Programmable Logic Controllers (PLC's) and industrial robots. PLC's and the programming flexibilities gives a manufacturer the possibility to manufacture a certain range of similar products [3, 4]. Since these systems are bulky and have low utilization of their functionalities due to certain products, it leads to high initial cost and less production rate compared to DMS. Even though researchers are working with many variants to improve the production rate in FMS it is limited due to its basic design. To overcome these constraints in the manufacturing system, researchers are working towards Reconfigurable Manufacturing Systems (RMS) [5]. RMS can prove to have a combination of a higher production rate closer to DMS but still with the flexibility of FMS. Reconfigurability can be achieved by the application software, design variables and Human-machine interface (HMI) [6]. Reconfigurability of RMS includes its wide range of subsystems to make a complete automation. Here the subsystems are for example machining, machine tools, fixturing, assembling, material handling, and control. To achieve reconfigurable production systems in the industry, often knowledge of external competence is required which most of the manufacturers are lacking.

To overcome these issues, the implementation of the Plug & Produce (P&P) concept using Multi-agent System (MAS) is proposed. In the manufacturing process when mass customization is required, P&P concept is adopted to add or remove a module in a quick and easy manner. P&P concept means it addresses the physical aspect of flexibility and allows a quick reconfiguration of the process modules. Hence, the system can easily adapt to the new production situation based on the new product variants and new customization. To enable this P&P concept, a multi-agent system concept is proposed [7]. The use of MAS technology addresses the flexibility in terms of software that can be logically controlled [8]. It contains distributed individual intelligent agents throughout the manufacturing process which can negotiate (interact) with each other to make decisions [9, 10]. Thereby these agents can solve more complex tasks by negotiating new solutions, which a single agent cannot solve.

The agents can be represented as parts to be processed in some manufacturing process and resources, such as a robot or a gripper. A simplified example of the Plug & Produce concept using MAS is shown in Fig. 1, here the agents can communicate to other agents to perform the transporting task. For example, the robot would first attach the gripper tool that in turn grips the part at station 1, then move the part to station 2. Within this type of solution, the agents representing robots can request a robot path in order to fulfil a goal of transportation [11]. These paths must be automatically generated and deployed online to facilitate a fully autonomous system, otherwise, all paths must be manually created and stored in advance. Online meaning is that it can generate a path instantly

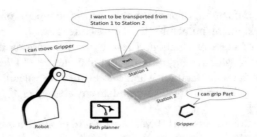

Fig. 1. Example of a Plug & Produce system using agents to perform an operation.

or during the production when the agent requests it. In this article, a path planner agent is described that automatically deploys a path when a start and goal point is given. For path planning, it uses the sampling-based algorithm Rapidly exploring Random Tree (RRT) [12]. The RRT algorithm is validated by Ramasamy et al. previously to be used as a path planning algorithm in Robotstudio simulation environment [13, 14] and tested with P&P environment [15] and a sample of generated path is shown in Fig. 2.

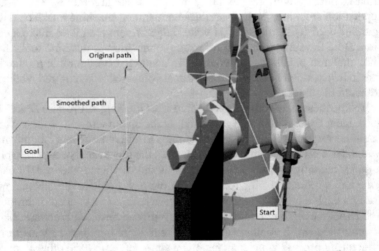

Fig. 2. A sample of a generated path from the path planner [11].

In this paper, a configurable multi-agent system is used (C-MAS). This means that agents are configured rather than programmed. The configurations contain the configuration classes and configuration entities defined in a configuration tool. Figure 3 shows a simplified example of how an agent configuration can be described in a configuration tool.

Parts and resources can have variables; in this Fig. 3 one variable is defined that identify the position of the parts and resources. Also, more complex variables such as CAD geometries could be also included. Interfaces are defined to identify the compatibility for connecting agents together, in this case: the part, gripper and robot. The part has a goal i.e., part has to be moved from station 1 to station 2. The goal needs a process plan to be reached. Process plan means the list of skills to be executed to achieve the goal. The process plan contains demands for interfaces, for example to grip a part, it needs a gripper interface. The interface has an instruction to run a skill that must be executed to perform that gripping operation. Again, the skill has a process plan to complete the gripping action. The motivation towards this project is achieving an agent-based path planner in the P&P environment. The reason for the path planner in the P&P environment is, we move the process modules around the cell depends on the needs and applications. For that purpose, we need an automatic path generator to generate the path instantly based on the agent requirements. This requires updated model of the P&P environment. In a future approach the simulation could be updated online based on incremental changes in the layout of the cell. In this article, an agent-based path planner is implemented as one of the agents in a multiagent system. For generating the path RRT path planning algorithm is employed. When a robot agent requests for a path, the path planner agent will fetch a path from the database if it is already generated for the same location with same environment, otherwise the path planner will generate a collision free path from the implemented RRT algorithm.

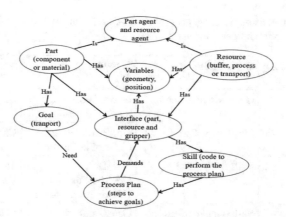

Fig. 3. Agent configuration classes

2   Manufacturing Scenario

A manufacturing scenario is created for a simple task of transportation with four agents. One agent represents the robot (it is a robot), the second one is a part agent (the part to be transported), the third one is a path planner agent (the path to be generated based on the start and goals points), and the fourth one is the gripper agent to grip the part. There are two stations, and it is stationary where the part can be moved from one station to the other station. If it is not a stationary station, these two stations may also be represented as an agent in MAS. But here it is assumed as a stationary object, so it has the same coordinate system as the robot. All the four mentioned agents can be configured to have their own goals for parts, skills for resources and all the agents have variables and interfaces. For example, when configuring the gripper agent, it is initialized with variables that define the coordinates for pick and place positions for the gripper to perform any transporting operation. Interfaces are where the agents connect and interact with each other. Skills for resources and their variables for interacting with other agents are also located on the interface. When we use a part with a resource, then the part agent and resource agent should have a matching interface, so that these two agents can interact with each other. Here the part is having a goal for transporting operation. To achieve this transporting operation, part search for a process plan. The process plan contains the set of instructions to perform the goal. The first instruction must call a robot gripper to perform the movement skill. From the move skill the gripper needs a robot to move the gripper. It finds the robot and lets it move the gripper. The robot realizes that it needs a path to move from station1 to station 2. Now the robot calls the path planner that has a skill of path generation, when the path generation skill is performed, a path is generated in the path planner agent, and it is sent to the robot agent where the robot performs the skill of transportation. When the robot agent calls the skills, it sends with a variable called *PathID*.

Table 1. A simplified example of agent configuration to implement the manufacturing scenario with skills for resources and goals for parts.

Agent (r = resource, p = part)	Goals/Skills	Variables	Interfaces
Part (p)	Goal: Move to station 2	Station1Position, Station2Position	GripInterface
Gripper (r)	Skill: MovePart	GripPartPosition, UnGripPartPosition	GripInterface TransportInterface
Robot (r)	Skill: MoveTool	PathID, GripToolPosition, UnGripToolPosition	TransportInterface PlannerInterface
Path planner (r)	Skill: GeneratePath	GeneratedPath, PathID, StartPoint, GoalPoint	PlannerInterface

The path planner saves all the generated paths with the corresponding *PathID*. Each time a new path is requested the path generator checks with the *PathID* information along with its environment in the database. If already a path is generated for this environment, it will retrieve the same path from the database otherwise path planner generates a new path. The process plan for transportation must find the position to grip the object. Table 1 shows a simplified example of how the configuration of each agent could be configured for this scenario. Here the coordinates of two stationary stations are given as variables in the robot agent.

The Part agent will communicate to the GripInterface on the gripper agent so that it can grip the part. This is done by setting the inputs *GripPartPosition* and *UnGripPart-Position* on the gripper based on the *Station1Position* and *Station2Position*. Any needed translation of coordinates can be defined on the process plans for goals and skills in each agent. The gripper finds the robot on the interface *TransportInterface* and forwards the positions to the *GripToolPosition* and *UnGripToolPosition*. The robot needs a path and connects to the path planner on interface *PlannerInterface* and calls the skill *GeneratePath* with *PathID*, *StartPoint* and *GoalPoint*. If needed, the gripper could also hold its tool data and each agent could have attached CAD geometry data to be used in the path planner. When a path is generated, the robot can access the *GeneratedPath* on the path planner through the connected *PlannerInterface*.

3   Design and Implementation

The path planner is implemented as an add-in software for ABB RobotStudio, implemented in C# using Microsoft Visual Studio. Figure 4 shows the overview of path planner code structure. The path planner was divided into three parts: "Extensible Markup Language (XML) files", "Main functions", and "Supporting functions and classes". The XML files are used to create add-in control buttons in RobotStudio, where these buttons can easily be used by humans. The main functions contain "PathGeneratorAddIn" and "PathPlanner". The PathGeneratorAddIn executes commands triggered by the controls defined in XML files and the PathPlanner contain the path planning algorithm for transportation as the main function. The supporting functions and classes contain a function for creating a range of workspace, a collision detection module, a path smoother and a function to save the program. The collision detection function was enabled by simulating the robot's movement in its configuration space which is created in the RobotStudio simulation environment.

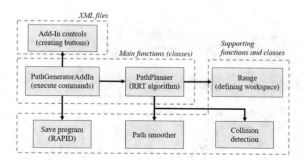

Fig. 4. Overview of the path planner code structure.

A sampling-based RRT algorithm is used to generate a collision-free path for transportation. The path planner is implemented as an agent. The path planner is in this paper connected to a multi-agent system and the requirements for this path planner agent are:

- Start and goal points should be fetched automatically from other agents.
- Each path generated should be saved automatically.
- Expecting a feasible path solution always if the one exists.

- Path planners can have any path planning algorithm such as RRT, since we use path planners as an agent.

Based on these requirements the proposed system is shown in Fig. 5. In the agent system, the path planner agent has a skill which is called as *GeneratePath* and robot agent has a skill which is called as *MoveTool*. The path planner service input and output are written in JavaScript Object Notation (JSON) [16] format. An agent can request a path by sending start and goal points to the path generator. The path generator generates a path, forwards the path to a requesting agent and saves the generated path in the path database along with start and goal points and information about the environment. The information about the environment means it contains robot position, tools, modules, and fences. When a new path is requested always the path generator compares checks a path that is saved already in the database in comparing with start and goals points and information about the environment. If everything matches it will retrieve the stored path and send it to the agent. If any one of the parameters is not matching, the path generator will generate a new path based on the data provided and send it. Thereby if it needs to generate the same path again it saves the computational time of the path planning algorithm. Any type of path planning algorithm can be easily exchanged or modified in the path generator, it doesn't require to change any communication between the agents and path planner or between the agents. Here the limitation is the path generator is implemented in RobotStudio, so it can accommodate only ABB robots. Agent based Path planner service contains a path generator (the main program), inputs in JSON format and output (generated path) in text format which is shown in Fig. 6. Path database contains the path folder and path input storage. The path folder contains the module file and program (PRG) file. Module file contains tool data, robot targets and move instructions.

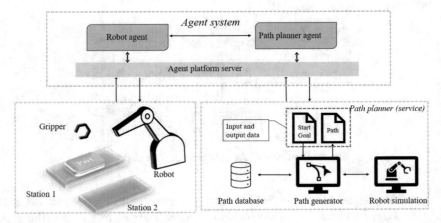

Fig. 5. Proposed system design where the agent system is used to communicate between a robot (on the left) and a path planner service (on the right).

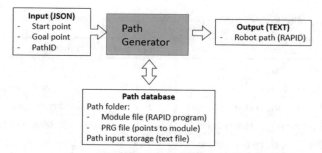

Fig. 6. Agent-based path planner service

As long as we have a module file in the path folder the content of PRG file will not change. It means that the module file only needs to be sent as output to the agent (path in the text format). Input consists of start point, goal point, path identifier (Path ID) and it is sent to the agent system in the format of JSON data strings. The start and goal points are represented as position (x, y, z) and orientation (r_x, r_y, r_z) of the points. The path ID is a string value and output is in the form of RAPID (ABB robot's programming language) program which can be sent as a text file to the agent system. When a path is generated, it will be saved in the path database with the following structure shown in Table 2. Each path input corresponds to 13 lines of text. Those are the start point comprises of 6 lines i.e., 3 for position and 3 for orientation, similarly goal point comprises of 6 lines and the last line carries the path ID. Then the new path information will be appended and so on. Use of visual studio project properties "Post build event command line" enables the path generator code will be automatically saved in to RobotStudio API reference. This procedure copies the path generator add-in's DLL file to RobotStudio's add-in directory whenever the solution is rebuilt in Visual Studio. After configuring the agents, the communication between the path planner and the agent system is tested with a simplified scenario. The scenario is created in RobotStudio with robot installed in it, not with the actual robot connected to it. So, the part variables and robot variables are not needed for this simplified test. The goal is, path planner must generate a path upon the request from the part agent. For testing, part agent, path panner agent and the robot agent must be online in the agent platform server. Steps to perform the test is given below.

- Part agent finds its goal and search for process plan (goal is transportation).
- Process plan demands an interface with a skill (skill is to move part from station 1 to station 2).
- Gripper requests assistance from the robot.
- Robot agent requests for a path from path planner.
- Path planner needs data (start and goal points with path ID) to generate a path.
- Robot agent retrieves data from part agent and sends to path planner.
- Path planner checks for a path from database, if exists sends it otherwise generate and sends to robot agent.
- The robot agent confirms that it received a path by printing a short message.

Table 2. Generated path storage structure in path database

Line no	Value name	Line no	Value name
1	Start point1, x-position	10	Goal point1, rx-orientation
2	Start point1, y-position	11	Goal point1, ry-orientation
3	Start point1, z-position	12	Goal point1, rz-orientation
4	Start point1, rx-orientation	13	Path1 identifier (ID)
5	Start point1, ry-orientation	14	Start point2, x-position
6	Start point1, rz-orientation	15	Start point2,y-position
7	Goal point1, x-position	...	...
8	Goal point1, y-position	26	Path2 identifier (ID)
9	Goal point1, z-position	...	...

When the agents can communicate, the part agent can request the gripper agent for transportation, the gripper can get assistance from the robot to move the gripper, the robot agent can request the path planner for a path to perform transportation, the path planner can generate a path based on the algorithm implemented in it and sends to the robot agent. The above steps i.e., the communication and path generation between the agents was established and is validated by looking at the output from the path planner to see if it was corresponding to the data that was expected by the agent system. Thus, this paper has designed and implemented a functioning path planner agent.

4   Conclusion

This paper presents a path planner agent that can act as a service in a manufacturing cell. The path planner agent is developed and implemented successfully with an agent system for the simple manufacturing scenario presented. The agent system presented in this paper was used for communication and to store configured data such as the positions for stations 1 and 2. A robot agent was designed that can communicate with the path planner through the agent network and request a path to be created. The path planner is reusing paths by saving any previously generated path along with its environmental data and path id. This reduces the computational time of path generation and in some cases eliminated, since the path planner agent is fetching a path for the environment if the path exists. We found that the implemented path planner agent can negotiate, interact and communicate with other agents to generate a path, thereby it completes the goal of transportation.

As of now, we have modelled a robot in robot studio to test this agent concept and implemented a sampling based RRT algorithm as a path planning algorithm in the path generator. Currently, working on establishing online communication between the robot agent and a physical robot. For future work, we will investigate adding other sampling-based optimized path planning algorithms in the path planner agent. When a robot agent requests to generate a path, the path planner agent will be able to choose the algorithm based on the needs for example less computational time or a less energy-consuming path.

Acknowledgements. The work was carried out at the Production Technology Centre at University West, Sweden and funded for PoPCoRN project, Dnr. 20200036 by KK-stiftelsen, Sweden. Their support is gratefully acknowledged.

References

1. Pedersen, M.R., et al.: Robot skills for manufacturing: from concept to industrial deployment. Robot. Comput.-Integr. Manuf. **37**, 282–291 (2016)
2. Zhang, G., Liu, R., Gong, L., Huang, Q.: An analytical comparison on cost and performance among DMS, AMS, FMS and RMS. In: Dashchenko, A.I. (eds) Reconfigurable Manufacturing Systems and Transformable Factories. Springer, Heidelberg (2006). https://doi.org/10.1007/3-540-29397-3_33
3. ElMaraghy, H.A.: Flexible and reconfigurable manufacturing systems paradigms. Int. J. Flex. Manuf. Syst. **17**, 261–276 (2005)
4. Zeballos, L.J.: A constraint programming approach to tool allocation and production scheduling in flexible manufacturing systems. Robot. Comput.-Integr. Manuf. **26**, 725–743 (2010)
5. Galan, R., Racero, J., Eguia, I., Garcia, J.M.: A systematic approach for product families formation in Reconfigurable Manufacturing Systems. Robot. Comput.-Integr. Manuf. **23**, 489–502 (2007)
6. Papcun, P., Kajáti, E., Koziorek, J.: Human machine interface in concept of Industry 4.0. In: 2018 World Symposium on Digital Intelligence for Systems and Machines (DISA), pp. 289–296 (2018)
7. Bennulf, M., Danielsson, F., Svensson, B., Lennartson, B.: Goal-oriented process plans in a multiagent system for plug & produce. IEEE Trans. Ind. Inform. **17**, 2411–2421 (2021)
8. Karnouskos, S., Leitão, P.: Key contributing factors to the acceptance of agents in industrial environments. IEEE Trans. Ind. Inform. **13**, 696–703 (2017)
9. Rocha, A.D., Tripa, J., Alemão, D., Peres, R.S., Barata, J.: Agent-based plug and produce cyber-physical production system – test case. In: 2019 IEEE 17th International Conference on Industrial Informatics (INDIN), pp. 1545–1551 (2019)
10. Wooldridge, M., Jennings, N.R.: Intelligent agents: theory and practice. Knowl. Eng. Rev. **10**, 115–152 (1995)
11. Hammar, S.: Automated path planning for supporting autonomous industrial robots in multi-agent systems (2021)
12. Karaman, S., Frazzoli, E.: Sampling-Based Algorithms for Optimal Motion Planning (2011)
13. Operating manual RobotStudio. https://library.e.abb.com/public/244a8a5c10ef8875c1257b4b0052193c/3HAC032104-001_revD_en.pdf. Accessed 04 Oct 2021
14. Abreu, P., Barbosa, M.R., Lopes, A.M.: Virtual experiment for teaching robot programming. In: 2014 11th International Conference on Remote Engineering and Virtual Instrumentation (REV), pp. 395–396 (2014)

15. Ramasamy, S., Zhang, X., Bennulf, M., Danielsson, F.: Automated path planning for plug produce in a cutting-tool changing application. In: Presented at the 24th IEEE International Conference on Emerging Technologies and Factory Automation (ETFA) (2019)
16. Bassett, L.: Introduction to JavaScript Object Notation: A To-the-Point Guide to JSON. O'Reilly Media, Inc. (2015)

Assessing Visual Identification Challenges for Unmarked and Similar Aircraft Components

Daniel Schoepflin[✉], Johann Gierecker, and Thorsten Schüppstuhl

Institute of Aircraft Production Technology, Hamburg University of Technology, TUHH, Denickestr. 17, 21073 Hamburg, Germany
Daniel.Schoepflin@tuhh.de
https://www.ifpt-tuhh.de

Abstract. Highest demands for complete traceability and quality control of each component, require thorough identification of each produced, replaced, and (dis-)assembled aircraft component. As many production and MRO-processes for modern aircraft remain to be carried out manually, this poses a great challenge. Many small components either do not feature a Part Number or in MRO-processes their Part Number is occluded or not readable due to dirt and wear. Considering unmarked components with a high resemblance to one another and few characteristics, e.g. standard parts such as bushings and pipes, manual identification is an error-prone task. Avoiding errors through digitalized procedures has the potential to significantly reduce error rates and costs for a typical manual dual control. However, automated identification of components has to overcome the high classification complexity that originates in the manifold of aircraft components and is additionally increased by individualistic MRO modifications for specific aircraft. This work presents a methodological approach to reveal possible challenges for identification procedures and gives special focus to the assessment of similarities between components. Two similarity metrics are introduced that are calculated either through feature-based analysis or through 3D-shape similarity assessment. The methodology is demonstrated with two to this date unsolved Use-Cases that represent different challenges of visual identification systems for similar and unmarked components.

Keywords: Visual sensor applications · Similarity of objects · Identification challenges · Object classification

1 Introduction

Despite recent advantages, modern aircraft production and maintenance are mainly performed by manual assembly [1]. Highest demands for complete traceability and quality control of each assembly step, require high efforts in process supervision. One of those supervision necessities is checking, whether the correct component is chosen for the next assembly step. Typically, such verifications can

© The Author(s) 2023
K.-Y. Kim et al. (Eds.): FAIM 2022, LNME, pp. 135–145, 2023.
https://doi.org/10.1007/978-3-031-18326-3_14

be digitized through the help of markers, e.g. RFID or 2D/1D-Codes. However, many components in the aviation industry either are not permitted to wear such codes or cannot bear such codes due to their surface area being a functional area. Therefore, manual identification of components is necessary for various stages of a (dis-)assembly. With manual identification being error-prone, this poses a great factor for process instability. This applies even more so if the components have a high degree of similarity to each other.

Due to the necessary waiver of markings, visual and markerless identification have to be employed. Such approaches are feasible for distinctive and feature-rich aviation components [3]. Within this work, we focus on components that wear little to no features, and have high similarities. Due to the highly individualistic nature of such identification problems, solutions are hardly transferable between problems. We, therefore, aim to contribute a transferable baseline for analysis of those problems. A methodology is presented that guides through the assessment of challenges for such identification systems. Special focus is given to the assessment of similarity and task identification complexity.

2   Related Work and State of the Art

Designing visual sensor applications for industrial processes is a task that requires expert knowledge both about sensor systems as well as the application domain [1,6,9]. To assist in the design process, methodologies have been developed that guide in the selection of sensors: The approaches of sensor planning methodologies augment the selection process towards configuration parameters such as the extrinsic pose of the camera and illumination of the scene. A sensor planning system outputs to the user camera pose, optical settings, and illuminator settings [4]. Sensor planning approaches have been developed for specific visual applications such as surface inspection [5]. They contribute a flexible yet automated planning pipeline that is motivated by the trend of individualistic customization of products.

In the domain of aircraft production [1] proposed an assistance approach to design visual sensor applications for assembly supervision. This specifically considers the assembly task-specific generation of viewpoint candidates that allow detecting the successful assembly operation. They augmented that approach by adding an automated enablement through AI-based processing trained with synthetic data. This synthetic data is generated with the same 3D models and camera-specifics used to configure the system. As they targeted assembly situations that follow pre-defined assembly patterns, the object recognition and localization tasks are well defined. Focusing on the distinction of different assembly situations [13] introduced a geometric analysis of assembly situations to derive view points for assembly supervision. Such approaches can be further improved by extending the metric for suitability of vision view-points towards similarity of considered components.

Focusing on the ability to detect aircraft components in production supplying logistic operations with delivery units, [2,3] provide the capability to enable

AI-based visual sensor applications with the help of synthetic training data. Such an approach can be incorporated into the design flow of this paper, however, is limited to components that can be differentiated through means of object detection and a top-view sensor configuration.

Addressing the challenges of identification of similar appearing car parts, [11] introduced a classification box that utilizes Deep-Learning based image processing, enabled by synthetic data. Focusing on the similarity of objects, [12] found the use of CNN-based image processing in principle applicable to such challenging situations. However, it is necessary to determine for which identification tasks special considerations have to be given. We, therefore, contribute an analysis methodology that focuses on similarity analysis. Assessment of similarities between 3D-shapes can be done by various methods [8]. We incorporate two approaches and derive a similarity metric.

3   Assessing Identification Challenges - Methodology

As shown in the previous sections, the selection of suitable technologies for identification tasks has to consider multiple parameters of each sensor and algorithm type. Mapping those parameters and the resulting abilities to individualistic identification problems, requires revealing the individualistic facets of each identification problem. For this analysis, a methodological approach is proposed that assesses requirements based on the component spectrum, reveals challenges, and defines the identification task complexity:

1. Revealing component features
2. Scale of geometric features
3. Assessing the similarity of components and their features
4. Using iterative subdivisioning to reduce the identification task complexity

The following sub-sections detail the steps of this methodical analysis approach. For each step several to be analyzed aspects are explained, and it is discussed how and when these aspects affect other steps or technology selection criteria.

Step 1: Revealing Component Features and their Value. Analyzing the component spectrum is the main driver of this methodological approach. As such the output of this step is the main input for subsequent analysis steps and may be re-visited whenever necessary information is missing in follow-up steps. With this step, highly individual outputs for each identification task are expected. It is therefore not possible to provide a comprehensive list that can be applied universally. In general, it can be differentiated between qualitative and quantitative or measurable features. Most of the qualitative features can be translated into one or more of several quantitative features (e.g. a button can be attributed to geometric and color features). Geometric features are mainly described by (1) surface features: planes, spheres, cylinders, cones, and free-form surfaces, and (2) curve features: lines, circles, ellipses, paraboles, splines [13].

This work addresses visual identification. The appearance of objects concerning features such as texture and colore schemes are mainly attributed to the surface of the object. Such features can be quantified through distribution maps and histogram analysis.

After features for each component are extracted, it may be beneficial to reflect those features against the entire component spectrum. This yields information with respect to the proper rating of suitability for different applications as well as the subsequent analysis. The main challenge, which is to be addressed in the following step, is whether these features can be detected by sensors and assessing the similarity of components.

Step 2: Scale of Geometric Features. After geometric features that allow for possible unique differentiation between components are identified, it has to be assessed whether those features can be detected and measured by sensors. Considering mainly geometric features, two parameters are relevant for that assessment, first the resolution r of the sensor and second the geometrical manufacturing tolerances of the component feature m. If the extreme values of the manufacturing tolerances and sensor resolution combined are greater than the range between two adjacent components feature values d, unique identification is not possible. Therefore the following has to hold:

$$\max(r) + \max(m) \leq \frac{d}{2} \tag{1}$$

The criterion to evaluate the suitability of sensors, therefore, is, whether this inequality holds.

Step 3: Assessing the Similarity of Components. The similarity of components is the main challenge for successful and unambiguous identification tasks. Considering the domain of the aviation industry, this poses a considerable challenge for modularized components. Choosing appropriate sensors and identification algorithms is a key factor to ensure that distinctive features are not only detected but also accordingly processed. It is, therefore, necessary to assess the distinctivity of each component with respect to a different component of the same spectrum. For this similarity measures may be applied. Two different methods for the assessment of component similarities are presented.

1. **3D-Model-based analysis:** under assumption of available 3D-models geometric analysis can be directly applied. Multiple approaches are viable for assessment of shape similarity [14]. Depending on requirements with respect to pose-invariance or scale suitable metrics have to be chosen. In the following examples, the *Hausdorff*-distance is used:

$$\begin{aligned} \hat{d}_H(X,Y) &= \max_{x \in X}\{\min_{y \in Y}\{\|x,y\|\}\} \\ d_H(X,Y) &= \max\{\hat{d}_H(X,Y), \hat{d}_H(Y,X)\}. \end{aligned} \tag{2}$$

With the euclidean distance $\| \cdot \|$, small values of the *Hausdorff*-distance denote that each of the elements x in a set X has an element $y \in Y$ for which the distance is small. We use this distance to assess the similarity between two shapes:

$$S = 1 - d_H(X, Y) \in [0 \dots 1], \tag{3}$$

where 0 denotes no similarity, and 1 denotes that both shapes are identical.

2. Feature-based analysis: under the assumption of a feature set that can be applied to describe the component spectrum, those feature-sets can be analyzed and similarities can be identified. In accordance with the previous step, both qualitative features can be analysed through histogram distributions. Consider a set $\mathcal{F}$ of n features:

$$\mathcal{F} = \{F_l \,|\, l = 1 \dots n\}. \tag{4}$$

For each feature F_l chose an appropriate bin width h. A narrower bin-width allows for good distinctivity but requires sufficiently accurate measurement capabilities. With the chosen bin width, calculate the histogram distribution $\mathcal{H}$, which can be represented as set of bins:

$$\mathcal{H}_{F_l} = \{B_i^l \,|\, i = 1 \dots k\}, \tag{5}$$

with the number of bins calculated through the bin width h and max/min values x in each Feature range:

$$k = \left\lceil \frac{\max(x \in F_l) - \min(x \in F_l)}{h} \right\rceil. \tag{6}$$

The similarity between two components X and Y, is calculated with the number of occurrences the two components are listed in the same bin. This is denoted through the Kronecker delta $\delta_{(X \in B_i^l),(Y \in B_i^l)}$:

$$S = \frac{\sum_l^n \sum_i^k \delta_{(X \in B_i^l),(Y \in B_i^l)}}{n}. \tag{7}$$

Again the similarity is ranged between 0 and 1, with the latter indicating that none of the features can be utilized to distinguish both components from another.

Step 4: Using Iterative Subdivisioning to Reduce the Identification Task Complexity. Within each identification task, the identification task complexity $\mathcal{C}$ denotes from how many different components a specific component one has to be differentiated. An index notation is introduced $\mathcal{C}_{i,j}$ that denotes the complexity for different subsets of the component spectrum.

Based on the previous step of feature analysis, similarity assessment, and sensor applicability assessment, suitable distinctive features can be chosen to subdivide the component spectrum alongside that feature (s. Fig. 1). Each subdivisionend spectrum contains very similar or same values of that specific feature, which therefore can be considered no longer useful for identification purposes within this smaller spectrum. This procedure may be applied iteratively until,

a) spawning branches (s. Fig. 1) result in a complexity of $\mathcal{C}_{i,j} = 1$ through feature measurement.

b) AI-based or Template-Matching-based are deemed applied.

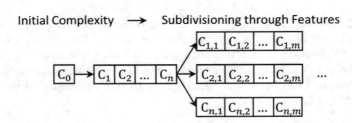

Fig. 1. Index notation of the identification task complexity value. Each subdivision step, reduces the complexity by utilizing one feature to subdivide the component spectrum into further sub-spectrums.

4   Application of Methodology to Use-Cases

Two industrial application scenarios are used to demonstrate the presented analysis approach: (1) Identification of bushings and (2) Identification of tubes. Both Use-Cases originate from Aircraft Industry and are considered to this date unsolved.

4.1   Use-Case 1 - Type-Identification of Bushings

Aircraft systems that contain actuators or moving parts, often include bushings. For E.g. landing gear systems contain up to several hundred different bushings. Since the bearings consist mainly of functional areas, attaching an identification marker is impossible and visual identification is necessary. In the analysed component spectrum, 23 different bushings are considered.

Step 1 - Component Features: Based on the construction data, four features are relevant for the identification of bushings: (1) length overall [mm], (2) outer diameter of the main bushing body [mm], (3) inner diameter [mm], and (4) the collar outer diameter. The inner diameter is not considered a suitable identification feature, since the considered bushings share the same feature value. Value distributions for the other features are shown in Fig. 2.

Step 2 - Scale of Geometric Features: Besides the presence of a collar, which can be considered a binary feature, lengths and diameters are quantitative measurable features. Therefore it has to be assessed, how narrowly two entries of two geometric measurements are located. Considering a dimensional manufacturing tolerance of 0.01 mm and the minimal distance d between two feature measurements (0.03 mm), the inequation 1 can be converted to yield the necessary resolution of $r = 0.005$ mm or 5 m.

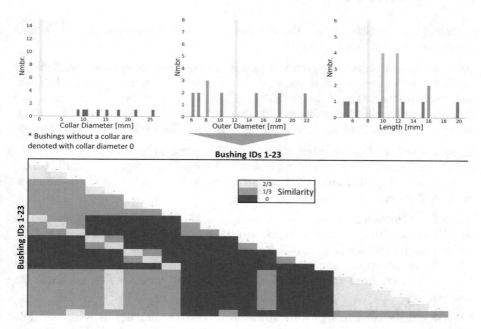

Fig. 2. Heat-map representation of bushing similarities. Due to the three distinguishing features, three similarity steps are calculated. Bushings without a collar, are represented in the histogram analysis with a collar diameter of 0.

Step 3 - Similarities: Although the design of bushings is not limited to the above parameters, bushings share a distinct similarity. In the considered use-cases no oil outlets, inlet, or similar feature is considered on the bushings. However, each bushing is uniquely described by the geometric features length as well as the inner and outer diameter, and the presence of a collar and its outer diameter. The inner diameter is for all bushings equal. Assessing the similarity of the bushings through the presented feature-based approach yields the in Fig. 2 shown results. With the three resulting features, all bushings can be distinguished.

Step 4 - Subdivisioning: The entire component spectrum of 23 bushings can be divided by the binary feature of a collar. Assessing the subset of collared bushings, 8 bushings have to be uniquely identified. Considering the subset of bushings without a collar, 15 bushings have to be differentiated.

4.2 Use-Case 2 - Identification of Tubes

Tubes are omnipresent throughout aircraft systems, being present in actuator systems, fueling, lavatories, engines, and similar systems. With flown systems, handled tubes are often soiled when going into MRO procedures and MSN plaques are often unreadable. Visual identification is carried out by manual matching of components to reference images. In the considered Use-Case, nine tubes are to be differentiated.

Step 1 - Component Features: The considered tubes are of the same diameter and have no flanges or similar mounted features. Thus, only geometric features may be extracted. Out of the possible features, three describe the tubes best: (1) End-to-End length of each tube, (2) Bending radii over the curve of the tube, and (3) Distances between bends but also distances between a bend and the end.

Step 2 - Scale of Geometric Features: While the above features can be measured, adjacent feature values are above $> 1\,\mathrm{mm}$. Thus, regular manufacturing tolerances and typical industry-standard sensors do not affect the measurement of these features.

Step 3 - Similarities: For this Use-Case the 3D Shape Similarity approach is used. The resulting similarity values are shown in Fig. 3. As intuitively visible, tubes 1 and 2 share a high similarity alongside tubes 2 and 3. Pairings 7 and 8 may also be identified as cumbersome. However, since the similarity with other tubes is low for both tubes 7 and 8, subdivisioning is a suitable approach for this pairing.

Step 4 - Subdivisioning: With the above-shown similarities between multiple pairs of tubes, it is recommendable to subdivide the tube identification. As such At least two sub-spectrums should be formed that mix tubes 1, 2, 3 accordingly to their similarity value.

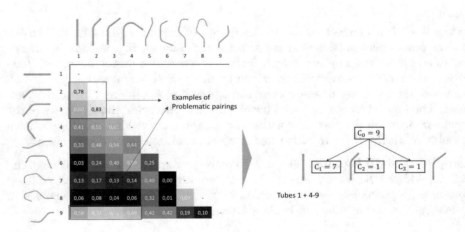

Fig. 3. Representation of the nine considered tubes and their respective similarity values.

5 Discussion

The above approach infers possible identification challenges. By Revealing those challenges it is possible to address those specific challenges in the selection of suitable identification algorithms and procedures. Considering the first presented use-case, the identification of bushings, feature selection allows presenting a set of distinctive bushing parameters. Calculating the similarity based on those features allows for selecting suitable identification strategies. Since the distinction between certain similar bushings is only possible based on a quantitative feature, measurement processes have to apply to accurately measure this specific feature.

This is in contrast to the second use case. There, the geometric similarity was not described by features but by the continuous similarity score. The derived subdivisions of the component spectrum can be used to derive classification tasks for AI-based object classification. For each subdivision, classification should be easier to train than for the component spectrum in its entirety. Utilizing this reduction in task complexity may yield more robust differentiation between those similar components.

Nevertheless, the presented approach relies on profound knowledge and understanding of the component spectrum. The limitation of the first similarity analysis is the resulting dependence on the quantified parameters of the component spectrum. Nevertheless, if distinctive parameters are found, the analysis can reveal whether the robust distinction between components is possible. If no such parameters are derived, the second similarity approach can be used. However, the geometric-based similarity analysis does not reveal which features can be measured or used to distinguish components from one another.

6 Conclusion and Outlook

This work presents a methodology to analyze similar and unmarked component spectrums with respect to identification challenges. Particular focus is given to similarities between components and revealing those similarities. A four-step methodology is presented that includes feature analysis, feature applicability assessment for sensors, similarity revealing, and identification of task complexity subdivision. The two presented methods for similarity revealing allow early discovery of possible mix-ups that pose threat to identification systems. By subdivision of identification task and reduction of the task complexity, this problem can be addressed in the design phase of the identification system.

Future work will address a combination of this approach with sensor selection and planning phases as well as further identification-algorithmic discussions.

Acknowledgements. Work was funded by IFB Hamburg, Germany under grant number 51161730.

References

1. Gierecker, J., Schoepflin, D., Schmedemann, O., Schüppstuhl, T.: Configuration and enablement of vision sensor solutions through a combined simulation based process chain. In: Schüppstuhl, T., Tracht, K., Raatz, A. (eds.) Annals of Scientific Society for Assembly, Handling and Industrial Robotics 2021, pp. 313–324. Springer, Cham (2022). https://doi.org/10.1007/978-3-030-74032-0_26

2. Schoepflin, D., Iyer, K., Gomse, M., Schüppstuhl, T.: Towards synthetic AI training data for image classification in intralogistic settings. In: Schüppstuhl, T., Tracht, K., Raatz, A. (eds.) Annals of Scientific Society for Assembly, Handling and Industrial Robotics 2021, pp. 325–336. Springer, Cham (2022). https://doi.org/10.1007/978-3-030-74032-0_27

3. Schoepflin, D., Holst, D., Gomse, M., Schüppstuhl, T.: Synthetic training data generation for visual object identification on load carriers. Proc. CIRP **104**, 1257–1262 (2021). https://doi.org/10.1016/j.procir.2021.11.211

4. Tarabanis, K.A., Allen, P.K., Tsai, R.Y.: A survey of sensor planning in computer vision. IEEE Trans. Robot. Autom. **11**(1), 86–104 (1995)

5. Gospodnetic, P., Mosbach, D., Rauhut, M., Hagen, H.: Flexible surface inspection planning pipeline. In: 2020 6th International Conference on Control, Automation and Robotics (ICCAR), pp. 644–652. IEEE, April 2020

6. Burla, A., Haist, T., Lyda, W., Osten, W.: An assistance system for the selection of sensors in multi-scale measurement systems. In: Interferometry XV: Applications, vol. 7791, p. 77910I. International Society for Optics and Photonics, August 2010

7. Jing, W., et al.: Sampling-based view planning for 3D visual coverage task with unmanned aerial vehicle. In: 2016 IEEE/RSJ International Conference on Intelligent Robots and Systems (IROS), pp. 1808–1815. IEEE, October 2016

8. Biasotti, S., Cerri, A., Bronstein, A., Bronstein, M.: Recent trends, applications, and perspectives in 3D shape similarity assessment. In: Computer Graphics Forum, vol. 35, no. 6, pp. 87–119, September 2016

9. Ma, J., et al.: Image matching from handcrafted to deep features: a survey. Int. J. Comput. Vis. **129**(1), 23–79 (2021). https://doi.org/10.1007/s11263-020-01359-2

10. Torrente, M.L., et al.: Recognition of feature curves on 3D shapes using an algebraic approach to Hough transforms. Pattern Recogn. **73**, 111–130 (2018)

11. Börold, A., Teucke, M., Rust, J., Freitag, M.: Recognition of car parts in automotive supply chains by combining synthetically generated training data with classical and deep learning based image processing. Proc. CIRP **93**, 377–382 (2020)

12. Lehr, J., Schlüter, M., Krüger, J.: Classification of similar objects of different sizes using a reference object by means of convolutional neural networks. In: 2019 24th IEEE International Conference on Emerging Technologies and Factory Automation (ETFA), pp. 1519–1522. IEEE, September 2019

13. Gierecker, J., Schüppstuhl, T.: Assembly specific viewpoint generation as part of a simulation based sensor planning pipeline. Proc. CIRP **104**, 981–986 (2021)

14. Cardone, A., Gupta, S.K., Karnik, M.: A survey of shape similarity assessment algorithms for product design and manufacturing applications. J. Comput. Inf. Sci. Eng. **3**(2), 109–118 (2003)

Projecting Product-Aware Cues as Assembly Intentions for Human-Robot Collaboration

Joe David[1,2(✉)], Eric Coatanéa[1], and Andrei Lobov[2]

[1] Tampere University, 33720 Tampere, Finland
{Joe.David,Eric.Coatanea}@tuni.fi
[2] Norwegian University of Science and Technology, Verkstedteknisk, 213, Gløshaugen, Norway
joesd@stud.ntnu.no, andrei.lobov@ntnu.no

Abstract. Collaborative environments between humans and robots are often characterized by simultaneous tasks carried out in close proximity. Recognizing robot intent in such circumstances can be crucial for operator safety and cannot be determined from robot motion alone. Projecting robot intentions on the product or the part the operator is collaborating on has the advantage that it is in the operator's field of view and has the operator's undivided attention. However, intention projection methods in literature use manual techniques for this purpose which can be prohibitively time consuming and unscalable to different part geometries. This problem is only more relevant in today's manufacturing scenario that is characterized by part variety and volume. To this end, this study proposes (oriented) bounding boxes as a generalizable information construct for projecting assembly intentions that is capable of coping with different part geometries. The approach makes use of a digital thread framework for on-demand, run-time computation and retrieval of these bounding boxes from product CAD models and does so automatically without human intervention. A case-study with a real diesel engine assembly informs appreciable results and preliminary observations are discussed before presenting future directions for research.

Keywords: Intention · Human-robot collaboration · Multi-agent systems · Digital thread · Product-aware · Knowledge-based engineering · CAD

1 Introduction

The transition of manufacturing from the fourth to fifth industrial revolution places the well-being of the human workforce at its core and leverages the synergy between them and autonomous machines [17]. In a human-robot collaborative environment, this means that humans will work alongside fence-less robots that exchange intentions and desires in a seamless and safe fashion between them. This would enable flourishing a trusted autonomy between interacting agents that would contribute towards an overall efficient manufacturing process [17].

© The Author(s) 2023
K.-Y. Kim et al. (Eds.): FAIM 2022, LNME, pp. 146–159, 2023.
https://doi.org/10.1007/978-3-031-18326-3_15

When working in close proximity with a robot as in the case of a collaborative assembly, the operator must be aware of robot intentions as they directly translate to the operator's safety. One way to do this is for the robot to express its intentions with the part it is interacting with and a popular approach for the same has been by augmenting the operator's reality by projecting intentions [3,22,26]. Using such augmented reality cues has demonstrated benefits in real manufacturing scenarios [8].

However, augmenting reality with head worn displays has proven not to be suitable for industrial environments due to bad ergonomics among others [12,14]. Using only a projector to do so has the advantage that it requires no equipment that the operator needs to wear, supports easy operator switching and supports simultaneous usage by multiple operators [25]. Further, spatially augmenting these projections on the product is said to reduce ambiguities and miscommunication as the operator is not required to divide attention between the task and an externally projected display [3].

To this end, this paper contributes with an approach to project spatially augmented product-aware assembly intentions. The novelty in the approach is twofold. First, bounding boxes are introduced as a novel information construct that approximates effectively and efficiently regions of interactions and are purposed to convey intentions associated with assembly and sub-assembly parts. Second, the approach taken to obtain them entails using the ubiquitous assembly design software that uses a digital thread framework for on-demand, online and dynamic computation of data that defines the bounding boxes and spatially augments the operator's reality. This is in contrast with most works in literature that manually extracts required information (e.g. wireframes [3] or reference geometries [27]) from the product's CAD model for the said purpose, sometimes using specialized software. Our approach, once completed, automates its extraction without human intervention and albeit simple, scales well with product sub-assembly parts of different sizes at any position within the assembly without the need for any reprogramming. Thus, it realizes a scalable approach that robots (or human operators) can use to project intentions about product assembly.

Such flexible approaches were deemed necessary in a recent study that reported interviews with automation and shop-floor operators [11]. Specifically, they expressed the need for intelligent robots that are updated automatically and aware of the product type it should work with in the context of human-robot collaboration. According to them, switching smoothly between products would better the efficiency of manufacturing processes. The work presented herein addresses these in the context of intent communication in human-robot collaborative product assembly.

The next section reviews existing works in literature related to projecting product-aware intentions. This alludes to research objectives of this study and a description of the research setting and scope in Sect. 3. Section 4 presents theoretical background for the approach described in Sect. 5. Section 6 presents results of the study and preliminary observations are discussed. The paper is summarized in a conclusion section in Sect. 7 that also presents future directions of this research.

2   Related Work

Projecting and consequently communicating intentions associated with a product between humans and robots is not a new idea in itself with many originating circa 1993 at least [16]. Notable works include that of Terashima and Sakane [25] that prototypes a 'digital desk' that allows two levels of interaction via a virtual operational panel (VOP) and an interactive image panel (IIP). While the VOP is used to communicate task dependent operations, the IIP streams the robot's workspace by a separate vision system with which the operator is able to convey target object intentions by touching with his/her hands. Thus, it did not capture the physical aspects of a collocated setup as collaborative relationships with robots were not necessarily the goal back then and such systems where used for "guiding and teaching robot tasks" as opposed to true intent communication and also worked in a single direction from the operator to the robot. Later around the same period, Sato and Sakane [21] added a third subsystem in addition to the VOP and IIP called the 'interactive hand pointer (IHP)' that allowed the operator to point directly at an object in the robot's workspace to convey his/her intentions thereby removing the need for separate a workspace or display. However, this too worked in the direction of the operator to the robot. Needless to say, the ability for the robot to convey intentions is crucial for operator safety.

More recent works include that of Schwerdtfeger et al. [22] that explores the use of a mobile head-mounted laser projector on a helmet in an attempt to do away with the discomfort of conventional head-mounted displays to display product-aware intentions. The device projects simple 3D aligned augmentations for welding points on the surface of the part the operator interacts with (a car door) while instructions are provided on a standard stationary computer monitor. However, the position of the weld points were defined off-line by a tracked pointer. Further, the device was later reported as "too heavy and big" for use as a head-mounted device [23]. A subsequent developed hybrid solution entails a tripod mounted projector that benefits from partially mobile - partially stationary degrees of freedom but requires careful pose estimation each time the tripod is moved [23]. Sand et al. [20] present 'smARt.assembly' which is a projection-based AR system to guide the operator to pick parts from an assembly rack during manufacturing assembly. The projector projects nested rectangles as an animation on the label of the part the operator is supposed to pick. The projection also entails a 2D image of the digital 3D model of the corresponding step that is presented on a panel on one side of the assembly station which is separated from the assembly workspace.

Uva et al. [27] present a spatial augmented reality solution that bears a close resemblance to the work reported in this paper. As the system was built with the goal of projecting technical guidance instructions, they use the reference geometries in the CAD model of only the base (fixed) part as an occlusion model and not the assembled part. This, as they say, was to reduce effort in the authoring phase which is evidence to the difficulties involved in extracting required geometry data for projection by manual means. Further, to use reference geometries of only the base part to project intent in a collaborative human-robot scenario would be a problem as the assembled part can take any form factor that is not accounted for while projecting intent and can be dangerous, say for example when a robot is placing a component that spans to an area where the operator is simultaneously working. Also, the solution makes use of a third party software, Unity, that requires specialized expertise for development not commonly found in run-of-the-mill manufacturing enterprises. Andersen et al. [3] present a product aware intention projection solution that tracks the object in real-time and projects wireframes of the object. However, the approach involves generating "a large" number of edge maps offline from the CAD model of the object. Further, it is not clear how the they manage to illuminate the parts of the door the robot works with. As noted previously, manual approaches can be prohibitively time consuming and difficult to scale to different object parts at poses not determined beforehand.

3   Research Objectives, Setting and Scope

Existing approaches reviewed in the previous section either require pre-processing of the CAD model of the product, uses manual techniques, requires special developer expertise or are unsuitable for use in collaborative environments. Thus, we identify a gap for a simple, scalable solution in the automation of intent projection methods for use in collaborative environments between humans and robots. To this end, this research sets the following as its objectives:

1. Realize a generalizable intent information construct that can be used to project product-aware intentions for product assembly.
2. Use it to do so in a manner with minimal human intervention, preferably with *in-situ* software.

The research is carried in a laboratory environment shown in Fig. 1a. It consists of a DLP projector (1920 × 1080) and a Kinect Camera (RGB-D) mounted atop a height adjustable table that acts as a collaborative working space between a table-mounted UR5 collaborative robot and a human operator. The experiments are conducted with respect to an assembly of a real diesel engine.

Although, we consider only pick and place tasks in this study, the presented concept may be used for other tasks that require the representation of part geometries in the manner proposed herein (e.g. screwing). Also, the scope of this paper does not extend beyond the identification of the information construct and a reflection on its use in the case of diesel engine assembly. Consequently, details of the digital thread framework that accesses and processes the CAD model have been intentionally left out after a brief overview. However, the reader is provided references to our previous work [6,7] that has them as the focus of the study. Further, we do not deal with the pose estimation problem of the part and assume that its pose is known. Pose estimation based on the CAD model has been the subject of several focused studies [10,18,19] and the approach presented herein is expected to be built upon any of them.

(a) Physical Setup   (b) Mixed-reality interface

Fig. 1. Human-robot collaboration setting

4   Theoretical Background

4.1   Oriented Minimum Bounding Boxes

In three dimensional euclidean space ($\mathbb{R}^3$), a minimum[1] bounding box around a 3D object is the minimum or smallest cuboid (by volume) that completely encloses it. If the edges of the bounding box are parallel to a coordinate axes of a Cartesian coordinate system, it is an axis-aligned bounding box (AABB) with respect to that coordinate system. On the other hand, if its edges are inclined at an angle to the coordinate axes of a Cartesian coordinate system, then they are oriented-bounding boxes (OBB) with respect to that coordinate system.

In this paper, we use bounding boxes to compute the minimum enclosing cuboid of a sub-assembly part geometry. Depending on the geometry of the sub-assembly part and how it is aligned within the entire assembly, the AABB may or may not be the smallest enclosing cuboid, but an OBB will always be. For this reason, in this study, we only refer to OBBs. AABBs will be the smallest cuboid and same as the OBB when the part is positioned such that its OBB is aligned with the coordinate system of the assembly part. Bounding boxes are further discussed in the Sect. 5 with examples (Fig. 3).

4.2   Camera and Projector Model

A camera can be modelled using the pin-hole camera model [24] that describes how a point in the 3D world is mapped onto its image plane. The projector too

---

[1] In this paper, we deal with only 'minimum' bounding boxes and we omit the word 'minimum' henceforth for brevity.

can be considered as an inverse camera where the rays of light are reversed, i.e. the light is projected instead of being captured [9]. Hence, the ideas underlying the calibration techniques that determine its intrinsic parameters used for a camera such as Zhang's [28], can be used for a projector as well [9].

The homogeneous transformation for a point X in the world coordinate system $\{\mathbb{W}\}$ ($\mathbb{R}^3$), to a point x in the pixel coordinate system $\{I_K\}$ ($\mathbb{R}^2$) of the image plane whose coordinate system origin is located at X_O is given by equation:

$$\mathrm{x}_{\{I_K\}} = \mathrm{P}\mathrm{X}_{\{\mathbb{W}\}} \tag{1}$$

where P is the direct linear transform (DLT)

$$\mathrm{P} = \mathrm{K}R[I_3| - X_O] \tag{2}$$

where K is a 3×3 matrix and defines five intrinsic parameters obtained through calibration and $R[I_3|-X_o]$ defines the 6 (3 translational + 3 rotational) extrinsic parameters or the rigid body transformation in a 3×4 matrix.

$$R[I_3| - X_o] = \begin{bmatrix} r_{11} & r_{12} & r_{13} & t_1 \\ r_{21} & r_{22} & r_{23} & t_2 \\ r_{31} & r_{32} & r_{33} & t_3 \end{bmatrix} \tag{3}$$

K and Eq. 3 can be multiplied together to realize the transform as a 3×4 matrix substituted in Eq. 1 as

$$\begin{bmatrix} x \\ y \\ 1 \end{bmatrix} = \begin{bmatrix} p_{11} & p_{12} & p_{13} & p_{14} \\ p_{21} & p_{22} & p_{23} & p_{24} \\ p_{31} & p_{32} & p_{33} & 1 \end{bmatrix} \begin{bmatrix} X \\ Y \\ Z \\ 1 \end{bmatrix} \tag{4}$$

where x, y is the pixel coordinate on the image plane of the projector of a point in the real world with coordinates X, Y, Z, both expressed in homogeneous coordinates and defined up to a scale factor.

5   Methods

5.1   Estimating Intrinsic and Extrinsic Parameters

We used a manual approach to establish correspondences between the projector pixels and the calibration landmarks (a printed planar checkerboard pattern) for calibration using the OpenCV library [4] to estimate the projector intrinsic matrix, K. To determine projector extrinsics, we used the Perspective-n-Point (PnP) pose computation method using similar correspondences, again using the OpenCV library.

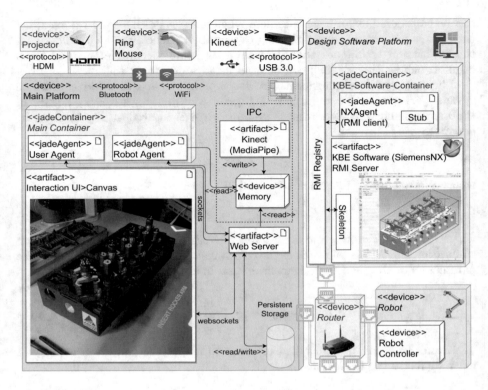

Fig. 2. Deployment diagram representing the architecture of the system

5.2 Digital Thread Framework

Figure 2 shows a condensed deployment diagram of the architecture of the digital thread framework employed in the HRC environment. The digital thread framework uses an agent-based framework (JADE) that integrates the assembly design environment (Design Software Platform) that exposes the product model via an API that provides with the data needed to project intentions from the product CAD model. It also consists of a purpose-built web application (Interaction UI) that is projected onto the shared work table that acts as a real-world canvas to project intentions and which the operator interacts with (Fig. 1b) to facilitate bi-directional communication with the robot. The operator's hand is tracked via an open-source hand recognition framework, MediaPipe (Kinect), while input is received from a ring mouse (Ring Mouse) worn by the operator. Further details pertaining to the digital thread framework, associated components and the interaction model can be found in our earlier works focused on the framework [7] and the web-based interaction model [6].

5.3   Product Design Environment

The digital thread framework maintains an online connection with the product design environment, Siemens NX. As a software also built for knowledge-based engineering (KBE), Siemens NX has a rich set of API, that allows to interact with the product geometry via the NXOpen API [2] and permits building digital thread applications [5] that help integrate product lifecycle information. It is with this API that the core functionality of our approach is realized.

When the robot is interacting with any of the sub-assembly parts, it requests the design software for the information it needs pertaining to the interacting part to project its intent. For a task such as that for an incoming placing operation which requires, conveying as intention, the relative position of a part in a base part (the part it is assembled into) but not the whole part itself, the coordinates for only the lower face of the bounding box is requested. For tasks that require, conveying as intention, the geometry of the whole part (as guidance instruction for example), the coordinates that defines the entire bounding box is requested. Thus, necessary information of any sub-assembly part can be obtained from the CAD file of the assembly and communicated to agents that require it dynamically at system run-time on request.

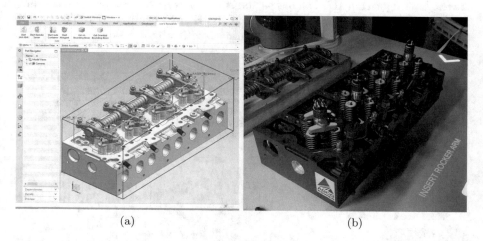

(a)   (b)

Fig. 3. (a) Computed bounding boxes viewed in the assembly design environment (b) Intent projection for the rocker arm

Figure 3a shows the CAD model of a real diesel engine loaded in the design software, NX. The diesel engine consists of many other parts but here only two sub-assembly parts (besides the fixed base part, engine block), a rocker arm shaft and rocker arm (8 nos), are shown to keep the demonstration concise and clear. However, the method scales well for parts of different sizes at any sub-assembly pose with the base part. Bounding boxes that the robot estimates are superimposed on the respective parts in NX and shown in Fig. 3a. Note the orientation of the absolute coordinate system $\{\mathbb{P}\}$ (ACS) located at the bottom left of the viewport. The bounding boxes of the engine block (in red) and the rocker arm shaft (in pink) have their edges aligned along the axes of the ACS. Hence both their bounding boxes are axis aligned. However, the rocker arms are positioned such that they are inclined with respect to the ACS. Hence the computed bounding boxes are oriented bounding boxes (with respect to ACS).

5.4 Intent Projection

The robot agent uses the data it receives from the assembly design software to project intentions through a web-based mixed reality user interface [6] projected onto the shared environment (Interaction UI in Fig. 2 and Mixed Reality Interface in Fig. 4). Specifically, the robot agent uses the HTML5 Canvas API [1] to draw shapes and to write text to reveal its intent. As earlier mentioned, the current iteration of the development works on known poses of parts with respect to the real world external coordinate system $\{\mathbb{W}\}$. The coordinates that define the bounding box are computed from the CAD model and is transformed to $\{\mathbb{W}\}$ using the known pose and subsequently to the projector's image plane using the DLT (Eq. 2 & Eq. 4). Once the coordinates are mapped to the projector plane, a convex hull algorithm [13] calculates the smallest convex polygon that contains these points and fills it with colour using the HTML Canvas API. The overall steps taken for intent projection are summarized in Fig. 4:

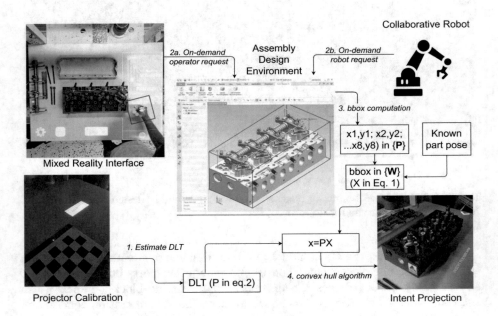

Fig. 4. Summary of steps involved in intent projection

6   Experimental Results and Observations

The approach described in the previous section is used to project intentions in a real diesel engine assembly in a laboratory environment. The results are shown for the two sub-assembly parts, namely the rocker arms and the rocker arm shaft in Fig. 3b and Fig. 5a respectively. While a comprehensive user study is not in the objectives of this paper, this section documents some preliminary observations made during the development and experimental process along with a general discussion.

6.1   Occlusions

Inherent to the single projector setup, the projections suffer from occlusions both from the operator and the robot. This can be seen in Fig. 5a where the shadow of the robot arm is cast onto the middle portion of the rocker arm shaft. Further protruding parts too cause occlusion. In Fig. 3b, it can be seen that the ignition coil occludes portions of the green bounding boxes projected on the rocker arm. While for small assemblies, such as the one presented here, this can be solved by skewing the projector or placing it vertically on top, larger assemblies are bound to suffer from occlusions from part geometry. However, considering the objectives of the study, this is not a limitation of the presented approach but that of the hardware setup. Depending on part geometry, multi-projector systems or a mobile projector setup [15] are two ways literature have minimized occlusions and with such additional hardware occlusions may be minimized.

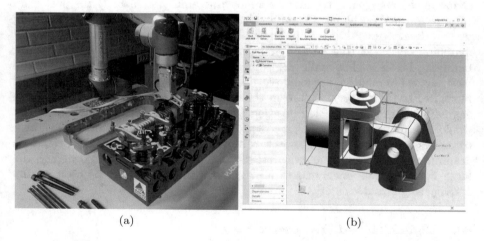

(a)   (b)

Fig. 5. (a) Intentions projected for the rocker arm shaft (b) Axis-aligned bounding boxes for a universal joint assembly and their intersection (cyan)

6.2   Bounding Boxes for Intent Projection

Bounding boxes generalize the problem of projecting intentions by approximating its shape quite well and do so efficiently. A 3D bounding box can be defined completely by six floating point numbers (minX, minY, minZ; maxX, maxY, maxZ) rather than, wireframes for example, that are defined by a series of points that define the boundary. Computing their intersection is a common and efficient method of collision detection that most if not all 3D software come with built-in functions for its computation. If not, it is possible to iterate through the geometry to compute them manually using an algorithm. Further, intersection of bounding boxes of assembly components can be used to locate the mating positions between sub-assembly components within an assembly when there is no third part involved (e.g. welded joints). Figure 5b shows the intersection of two axis-aligned bounding boxes of a universal joint in cyan to illustrate this. Note that the same code that was used in the diesel engine assembly case presented earlier was used, which demonstrates the scalability of the approach. In our work, while the robot uses bounding boxes to project its intentions to the operator, in a similar way the operator can use the mixed-reality interface (Fig. 1b) to request similar projection cues of sub-assembly parts. However, these projection cues projected at the behest of the operator are not used by the robot to perceive the operator's intentions. Rather, it is to reduce the cognitive load on the operator and to assist with the assembly process in general. To the best of our knowledge, such flexible functionality have not been implemented for intention projection purposes.

However, it can be argued that using bounding boxes can cause to loose the shape of the part geometry and thus loose the ability to identify the part from the box projection alone. In our work, we compensate for this by presenting textual descriptions that are automatically loaded from the design software with matching colors as that of the bounding box. Another issue with projecting bounding boxes is that while projecting intentions for a large part, the entire base part is illuminated. For example, the engine frame that lies beside the engine block in Fig. 5a is largely hollow in the center but spans along the edges of the engine block in its assembled position (not shown). As such, its bounding box would illuminate the entire engine block which can be difficult for the operator to understand given the box projection alone. In such cases, textual descriptions are important to prevent any confusions for the operator. On the contrary, the large bounding box encompasses all the areas that require to be clear of any activity to guarantee safety.

6.3   Digital Thread Framework

The digital thread framework provides important information necessary for correct projection of the bounding boxes from a (type of) software that is commonly used in manufacturing enterprises, i.e. the product design software. Thus, the requirement of a third party software that requires specialized development expertise is avoided, C# for Unity as an example. Rather, modern KBE software vendors, expose their CAD kernel with a rich set of API that can be then used to drive applications that use the product in the manufacturing processes. NX, in particular exposes that CAD kernel via a Common Object Model that has bindings in four general purpose languages, Java, C++, Python and .NET which considerably reduces the barrier for such application developers. However exposing the entire CAD kernel with APIs means an overwhelming amount of programming constructs and our experience is that finding the right constructs to perform simple operations can sometimes be time consuming. However, as we got acquainted with the API, we experienced this less. Lastly, since such software is already well integrated with a traditional manufacturing enterprise, interfacing it with related systems is expected to be easy as was in our case and such solutions could be expected to be well received by the involved stakeholders.

7   Conclusions and Future Work

Collaborative tasks between humans and robots are becoming commonplace and recognizing intentions of agents that behave autonomously is pivotal in guaranteeing operator safety. The work presented in this paper presents a generalizable information construct in the form of oriented bounding boxes that is expected to foster greater situational awareness between agents engaged in collaborative assembly. The approach uses only the ubiquitous assembly design software and exploits the flexibility of a KBE software API to realize a scalable solution for on-demand, online computation of the required information dynamically at system run-time.

As future work, we aim to develop an information model or vocabulary that semantically grounds agent interactions. The work presented herein is expected to support the notion of agent intentions during these interactions. Another possible direction of future research includes pose estimation of the parts. We would like to investigate if we could, in a similar manner, automate the extraction of sufficient information that could train models for run-time identification and pose detection of assembly parts.

References

1. Canvas API - Web APIs: MDN. https://developer.mozilla.org/en-US/docs/Web/API/Canvas_API
2. Siemens Documentation: NX Open Programmer's Guide. https://docs.plm.automation.siemens.com/tdoc/nx/12/nx_api#uid:xid1162445:index_nxopen_prog_guide
3. Andersen, R.S., Madsen, O., Moeslund, T.B., Amor, H.B.: Projecting robot intentions into human environments. In: 2016 25th IEEE International Symposium on Robot and Human Interactive Communication (RO-MAN), pp. 294–301 (2016). https://doi.org/10.1109/ROMAN.2016.7745145
4. Bradski, G.: The OpenCV library. Dr. Dobb's J. Softw. Tools **25**(11), 120–123 (2000)

5. David, J., Järvenpää, E., Lobov, A.: Digital threads via knowledge-based engineering systems. In: 2021 30th Conference of Open Innovations Association FRUCT, pp. 42–51 (2021). https://doi.org/10.23919/FRUCT53335.2021.9599986
6. David, J., Järvenpää, E., Lobov, A.: A web-based mixed reality interface facilitating explicit agent-oriented interactions for human-robot collaboration. In: 2022 8th International Conference on Mechatronics and Robotics Engineering (ICMRE), pp. 174–181 (2022). https://doi.org/10.1109/ICMRE54455.2022.9734094
7. David, J., Lobov, A., Järvenpää, E., Lanz, M.: Enabling the digital thread for product aware human and robot collaboration - an agent-oriented system architecture. In: 2021 20th International Conference on Advanced Robotics (ICAR), pp. 1011–1016 (2021). https://doi.org/10.1109/ICAR53236.2021.9659352
8. Doshi, A., Smith, R.T., Thomas, B.H., Bouras, C.: Use of projector based augmented reality to improve manual spot-welding precision and accuracy for automotive manufacturing. Int. J. Adv. Manuf. Technol. 89(5), 1279–1293 (2017). https://doi.org/10.1007/s00170-016-9164-5
9. Falcao, G., Hurtos, N., Massich, J.: Plane-based calibration of a projector-camera system. VIBOT Master 9, 1–12 (2008)
10. Fan, Z., Zhu, Y., He, Y., Sun, Q., Liu, H., He, J.: Deep learning on monocular object pose detection and tracking: a comprehensive overview. arXiv preprint arXiv:2105.14291 (2021)
11. Gustavsson, P., Syberfeldt, A.: The industry's perspective of suitable tasks for human-robot collaboration in assembly manufacturing. IOP Conf. Ser. Mater. Sci. Eng. 1063(1), 012010 (2021). https://doi.org/10.1088/1757-899x/1063/1/012010
12. Hietanen, A., Pieters, R., Lanz, M., Latokartano, J., Kämäräinen, J.K.: AR-based interaction for human-robot collaborative manufacturing. Robot. Comput.-Integr. Manuf. 63, 101891 (2020). https://doi.org/10.1016/j.rcim.2019.101891. https://www.sciencedirect.com/science/article/pii/S0736584519307355
13. Jarvis, R.: On the identification of the convex hull of a finite set of points in the plane. Inf. Process. Lett. 2(1), 18–21 (1973)
14. van Krevelen, D., Poelman, R.: A survey of augmented reality technologies, applications and limitations. Int. J. Virtual Reality 9(2), 1–20 (2010). https://doi.org/10.20870/IJVR.2010.9.2.2767. https://ijvr.eu/article/view/2767
15. Leutert, F., Herrmann, C., Schilling, K.: A spatial augmented reality system for intuitive display of robotic data. In: 2013 8th ACM/IEEE International Conference on Human-Robot Interaction (HRI), pp. 179–180 (2013). https://doi.org/10.1109/HRI.2013.6483560
16. Milgram, P., Zhai, S., Drascic, D., Grodski, J.: Applications of augmented reality for human-robot communication. In: Proceedings of 1993 IEEE/RSJ International Conference on Intelligent Robots and Systems (IROS 1993), vol. 3, pp. 1467–1472. IEEE (1993)
17. Nahavandi, S.: Industry 5.0-a human-centric solution. Sustainability 11(16) (2019). https://doi.org/10.3390/su11164371. https://www.mdpi.com/2071-1050/11/16/4371
18. Nguyen, D.D., Ko, J.P., Jeon, J.W.: Determination of 3D object pose in point cloud with CAD model. In: 2015 21st Korea-Japan Joint Workshop on Frontiers of Computer Vision (FCV), pp. 1–6 (2015). https://doi.org/10.1109/FCV.2015.7103725
19. ten Pas, A., Gualtieri, M., Saenko, K., Platt, R.: Grasp pose detection in point clouds. Int. J. Robot. Res. 36(13–14), 1455–1473 (2017)

20. Sand, O., Büttner, S., Paelke, V., Röcker, C.: smARt.Assembly – projection-based augmented reality for supporting assembly workers. In: Lackey, S., Shumaker, R. (eds.) VAMR 2016. LNCS, vol. 9740, pp. 643–652. Springer, Cham (2016). https://doi.org/10.1007/978-3-319-39907-2_61

21. Sato, S., Sakane, S.: A human-robot interface using an interactive hand pointer that projects a mark in the real work space. In: Proceedings 2000 ICRA. Millennium Conference. IEEE International Conference on Robotics and Automation. Symposia Proceedings (Cat. No. 00CH37065), vol. 1, pp. 589–595 (2000). https://doi.org/10.1109/ROBOT.2000.844117

22. Schwerdtfeger, B., Klinker, G.: Hybrid information presentation: combining a portable augmented reality laser projector and a conventional computer display. In: Froehlich, B., Blach, R., van Liere, R. (eds.) Eurographics Symposium on Virtual Environments, Short Papers and Posters. The Eurographics Association (2007). https://doi.org/10.2312/PE/VE2007Short/027-032

23. Schwerdtfeger, B., Pustka, D., Hofhauser, A., Klinker, G.: Using laser projectors for augmented reality. In: Proceedings of the 2008 ACM Symposium on Virtual Reality Software and Technology, VRST 2008, pp. 134–137. Association for Computing Machinery, New York (2008). https://doi.org/10.1145/1450579.1450608

24. Sturm, P.: Pinhole camera model. In: Ikeuchi, K. (ed.) Computer Vision, pp. 610–613. Springer, Boston (2014). https://doi.org/10.1007/978-0-387-31439-6_472

25. Terashima, M., Sakane, S.: A human-robot interface using an extended digital desk. In: Proceedings 1999 IEEE International Conference on Robotics and Automation (Cat. No. 99CH36288C), vol. 4, pp. 2874–2880. IEEE (1999)

26. Uva, A.E., et al.: Design of a projective AR workbench for manual working stations. In: De Paolis, L.T., Mongelli, A. (eds.) AVR 2016. LNCS, vol. 9768, pp. 358–367. Springer, Cham (2016). https://doi.org/10.1007/978-3-319-40621-3_25

27. Uva, A.E., Gattullo, M., Manghisi, V.M., Spagnulo, D., Cascella, G.L., Fiorentino, M.: Evaluating the effectiveness of spatial augmented reality in smart manufacturing: a solution for manual working stations. Int. J. Adv. Manuf. Technol. **94**(1), 509–521 (2018). https://doi.org/10.1007/s00170-017-0846-4

28. Zhang, Z.: A flexible new technique for camera calibration. IEEE Trans. Pattern Anal. Mach. Intell. **22**(11), 1330–1334 (2000). https://doi.org/10.1109/34.888718

Online Quality Inspection Approach for Submerged Arc Welding (SAW) by Utilizing IR-RGB Multimodal Monitoring and Deep Learning

Panagiotis Stavropoulos[✉], Alexios Papacharalampopoulos, and Kyriakos Sabatakakis

Laboratory for Manufacturing Systems and Automation (LMS), Department of Mechanical Engineering and Aeronautics, University of Patras, 26504 Rio, Patras, Greece
pstavr@lms.mech.upatras.gr

Abstract. Online, Image-based monitoring of arc welding requires direct visual contact with the seam or the melt pool. During SAW, these regions are covered with flux, making it difficult to correlate temperature and spatial related features with the weld quality. In this study, by using a dual-camera setup, IR and RGB images depicting the irradiated flux during fillet welding of S335 structural steel beams are captured and utilized to develop a Deep Learning model capable of assessing the quality of the seam, according to four classes namely "no weld", "good weld", "porosity" and "undercut/overlap", as they've emerged from visual offline inspection. The results proved that the camera-based monitoring could be a feasible online solution for defect classification in SAW with exceptional performance especially when a dual-modality setup is utilized. However, they've also pointed out that such a monitoring setup does not grand any real-world advantage when it comes to the classification of relatively large, defective seam regions.

Keywords: Intelligent Welding System (IWS) · Manufacturing process quality · Deep learning

1 Introduction

Submerged Arc Welding (SAW) is a fusion welding process which due to its high heat input is used to weld thick-section carbon steels. It is typically utilized in shipbuilding and in fabrication of pipes, pressure vessels and structural components for bridges and buildings due to its high productivity, low cost, and fully automated operation [1]. Despite though this high level of automation on the welding floor, today's SAW systems cannot be considered intelligent, as their autonomy and adaptability are limited and achieved by utilizing an entire ecosystem consisting of designers, skilled operators as well as auxiliary processes such as quality control [2]. This fact contradicts the case of other welding processes involved in light metal fabrication sectors like the automotive industry

© The Author(s) 2023
K.-Y. Kim et al. (Eds.): FAIM 2022, LNME, pp. 160–169, 2023.
https://doi.org/10.1007/978-3-031-18326-3_16

where key enabling technologies and frameworks such as Artificial Intelligence, Cyber-physical Systems, and the Internet of Things have been the moving force for introducing aspects of this intelligence to the welding systems [3].

From available solutions on the market [4–6] to advanced research approaches, systems or prototypes are integrating process monitoring, closed-loop control [7–9] and quality assessment functionalities [10–13] and thus implementing fractions of what is called Intelligent Welding System (IWS) [2]. On other hand, in case of SAW, the majority of the research studies are focused on process modeling [14–16] and process parameter optimization [17, 18] which although are enriching the existing knowledge base they are lacking a clear interface with the welding system and integrability. Nevertheless, holistic approaches are existing and focus mainly on the prediction and/or control of the seam's geometrical features by utilizing either the nominal input parameters [19, 20] or real-time measurements from retrofitted sensors [21]. To this end by looking at paradigms of other welding processes, integrating knowledge for defect detection into a welding system seems to be favored by the utilization of image sensors, as they can provide high-dimensional information of the process in a single instance [22]. Along with that, image classification, although until recently was carried out based on custom feature extraction algorithms and classic Machine Learning (ML) models, thanks to the establishment of the Convolutional Neural Networks (CNN), these two mechanisms were integrated into one optimizable structure which given the appropriate amount of training data can achieve high performance on image classification tasks [23]. Indicatively in [12] the authors by training a CNN using the data from a high-speed CMOS camera were able to achieve a 96.1% pore classification accuracy for the laser keyhole welding of 6061 aluminum. In [24] a monitoring system by utilizing a CNN was capable of determine the penetration state for the laser welding of tailor blanks with an accuracy of 94.6%, while in [25] with an end-to-end CNN achieved a classification accuracy of 96.35% on the penetration status during gas tungsten arc welding of stainless steel by using images depicting the weld pool.

With that been said, driven by 1) the lack of online defect detection approaches for the SAW 2) the recent advances on Image-based defect-detection for welding applications 3) a study which indicates that IR monitoring of SAW is feasible even though the thick layer of slag covers the seam [21] and 4) the need for integrating knowledge from the welding floor to the welding system, this study investigates the feasibility of image-based monitoring for SAW as a mean for online quality assessment. This is validated by introducing both an IR and an RGB camera which are feeding 3 different CNN models for classifying defective segments of the seam online. Their capabilities are assessed both in terms of classification and real-time inference performance while the need for dual-modality monitoring is determined afterward.

2 Quality Assessment Method

2.1 Model Selection and Architecture

When it comes to image classification many options are available [26, 27] however, CNNs' were selected in this study as they are incorporating beyond many well-known advantages [28] some unique features that are particularly suited to this application. As such the CNNs can be trained and operate directly with images without the need for additional feature extraction methods. A trained CNN can be re-trained with a lot less data, even for entirely different application [25, 29–31] while the development of end-to-end inference applications can be accelerated, as it comes standardized for the most part by the majority of Deep Learning tools and frameworks. Consequently, CNNs are presenting the required flexibility, adaptability, and ease of development that is expected by an IWS as regards its quality assurance capabilities.

On the other hand, while the CNN's non-linear nature is one of the key ingredients to its flexibility for learning complex relationships within the data among other image classification approaches it makes also sensitive to initial conditions. A high variance/ low bias remedy to this problem has emerged from the "Ensembled Learning [32] which although typically utilizes data from a single source, recently has been applied to multimodal scenarios [33, 34]. Herein findings on the improvement of the classification accuracy when combining the predictions from multiple neural networks (NN), seems to add the required bias that in turn counters the variance of a single trained NN. The results are predictions that are less sensitive to the specifics of the training data, the choice of the training scheme, and the serendipity of a single training run.

Based on the above the authors propose a 2-Branch CNN which uses the features of two trained CNN to fit a NN model as typically performed by applying the ensemble learning algorithms called "stacking". The top-level architecture of the 2-Branch CNN is depicted hereafter (Fig. 1). For each branch of the model 2 residual CNNs were user as feature extractors. The architecture of the CNN used to extract features from IR images (MRN) can be found in [35] while the architecture of the other one is the same with the ResNet18 [36], taking RGB images as input. The selection of residual architectures was made as the "shortcut" connection can "transform" a non-convex optimization problem (training of MRN and ResNet18) into a convex one (improve gradient flow) [37].

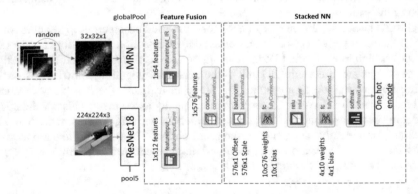

Fig. 1. 2-Branch CNN model architecture

2.2 Training of the CNNs

For the MRN and ResNet18 networks, data augmentation was performed during training, by apply a random rotational and scale transformation on the images of the training partition of the data (70%) on every epoch [38]. By that the model rarely encountered the same training example twice, as this is improbable given that the transformations are random. To update the networks' weights and biases and minimize the loss function the Stochastic Gradient Descent with Momentum (SGDM) [39] was used. Beyond the training algorithm, an early stopping criterion was used to avoid overfitting. This was implemented by calculating the loss of the model on the test (30%) partition of the data. If the current loss was greater than the current value and this situation persisted for more than 5 consecutive epochs, then the training was stopped. The model weights were the ones the model have from its last update before stopping.

The 2-branch CNN as depicted in Fig. 1 takes as inputs the features vectors emerged from the last average-pooling layers of the above networks. Considering that the two cameras are operating on different frame rates a single frame from the RGB corresponds to a batch of 33 IR frames. Thus, during training and inference, a single IR frame was selected randomly from this batch. Once again SGDM was used for optimization and cross-entropy as the loss function.

The development of all the models in this study was carried out using MATLAB's Deep Learning Toolbox [40]. The training was speeded up by utilizing a 4xGPU machine for linearly scaling the mini-batch size, distributing, and accelerating the computations.

3   Experimental Setup and Data Processing

Two stationary, cameras were used to capture the visible and IR electromagnetic emissions of a SAW process. The high-speed IR camera (32 by 32 pixels, 1000 fps, 1–5 μm) by NIT [41] and the RGB web-camera (1920 by 1080 pixel, 30 fps) by Logitech [42] both placed at a working distance of 140 cm from the target (Fig. 2) were monitoring the fillet welding of structural steel beam (S335, 28 mm thickness). This placement is also convenient for the online use of the monitoring system. A Tandem Arc SAW machine with a stationary head was used with a 4 mm filler wire, at 29 V, 680–700 A, and a welding speed of 45 cm/min.

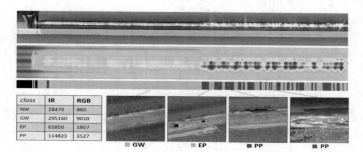

class	IR	RGB
NW	28470	860
GW	295160	9010
EP	65850	1957
PP	114823	3527

Fig. 2. Seam IR and RGB image concatenation & quality labels.

The defects were artificially introduced by applying to the surface of the beam's flange a grease-based layer for half of the beam's length. This had, as a result, the electrical connection with the workpiece to be compromised and the root and leg of the weld to be contaminated. The welded beam was inspected for defects across its length visually by authorized personnel and four classes were derived for segments of the seam and the corresponding image data (Fig. 2). These segments were labeled as: No Weld (NW), Good Weld (GW), Porosity (EP), and undercut/overlap (PP). Two of these classes were representing defective regions as defined in [1, 43]. The data were captured manually using the software provided by the camera manufacturer. As in a single frame, more than one defect may be present, the matching of a quality stamp with a frame was made based on the point that this frame's center was representing on the seam. For the IR camera due to the prolonged recording duration, the thermal drifting of the entire FPA was significant although linear [10]. Thus, to compensate that the minimum value of each frame was subtracted from the rest of the same frame. This step was integrated into the classification models' architecture. The last step regarding image processing, concerned the synchronization of the two videos as the recording was not triggered by a common signal. Thus, this was carried out by using a single synchronization point located at the center of each frame which corresponded to the start of the seam. Following that, as the framerates of the two cameras were different, a map was created matching a batch of 33 IR frames to a single RGB frame.

4   Results and Discussion

It is noted that, regarding training, the learning rate decayed over time with each iteration having as a result the algorithm to oscillate less as approached minima (see Fig. 3). The training process was carried out using an early stopping criterion with a patience value of 1 epoch and using the same test a train partition of the data as with the previous models. The network's parameters corresponding to the minimum loss observed during training were kept as the optimal ones. The classification performances of the MRN, ResNet18, and 2-Branch models on the test partition of the data are given in the Table 1 where the recall and precision metrics are calculated for each class [44]. The real-time inference performance of the models was evaluated by generating CUDA® [45] code for each model and using a MEX function as an entry point.

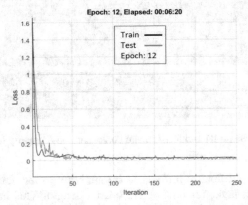

Fig. 3. Training progress of the 2-branch CNN

The networks after loaded to the memory were fed with the number of frames that would have been received for 10 s. The frames were passed as single instances to the models and the overall procedure was repeated 20 times. The average inference time for the MRN (10000 frames) was 7.4 s, for the ResNet18 (303 frames) was 1.7 s and for the 2-Branch (303 frames) was 2.3 s.

Table 1. Confusion matrices of the 3 models (test data).

	MRN		ResNet18		2-Branch	
	Precision	Recall	Precision	Recall	Precision	Recall
EP	97.1%	98.5%	99.7%	96.5%	99.7%	97.0%
GW	99.1%	99.2%	98.9%	100%	99.1%	99.8%
NW	98.9%	100%	100%	99.2%	100%	100%
PP	97.3%	96.1%	98.0%	97.3%	97.9%	97.6%

From Table 1 the ability of all the models to identify the GW and NW classes is superior compared to the EP and PP classes. This can be easily justified for the GW class as it includes the most instances and thus adding the most bias to the model's training. However, the same cannot be said for the NW class where the number of training instances was the smallest. Looking at the channels of the 1st convolutional layer it can be observed, especially for the RGB case, that the activations are more, and stronger when an image belongs to the NW class (see Fig. 4 - left).

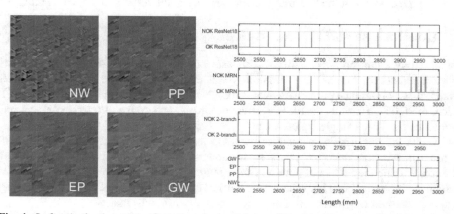

Fig. 4. Left - Activation of the first convolutional layer of ResNet18, right – Misclassification across the seam's length by the 3 models

In a real-world scenario, as well as in this approach the defects are the exception (minority) meaning that when one has occurred the system must be able to identify it. On the other hand, misfiring on such events and halting the production would raise the production cost. Hence, selecting the best model out of three comes down to calculating which has both the highest recall and precision scores for the PP and EP classes. This typically can be combined into a single metric namely F1 score [44]. As expected, the higher F1 score for these classes was achieved by the 2-Branch model (EP: 98.3%, PP: 97.7%) while to our surprise the ResNet18 model followed by slight margin (EP: 98.1%, PP: 97.6%). These findings highlighted that the main contribution regarding the identification of the EP and PP class are most probably due to the RGB images.

Before deriving a final conclusion, it must be taken into consideration that the models presented herein are meant to assess the quality of a seam across its length. In the figure above (see Fig. 4 - right) the target classes are plotted against the length of the seam. Additionally, the points where misclassification occurred by the models are plotted using a simple binary indicator. The misclassification of the EP and PP defects mainly occurred at the transition points and not along the area for which these defects persist. Considering that from the seam inspection results, the minimum length between two different quality labels was approximately 10mm and the maximum length for a misclassified area was a lot less than 10mm (Fig. 4 - right), the F1 score for these classes cannot affect the real-world performance of the model, as the process is fairly slow, and the models can offer real-time inference, below 1ms.

Coming to this point it is safe to conclude that all the models within the context of the current approach are capable of assessing the seam quality in real-time with 100% accuracy for all the classes. This implies that even with a cost-effective monitoring setup (RGB camera) the implementation of a quality assessment system for SAW is feasible without compromising the classification performance. Reaching the inference limits of the current approach as presented previously, would require on the one-hand a high spatial-resolution for the seam inspection procedure and on the other hand, a quite fast welding process given the fact that each camera is capable/configured for 1p/1mm.

5   Conclusions and Future Outlook

In this study, a non-invasive, online quality assessment approach was proposed for identifying defects for the SAW. Its feasibility was evaluated from the high-classification and real-time inference performance of the CNN models which achieved an average F1 score of 98.0% as regards the identification of porosity and undercut/overlap defects and inference time for a single frame <1 ms.

While the dual-modality monitoring setup increased the defect-identification performance, this was done by a very small amount. Additionally, it was observed that the misclassification was not persisted for a length greater than the smallest length for which the seam has been inspected. This implies that for that level of accuracy choosing a specific monitoring setup will not grand any real-world advantage.

From a system perspective, with the utilization of CNNs, the knowledge integration was proven once again that is moving towards standardization with the quality assessment functionalities to be as easy to integrate into a welding system as labeling some images. With such capacity both in terms of real-time utilization, assessment per length, future work would aim at the one hand to exhaust the capabilities of the system on faster welding scenarios (laser welding), test its adaptability, and introduce a standardized welding system development as regards quality assessment.

For future research, the maximum sampling rate for both cameras should be investigated, so that adequate information on the process is able to aggregated. Furthermore, there is still pending work to be done in multi-modal monitoring setup, as different types of manufacturing processes should be tested to check the complementarity of various sensors used.

Acknowledgements. This work is under the framework of EU Project AVANGARD, receiving funding from the European Union's Horizon 2020 research and innovation program (869986). The dissemination of results herein reflects only the authors' view and the Commission is not responsible for any use that may be made of the information it contains.

References

1. ASM Handbook committee: ASM Handbook: Welding, Brazing, and Soldering, 1st edn. ASM Int. (1993)
2. Wang, B., Hu, S.J., Sun, L., Freiheit, T.: Intelligent welding system technologies: state-of-the-art review and perspectives. J. Manuf. Syst. **56**, 373–391 (2020)
3. Mourtzis, D., Angelopoulos, J., Panopoulos, N.: Design and development of an IoT enabled platform for remote monitoring and predictive maintenance of industrial equipment. Procedia Manuf. **54**, 166–171 (2021)
4. CLAMIR. https://www.clamir.com/en/. Accessed 13 Feb 2022
5. Precitec Laser Welding Monitor LWM. https://www.precitec.com/laser-welding/products/process-monitoring/laser-welding-monitor/. Accessed 13 Feb 2022
6. 4D Photonics GmbH WeldWatcher®. https://4d-gmbh.de/how-is-process-monitoring-realized-by-the-weldwatcher/?lang=en. Accessed 13 Feb 2022
7. Günther, J., Pilarski, P.M., Helfrich, G., Shen, H., Diepold, K.: Intelligent laser welding through representation, prediction, and control learning: an architecture with deep neural networks and reinforcement learning. Mechatronics **34**, 1–11 (2019)
8. Masinelli, G., Le-Quang, T., Zanoli, S., Wasmer, K., Shevchik, S.A.: Adaptive laser welding control: a reinforcement learning approach. IEEE Access **8**, 103803–103814 (2020)
9. Franciosa, P., Sokolov, M., Sinha, S., Sun, T., Ceglarek, D.: Deep learning enhanced digital twin for closed-loop in-process quality improvement. CIRP Ann. **69**(1), 369–372 (2020)
10. Stavropoulos, P., Sabatakakis, K., Papacharalampopoulos, A., Mourtzis, D.: Infrared (IR) quality assessment of robotized resistance spot welding based on machine learning. Int. J. Adv. Manuf. Technol., 1–22 (2021). https://doi.org/10.1007/s00170-021-08320-8
11. Stavridis, J., Papacharalampopoulos, A., Stavropoulos, P.: Quality assessment in laser welding: a critical review. Int. J. Adv. Manuf. Technol. **94**(5–8), 1825–1847 (2018). https://doi.org/10.1007/s00170-017-0461-4
12. Zhang, B., Hong, K.M., Shin, Y.C.: Deep-learning-based porosity monitoring of laser welding process. Manuf. Lett. **23**, 62–66 (2020)
13. Zhang, Z., Wen, G., Chen, S.: Weld image deep learning-based on-line defects detection using convolutional neural networks for Al alloy in robotic arc welding. J. Manuf. Process. **45**, 208–216 (2019)
14. Cho, D.W., Song, W.H., Cho, M.H., Na, S.J.: Analysis of submerged arc welding process by three-dimensional computational fluid dynamics simulations. J. Mater. Process. Technol. **213**(12), 2278–2291 (2013)

15. Nezamdost, M.R., Esfahani, M.R.N., Hashemi, S.H., Mirbozorgi, S.A.: Investigation of temperature and residual stresses field of submerged arc welding by finite element method and experiments. Int. J. Adv. Manuf. Technol. **87**(1–4), 615–624 (2016). https://doi.org/10.1007/s00170-016-8509-4
16. Wen, S.W., Hilton, P., Farrugia, D.C.J.: Finite element modelling of a submerged arc welding process. J. Mater. Process. Technol. **119**(1–3), 203–209 (2001)
17. Karaoğlu, S., Secgin, A.: Sensitivity analysis of submerged arc welding process parameters. J. Mater. Process. Technol. **202**(1–3), 500–507 (2008)
18. Tarng, Y.S., Juang, S.C., Chang, C.H.: The use of grey-based Taguchi methods to determine submerged arc welding process parameters in hardfacing. J. Mater. Process. Technol. **128**(1–3), 1–6 (2002)
19. Gunaraj, V., Murugan, N.: Application of response surface methodology for predicting weld bead quality in submerged arc welding of pipes. J. Mater. Process. Technol. **88**(1–3), 266–275 (1999)
20. Murugan, N., Gunaraj, V.: Prediction and control of weld bead geometry and shape relationships in submerged arc welding of pipes. J. Mater. Process. Technol. **168**(3), 478–487 (2005)
21. Wikle III, H.C., Kottilingam, S., Zee, R.H., Chin, B.A.: Infrared sensing techniques for penetration depth control of the submerged arc welding process. J. Mater. Process. Technol. **113**(1–3), 228–233 (2001)
22. Knaak, C., Thombansen, U., Abels, P., Kröger, M.: Machine learning as a comparative tool to determine the relevance of signal features in laser welding. Procedia CIRP **74**, 623–627 (2018)
23. Cheon, S., Lee, H., Kim, C.O., Lee, S.H.: Convolutional neural network for wafer surface defect classification and the detection of unknown defect class. IEEE Trans. Semicond. Manuf. **32**(2), 163–170 (2019)
24. Zhang, Z., Li, B., Zhang, W., Lu, R., Wada, S., Zhang, Y.: Real-time penetration state monitoring using convolutional neural network for laser welding of tailor rolled blanks. J. Manuf. Syst. **54**, 348–360 (2020)
25. Jiao, W., Wang, Q., Cheng, Y., Zhang, Y.: End-to-end pre-diction of weld penetration: a deep learning and transfer learning based method. J. Manuf. Process. **63**, 191–197 (2021)
26. Liu, C., Law, A.C.C., Roberson, D., Kong, Z.J.: Image analysis-based closed loop quality control for additive manufacturing with fused filament fabrication. J. Manuf. Syst. **51**, 75–86 (2019)
27. Sudhagar, S., Sakthivel, M., Ganeshkumar, P.: Monitoring of friction stir welding based on vision system coupled with machine learning algorithm. Measurement **144**, 135–143 (2019)
28. Gu, J., Wang, Z., Kuen, J., Ma, L., Shahroudy, A., Shuai, B., et al.: Recent advances in convolutional neural networks. Pattern Recogn. **77**, 354–377 (2018)
29. Pan, H., Pang, Z., Wang, Y., Wang, Y., Chen, L.: A new image recognition and classification method combining transfer learning algorithm and MobileNet model for welding defects. IEEE Access **8**, 119951–119960 (2020)
30. Gellrich, S., et al.: Deep transfer learning for improved product quality prediction: a case study of aluminum gravity die casting. Procedia CIRP **104**, 912–917 (2021)
31. Papacharalampopoulos, A., Tzimanis, K., Sabatakakis, K., Stavropoulos, P.: Deep quality assessment of a solar reflector based on synthetic data: detecting surficial defects from manufacturing and use phase. Sensors **20**(19), 5481 (2020)
32. Friedman, J., Hastie, T., Tibshirani, R.: The Elements of Statistical Learning, 2nd edn. Springer, New York (2009). https://doi.org/10.1007/978-0-387-84858-7
33. Caggiano, A., Zhang, J., Alfieri, V., Caiazzo, F., Gao, R., Teti, R.: Machine learning-based image processing for on-line defect recognition in additive manufacturing. CIRP Ann. **64**(1), 451–454 (2019)

34. Yang, Z., Baraldi, P., Zio, E.: A multi-branch deep neural network model for failure prognostics based on multimodal data. J. Manuf. Syst. **59**, 42–50 (2021)
35. MATLAB Support Documentation. https://www.mathworks.com/help/deeplearning/ug/train-residual-network-for-image-classification.html. Accessed 13 Feb 2022
36. He, K., Zhang, X., Ren, S., Sun, J.: Deep residual learning for image recognition. In: 2016 Proceedings of the IEEE Conference on Computer Vision and Pattern Recognition, pp. 770–778 (2016)
37. Li, H., Xu, Z., Taylor, G., Studer, C., Goldstein, T.: Visualizing the loss landscape of neural nets. In: NeurIPS 2018, vol. 31 (2018)
38. Shorten, C., Khoshgoftaar, T.M.: A survey on image data augmentation for deep learning. J. Big Data **6**(1), 1–48 (2019)
39. Murphy, K.P.: Machine Learning: A Probabilistic Perspective, 1st edn. The MIT Press, London (2012)
40. MATLAB Deep Learning Toolbox. https://www.mathworks.com/products/deep-learning.html. Accessed 13 Feb 2022
41. NIT TACHYON 1024 microCAMERA. https://www.niteurope.com/en/tachyon-1024-microcamera/. Accessed 13 Feb 2022
42. LOGITECH C922 PRO HD STREAM WEBCAM. https://www.logitech.com/en-us/products/webcams/c922-pro-stream-webcam.960-001087.html. Accessed 13 Feb 2022
43. Kobe Steel Ltd.: The ABC's of Arc Welding and Inspection, 1st edn. Kobe Steel Ltd., Tokyo (2015)
44. Tharwat, A.: Classification assessment methods. Appl. Comput. Inform. **17**(1), 168–192 (2021)
45. NVIDIA CUDA Toolkit Documentation. https://docs.nvidia.com/cuda/. Accessed 13 Feb 2022

Detachable, Low-Cost Tool Holder for Grippers in Human-Robot Interaction

Christina Schmidbauer[✉], Hans Küffner-McCauley, Sebastian Schlund,
Marcus Ophoven, and Christian Clemenz

Institute of Management Science, TU Wien, Vienna, Austria
{christina.schmidbauer,hans.kueffner-mccauley,
sebastian.schlund}@tuwien.ac.at

Abstract. To hand over more than just pick & place tasks to an industrial collaborative robotic arm with a two-jaw gripper, the gripper must first be removed, and a new tool mounted. This tool change requires either human assistance or an expensive tool changer. The tools applied to the end-effector are often highly expensive and software system interfaces between different tools and robots are seldom available. Therefore, a holder was developed that allows the robot to pick up and operate a tool, such as an electric screwdriver, without having to demount the two-jaw gripper. Instead, the gripper's functionality is used to activate and deactivate the tool fixed to the holder. This paper presents the state-of-the-art of the underlying problem as well as the development process including simulations, the patented design, and the low-cost production of the tool holder. This detachable, low-cost tool holder enables a flexibilization of human-robot processes in manufacturing.

Keywords: Robot tools · Collaborative robots · Versatile production

1 Introduction

Industrial collaborative robotic arms (cobots) entered the market to enable physical human-robot interaction (HRI) in manufacturing. Due to their built-in force sensors, they allow safe interaction with human workers and employees on the shopfloor according to the applicable standards [10, 15]. This is a novelty in comparison to classical industrial robots, which must be separated from workers' workplaces by additional safety measures such as walls, fences or specific safety skins. It is important to note that not only the safety of the robot arm plays a significant role, but the entire application, including the end effector, tools, workpieces, peripherals, software, etc., must be considered when checking the conformity against safety of the application. This safety aspect enables new ways of human-robot collaboration.

Cobot manufacturers usually sell the robot arm without the end-effector, i.e. the tool, as this must be adapted to the application. The robot arm is therefore generically designed for a large number of applications, while the end-effector

© The Author(s) 2023
K.-Y. Kim et al. (Eds.): FAIM 2022, LNME, pp. 170–178, 2023.
https://doi.org/10.1007/978-3-031-18326-3_17

must be individually adapted to a specific task. These cobot tasks are e.g., assembly, inspection, kitting, joining, packing, and pick & place tasks as well as machine tending, screwing, gluing, soldering, and grinding [2,6].

In theory, this multitude of possible tasks allows a very high degree of flexibility with regard to the use of cobots. However, the mentioned safety considerations and either time-consuming manual or costly automatic tool changes limit this flexibility in practice. This particularly concerns manufacturers with small and medium batch sizes, where a high flexibility is required [8]. We present a new and patented solution to this problem of tool changing as we have designed and developed a detachable, low-cost tool holder which allows the cobot to switch autonomously between the function of a two-jaw gripper and a screwdriver at lower investment costs. The functionality of the holder has been patented (AT523914, WO2022000011, IPC B25J 015/02 [17]). Our design allows a flexible alternation between the human and cobot in the use of the tool such as a screwdriver, without removing the primary tool (two-jaw gripper) from the cobot. In addition, the tool (commercially available screwdriver) deployed can also be used by the human, enabling further cost savings.

In this paper, we present a short overview of available solutions (Sect. 2). In Sect. 3, we present the development process of our solution followed by a detailed description of our detachable, low-cost tool-holder for cobot grippers in terms of design and functions (Sect. 4). We finish with a discussion (Sect. 5), and a conclusion with outlook (Sect. 6).

2   State-of-the-Art

The integration of a tool to the cobotic arm consists of two steps: the hardware and the software integration. In terms of hardware integration there are existing tool holders and changers patented. Examples of prior art are activating a gun type tool through special end-effectors [5,16], special fixtures with electrical I/O capabilities [3], or form-fitting membranes enclosing the tool [4].

Tooling solutions for cobots on the market are solutions from cobot manufacturers, solutions from third party manufacturers, and manufacturers of automated and manual tool changing systems for cobots. Cobot manufacturers such as Franka Emika and ABB offer tools for their cobots. Newer third party tool specialists have established themselves on the market such as Robotiq and onRobot, but there are also established (automation) companies such as Schunk, Festo, and Zimmer selling various tools for cobots. Other available solutions focus on automated and manual tool changing systems such as those manufactured by SmartShift, TripleA-robotics, and Nordbo-Robotics, where tools are stored on a rack, similar to tool changing systems used in automated metrology systems. Due to their price positioning, these specific tools for cobots are suitable for high volumes, but less suitable for smaller lot sizes. Additionally, these systems cannot be operated by a human as these tools are controlled directly by the cobotic arm's software.

Software integration is a complex, individual topic: It depends on the cobot, tool and software environment that are used. To enable flexible HRI in manufacturing, easy software interfaces are used to quickly (re-)program the cobot without text-based programming knowledge. Best case, the cobot's software allows the integration of tools. This is offered, for example, by some manufacturers such as Universal Robots, but also by integrators such as Drag&Bot where tools only have to be integrated and configured once. Depending on the system, the integration and configuration can take up to one working day. If the software environment does not support the tool, then tools must be programmed via input and output signals. This is relatively simple regarding two-jaw grippers, but is complex or impossible with other tools. To find a solution to these challenges, a multi-stage approach was chosen which is detailed in the following section.

3   Development Process

As part of a student thesis project [11] in our learning and teaching factory, a low-cost solution was designed and developed to how a cobot, equipped with a standard two-jaw gripper, can also take over the task of screwing. A tool change should be avoided as far as possible in order to save time, costs and interfaces. The boundary conditions are given by the existing robot incl. gripper (Franka Emika Panda) and inline screwdriver (AEG SE 3.6), as well as the FDM (fused deposition modeling) 3D-printing process. The following requirements for the tool holder were identified using the factorization technique by Feldhusen and Grote [7]:

1. Enable picking up and setting down the entire holder.
2. Enable operating the screwdriver through the cobot's gripper.
3. Design for assembly and customization, including optimization of assembly connections, planning of assembly, possibilities for customization, and simple alignment of components.
4. Allow fixating and removing the screwdriver in and out of the holder.
5. Enable centering of the screwdriver.
6. Do not exceed the max. payload of robot (1kg for Franka Emika Panda): screwdriver (534g) + holder (x) $\leq$ payload of robot.

A morphological box was used to compare the different concepts and select a solution. The tool holder was designed using CAD software and finite element analysis, e.g., for most stressed parts such as the edge of the lock. The components were 3D-printed in polylactic acid (PLA) using FDM, assembled, and mounted to the cobot's gripper to validate its functionality. The weight of the first prototype was 418g. The successful validation showed that the holder was able to hold and operate the screwdriver. The initial design was then optimized considering two aspects: weight and 3D printing design. Topology optimization was used to reduce the weight of the holder by 10%. The optimized prototype was 3D printed in PLA using FDM and successfully validated in practice.

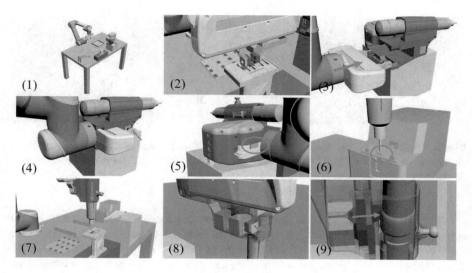

Fig. 1. (1) Full view of work table with cobot and latest holder prototype, (2) Pick & place function of the robot, (3) Screwdriver holder, (4) Holder fixation, (5) Holder locking, (6) Pick screw, (7) Positioning, (8) Screwdriver actuation, (9) Inside view of screwdriver actuation.

A process simulation illustrates the holder and its functionalities using the software "Autodesk 3ds Max®". The example scenario is a use case from the electronics industry, where a cobot assembles transistors on a heat sink. This requires the two-jaw gripper and the screwdriver holder. The operation consists of two parts, first a pick & place movement of the cobot in which it brings the heat sink and the transistors to a fixture, and then the tightening of bolts using the screwdriver. Therefore, the tool holder including the screwing tool has to be mounted to the gripper, then a bolt is magnetically attached to the tip of the screwdriver and screwed into the heat sink. Finally, the tool holder is removed from the gripper and set back into its initial resting position. Screenshots of the simulation (Fig. 1) illustrate the holder and its functionalities.

4   Detachable, Low-Cost Tool Holder for Robot Grippers

The design (screwing and manipulation module) and functions of the detachable, low-cost holder are described in the following. Figure 1 illustrates the simulations of the functions of the holder and Fig. 2 shows different views of the 3D-printed prototype. The screwing module was designed to incorporate a spring-loaded quick release system allowing for the fast and easy removal of the screwdriver. Pulling the release out, the screwdriver is removed. Letting go of the release causes the release to shoot back to its initial position. The design of the holster is individualized and only holds a specific screwdriver, but can be changed to accommodate different designs of inline tools. The cobot's gripper is fit into

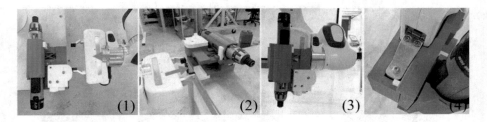

Fig. 2. (1–2) Tool holder fixation to the gripper, (3) side view of tool holder prototype mounted on a cobot, holding an inline screwdriver, and (4) screwdriver activation through two-jaw gripper.

a pocket in the manipulation module. The manipulation module contains the conversion mechanism, that converts the movement of the gripper's jaws to a force activating the screwdriver, consisting of two mirror-symmetric prisms and a counter-piece. When the jaws converge, the prisms converge toward each other causing the counter-piece to be pushed away, which in turn activates the initialization mechanism and subsequently the toggle switch of the screwdriver, see Fig. 1 (8–9). The gearbox houses the locking mechanism, which consists of a slider, gears, rack and pinion and the stop, responsible for blocking the locking mechanism. In its storage position, the slider is pushed inwards, until the gripper tool lifts the screwing tool, which causes the slider to move back into its initial position. The rack, which is mounted on the slider, causes the pinion to rotate, transferring force to the gears and subsequently to the lock, which is pushed forwards and thus secures the screwing tool to the gripper, see Fig. 1 (5).

To pick up the holder including the tool, the gripper's twin jaws are set to the maximum width position so that the jaws are guided into the holder (Fig. 1 (3)). In its initial position, the holder lies on an individualized stand, which through an extrusion keeps the locking mechanism deactivated by pushing the slider inwards. Once inside the screwing tool at the designated position, the jaws slowly converge until merely contact is made with the initialization mechanism (Fig. 1 (4)). The locking mechanism is activated to guarantee the fixation of the screwing tool to the gripper (Fig. 1 (5)). This mechanism is initiated by moving the gripper upwards and away from the stand, which causes the slider to fall outwards, and the locking mechanism to be activated, securely fixing the tool holder to the cobot gripper.

The operation of the screwdriver starts with picking up bolts through a magnetic tipped screwing bit and an automatic bolt feeding machine (Fig. 1 (6)). The cobot moves to the automatic bolt feeder, picks up a bolt and moves to the heat sink, placing the bolt into the threads of the previously placed transistor into the fixture (Fig. 1 (7)). In order for the bolt to twist itself into the threads, the screwdriver has to be activated. The gripper's jaws converge beyond the point of contact inside the pocket of the screwing tool, which causes the prisms inside the conversion mechanism to push against the counter-piece thus activating the toggle switch of the screwdriver (Fig. 1 (8–9)). Once the bolt has been tightened, the gripper's jaws diverge so that the screwdriver is deactivated. This process is repeated for the other bolts needed to complete the assembly.

In order to revert to pick & place sequences, the screwing tool has to be removed from the gripper of the cobot. This occurs analogue to the fixation of the tool. The tool is placed directly above the stand, so that the extrusion is directly underneath the slider. Once the holder is lowered so far that the slider is pushed upwards again, the lock is deactivated through the movement of the gears inside the locking mechanism. After the screwing tool has been set back onto the stand, the cobot can remove itself from the direct area of the screwing tool stand. The twin jaws diverge to the maximum width position, and the gripper backs out of the pocket of the screwing tool transversely. After successful removal, pick & place operations can be realized again.

5   Discussion

The described tool holder addresses the challenge of implementing cobot flexibility in terms of fast and flexible tool exchanges. The presented approach focuses especially on cobots as they are regarded to be used in very flexible environments with changing and rather small lot sizes. The cobot market volume is expected to grow up to 1B USD until 2023 and up to 8B USD until 2030 [9]. However, cobots are not seen as substitution to classical industrial robots, but as a new tool to open up new markets and applications. About 18,000 cobots were sold in 2019 in comparison to 363,000 industrial robots (5%) [9].

The potential price of the presented tool holder is positioned at the low-cost level of industrial tool holders. In terms of the envisaged costs of the system, the straightforward design, low-cost materials and the focus on off-the-shelf tools allows price ranges at about EUR 300–500, about 10% of the onRobot system which is considered as today's benchmark. The costs can be regarded as costs per function (such as screwdriving) as modules for the integration of further tools are under development. With the modular adaptability of the holster a similarly built tool could be used with an appropriate customized holster. Considering a growing market in robot usage in the industry grows naturally as the principal market for such a tool holder.

Against the mainstream of existing solutions, slight compromises of the needed availability and reliability are accepted at prototype level, but also regarding the final state. As the primary application area is HRI, this trade-off is actively accepted as long as a human co-worker may stand-in for the rapid problem-solving in low-threshold unforeseeable events (re-positioning, re-start, material replenishment for small lot sizes) [1].

As the solution today represents a technology readiness level of TRL5 (large scale prototype), further validation, integration and testing in different scenarios are necessary. The definition and approach of a considerable market poses a current two-sided challenge. First, the patented approach comprehensively covers robot/end-effector interfaces with a cutout/pocket such as the parallel jaw gripper (Fig. 2). Therefore, traditional interfaces are not covered by the presented solution. However, a significant share of cobot manufacturers offers solutions with a pre-assembled end-effector, usually a jaw gripper. Second, the presented solution stands in competition with existing tools and tool-changing systems. Therefore, regarding the commercialization, besides cobot and end-effector manufacturers, further potential partners such as automation companies, system integrators and tool manufacturers are investigated. In order to account for a significant market the prototype needs to be validated in terms of safety. Preliminary work on the conformity against safety of envisaged applications is considered within the SafeSecLab project [13].

6   Conclusion and Outlook

In this paper, we presented a new and patented tool holder that allows a cobot to switch autonomously between the function of a two-jaw gripper and a commercially available screwdriver, which can also be used by the human co-worker. This increases flexibility in terms of HRI and robot functionalities. The developed tool holder works around drawbacks of existing solutions such as automatic tool changers and specific tools for cobots by providing the robotic arm with a tool it can mount and remove independently, enabling it to switch between gripper and screwing tasks when required without manual tool installation through a worker. The existing prototypes were designed to operate in a controlled environment in the TU Wien Pilot Factory for Industry 4.0. For a broader use in industrial applications and environments the next step will include the transition from a prototype to a product. In this step different use cases will be validated and accordingly, business models will be created. Future work on the industrialization of the tool holder including an analysis of the production techniques is necessary. The general idea behind a tool holder is to improve the functionality of the robot in regard to different performable tasks. The more flexible a cobot can be used, the more different possible task allocations between human and machine become possible [14]. Thus, a direct positive impact on comprehensive work tasks and human factors, especially in terms of task load and adaptability is envisaged, contributing to more human-centered and even individualizeable work systems [12].

Acknowledgements. This research was supported by the Austrian Research Promotion Agency and the Austrian Ministry for Climate Action, Environment, Energy, Mobility, Innovation and Technology through the endowed professorship "HCCPPAS" (FFG-852789). The patent process was supported by the Research and Transfer Support, namely Karin Hofmann, at TU Wien. We thank Tudor B. Ionescu for his contribution in the development of the first idea of the holder, Johannes Hagenauer, who created the process simulation, Nemanja Miljevic for the further development of the holder, and Stefanie Bauer for her contribution to the market analysis.

References

1. Bauer, W., Ganschar, O., Pokorni, B., Schlund, S.: Concept of a failures management assistance system for the reaction on unforeseeable events during the ramp-up. Procedia CIRP **25**, 420 (2014)
2. Bauer, W.E., Bender, M., Braun, M., Rally, P., Scholtz, O.: Lightweight Robots in Manual Assembly - Best to Start Simply. Examining Companies' Initial Experiences with Lightweight Robots. Fraunhofer Institute for Industrial Engineering IAO, Stuttgart (2016)
3. DE102017120923A1 Verbindungseinrichtung und Installationssystem: (2022). https://worldwide.espacenet.com/patent/search/family/065441334/publication/DE102017120923A1?q=pn%3DDE102017120923A1
4. DE202014003133U1 Roboterarbeitsplatz: (2022). https://worldwide.espacenet.com/patent/search/family/052988036/publication/DE202014003133U1?q=pn%3DDE202014003133U1
5. EP3098031A1 Roboter mit Werkzeughalterträger: (2022). https://worldwide.espacenet.com/patent/search/family/053365474/publication/EP3098031A1?q=pn%3DEP3098031a1
6. Fast-Berglund, Å., Romero, D.: Strategies for implementing collaborative robot applications for the operator 4.0. In: Ameri, F., Stecke, K.E., von Cieminski, G., Kiritsis, D. (eds.) APMS 2019. IAICT, vol. 566, pp. 682–689. Springer, Cham (2019). https://doi.org/10.1007/978-3-030-30000-5_83
7. Feldhusen, J., Grote, K.H.: Pahl/Beitz Konstruktionslehre. Springer, Heidelberg (2013). https://doi.org/10.1007/978-3-642-29569-0
8. Gualtieri, L., Rojas, R.A., Ruiz Garcia, M.A., Rauch, E., Vidoni, R.: Implementation of a laboratory case study for intuitive collaboration between man and machine in SME assembly. In: Matt, D.T., Modrák, V., Zsifkovits, H. (eds.) Industry 4.0 for SMEs, pp. 335–382. Springer, Cham (2020). https://doi.org/10.1007/978-3-030-25425-4_12
9. IFR: Demystifying collaborative industrial robots: Positioning paper (2020)
10. ISO/TS 15066:2016(en) Robots and robotic devices - Collaborative robots: (2016)
11. Ophoven, M.A.N.: Konzeption und Konstruktion einer Schrauberhalterung für eine Mensch-Roboter-Kollaboration in der Montage und Validierung mittels Prototyping; Note: restricted until 10/2023. TU Wien (2019)
12. Rupprecht, P., Schlund, S.: Taxonomy for individualized and adaptive human-centered workplace design in industrial site assembly. In: Russo, D., Ahram, T., Karwowski, W., Di Bucchianico, G., Taiar, R. (eds.) IHSI 2021. AISC, vol. 1322, pp. 119–126. Springer, Cham (2021). https://doi.org/10.1007/978-3-030-68017-6_18
13. SafeSecLab: (2022). https://safeseclab.tuwien.ac.at/
14. Schmidbauer, C., Schlund, S., Ionescu, T.B., Hader, B.: Adaptive task sharing in human-robot interaction in assembly. In: IEEE International Conference on Industrial Engineering and Engineering Management (IEEM), pp. 546–550. IEEE (2020)
15. The European Parliament and the Council of the European Union: Directive 2006/42/EC of the European Parliament and of the Council of 17 May 2006 on machinery, and amending Directive 95/16/EC (recast) (2006)

16. US3,620,095 Mechanically Actuated Triggered Hand: (2022). https:// patentimages.storage.googleapis.com/pdfs/US3620095.pdf
17. WO2022000011A1 Tool holder for the detachable connection to a two-jaw gripper on a robot arm: (2022). https://worldwide.espacenet.com/publicationDetails/ originalDocument?FT=D&date=20220106&DB=&locale=de_EP&CC=WO& NR=2022000011A1&KC=A1&ND=2#

Intelligent Robotic Arm Path Planning (IRAP²) Framework to Improve Work Safety in Human-Robot Collaboration (HRC) Workspace Using Deep Deterministic Policy Gradient (DDPG) Algorithm

Xiangqian Wu, Li Yi$^{(\boxtimes)}$, Matthias Klar, Marco Hussong, Moritz Glatt, and Jan C. Aurich

TU Kaiserslautern, P.O. Box, 3049, 67653 Kaiserslautern, Germany
li.yi@mv.uni-kl.de

Abstract. Industrial robots are widely used in manufacturing systems. The places that humans share with robots are called human-robot collaboration (HRC) workspaces. To ensure the safety in HRC workspaces, a collision-avoidance system is required. In this paper, we regard the collision-avoidance as a problem during the robot action trajectory design and propose an intelligent robotic arm path planning (IRAP²) framework. The IRAP² framework is based on the deep deterministic policy gradient (DDPG) algorithm because the path planning is a typical continuous control problem in a dynamic environment, and DDPG is well suited for such problems. To test the IRAP² framework, we have studied a HRC workspace in which the robot size is larger than humans. At first, we have applied a physics engine to build a virtual HRC workspace including digital models of a robot and a human. Using this virtual HRC workspace as the environment model, we further trained an agent model using the DDPG algorithm. The trained model can optimize the motion path of the robot to avoid collision with the human.

Keywords: Human-robot collaboration · Path planning · Deep deterministic policy gradient · Reinforcement learning

1 Introduction

To handle more complex manufacturing tasks, industrial robots are widely used in manufacturing systems because robots can provide fast and precise executions in repetitive tasks [1]. Nevertheless, robots lack the flexibility and adaptability of humans, and therefore, recent robotics research has focused on human-robot collaboration (HRC), which ensures both precision and flexibility in manufacturing systems [2].

The places that humans share with robots are called HRC workspaces [2]. Whenever robots are working with humans in HRC workspaces, security concerns apply. For example, safety regulations are elaborated by numerous standards (e.g. ISO 10218). In

© The Author(s) 2023
K.-Y. Kim et al. (Eds.): FAIM 2022, LNME, pp. 179–187, 2023.
https://doi.org/10.1007/978-3-031-18326-3_18

conventional scenarios, robots need to be separated from humans by specific equipment, e.g. protective fences. As HRC workspaces require the coexistence of humans and robots in one place, new safety concerns are of importance, and former separation regulations systems cannot persist in HRC workspaces.

Aiming at the resulting safety problem in HRC workspaces, two major categories of measures are commonly applied [3]. The first category intends to minimize the injury risk when collisions between humans and robots cannot be avoided. Measures in this category include mechanical compliance systems (e.g., viscoelastic covering [4] or mechanical absorption systems [5]), lightweight robot structures (e.g. [6]) and safety strategies involving collision or contact detection respectively (e.g. [7]). The commercial robots applied in HRC workspaces usually comprise one or several of these features [3]. Another category includes measures that achieve an active collision avoidance. These measures incorporate information about the robot motion and the human operations using vision systems or other sensing modules. Based on this information, alternative trajectory paths are generated to avoid the forecasted collision [3]. The works related to the collision-avoidance based on different sensors can be found in [8–13].

In addition to sensorics, deep reinforcement learning (RL) is another important approach to realize the collision-free path planning in HRC workspaces. RL is a subclass of machine learning and consists of two main parts, the agent and the environment [14]. The agent receives a representation of the current state within the environment and selects an action based on a policy. Once the action is performed, the agent will receive a reward. The agent aims at learning a policy that maximizes the total discounted future reward [15]. RL has been used successfully in various application fields such as solving complex games [16], job shop scheduling [17], or factory layout planning [18]. In terms of the collision problems in HRC workspaces, the implementation of RL can be found in a number of studies, e.g. [19–21].

However, in approaches related to the RL-based collision-avoidance, the size of robots in their HRC workspaces is smaller than humans, and the case where the robot size is larger than human is not considered. When the robots are small, even if the collision cannot be avoided, the location of the collision is mostly in the human hands or arms, which does not lead to a high risk of fatal injuries. But when the size of the robot is larger than humans, the collision may occur in the head or torso, resulting in a higher risk of fatal injuries. Therefore, we are focusing on the case where the robot size is larger than a human and are proposing an intelligent robotic arm path planning (IRAP^2) framework. The IRAP^2 framework and its case are explained in remainder of the paper.

2   Problem Statement and Methodology

2.1   Problem Statement

In our case, we scaled the scenario down to the size of a desktop scenario, as depicted in Fig. 1. In our desktop-level HRC workspace, the height of the base of the robot is 138 mm, and the lengths of the first and second connecting links are 135 mm and 147 mm, respectively. Neglecting the degree of freedom (DoF) of the attached vacuum gripper, the robot has 3 DoF, as labeled from $J1$ to $J3$ in Fig. 1. The movement ranges of $J1$,

J2, and *J3* are $(-135°$ to $+135°)$, $(0°$ to $+85°)$, and $(-10°$ to $+95°)$, respectively. The maximum rotation speed of the joints is 320°/s. To make the human model compatible for the small robot, the height of the human is downscaled to 129 mm.

To make the training environment as similar as possible to the real physical environment, we have applied the 3D physics engine 'PyBullet' to build a 1:1 virtual model for training the RL model in the IRAP² framework. The virtual model consists of three parts: a virtual robot arm, a virtual human, and a virtual pickup object, as depicted in Fig. 1. The problem has been defined as to find out the shortest path to pick up the blue object without colliding with the human. In this work, we consider four different cases: (1) Pick up the target object with no humans (as comparison reference case); (2) Pick up the object with one human standing at one specific position; (3) Pick up the object with one human standing at one of two positions; and (4) Pick up with two humans standing at two specific positions, as depicted in Fig. 1.

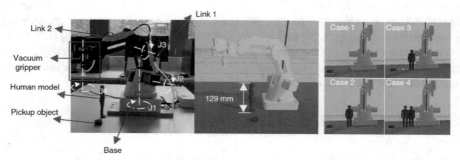

Fig. 1. Real and virtual scenario of a HRC workspace

2.2 The IRAP² Framework Based on DDPG

In HRC workspace, the motion path of the robot can be regarded as a sequence of decisions and can be planed using Deep Deterministic Policy Gradient (DDPG) algorithm.

Figure 2 illustrates the IRAP² framework based on the DDPG algorithm. In the virtual 3D physics environment, the virtual robot arm is allowed to explore various positions within the described ranges and obtains a reward according to its interaction with the environment. The action, current state, next state, and reward of the virtual robot can be denoted as a list of tuples $\{a_t, s_t, s_{t+1}, r_t\}$, which will be stored in the replay buffer and can be used as data for training the artificial deep neural network. For each training iteration, the replay buffer will sample 64 batches of $\{a_t, s_t, s_{t+1}, r_t\}$, and a critic network with the weight w_Q will calculate the state value $Q(s_t, a_t, w_Q)$ that determines the cumulated rewards of the state s_t. Furthermore, an actor network with the weight w_μ is used to obtain the behavioral policy $\mu(s_t, w_\mu)$, which is the action of the virtual robot for the next time step. For the stability of the training, two target networks are created for the critic and actor network, which are denoted as $Q'(s'_t, a'_t, w'_Q)$ and $\mu'(s'_t, w'_\mu)$, respectively. Weights of two target networks are updated slowly for each iteration (Soft Update). The update of weights of current critic and actor network in the DDPG algorithm is performed by minimizing the loss function through RMSProp optimizer. Loss functions for the actor and critic network (L_a and L_c) are expressed as follows.

$$L_a(w_\mu) = \nabla_\mu Q(s_t, a_t, w_Q) \nabla_{w_\mu} \mu(s_t, w_\mu) \tag{1}$$

$$L_c(w_Q) = \left[r + \gamma Q'\left(s'_t, \mu'(s'_t, w'_\mu), w'_Q\right) - Q(s_t, a_t, w_Q) \right]^2 \tag{2}$$

In Eq. (1) and (2), r denotes the reward, γ is the discount factor, and ∇ describes the gradient. The action, state, and reward functions of the four cases are summarized in Table 1, where φ_1, φ_2, φ_3 are the rotation angle of joints J_1, J_2, and J_3, respectively. The parameters α, β, ε, δ are the scaling constants to convert the distance values and reward values from environment model in PyBullet to the values that are suitable for training the DDPG neural networks. The parameters d_{ta}, d_{h_1}, d_{h_2} are the distance of the vacuum gripper to the target object, the first human, and the second human, respectively. Finally, the parameter i is an index to indicate whether the object has been successfully picked up or not. After the training, the optimal path of the agent can be exported to control the real robot in the HRC workspace.

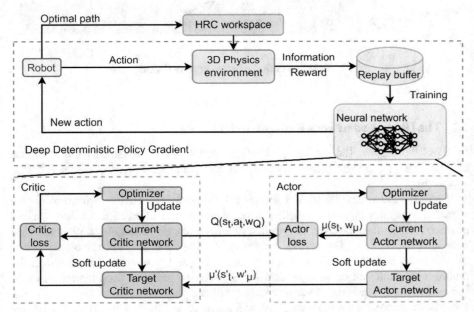

Fig. 2. Illustration of the IRAP^2 framework

Table 1. State and reward functions

Case	Action a	State s	Reward r	
1		$[\alpha p_r, \beta d_{ta}, i]$		Pick up the object on top
2, 3	$[\varphi_1, \varphi_2, \varphi_3]$	$[\alpha p_r, \beta d_{ta}, \varepsilon d_{h_1}, i]$	$r = \begin{cases} 150 \\ -80 \\ \delta d_{ta} \end{cases}$	Knock down human or object
4		$[\alpha p_r, \beta d_{ta}, \varepsilon d_{h_1}, \varepsilon d_{h_2}, i]$		otherwise

3   Results and Discussion

3.1   Evaluation of the Training Process

The first result is the training performance. Figure 3 shows the number of steps (one step implies one action by the agent) and rewards versus the training episodes. During each training episode, if the target object has been picked up, the episode will be closed. Moreover, the maximal steps are set to 300. In Fig. 3, it is seen that in the first case, the agent was not able to catch the target object until about 200 episodes. From 200 episodes to about 300 episodes, the number of steps is reduced to appr. 100, but the optimal path is not yet found. From the reward plot, it is seen that the agent in first case can always reach the optimal grasping path after about 300 episodes. In the second case, the step and reward plots clearly show that the agent can always reach the optimal path after about 200 episodes. In the third and fourth cases, the stable optimal path generation is not achieved until more than 500 episodes.

Comparing all cases, it can be concluded that the IRAP2 framework has a higher training efficiency when there is only one human standing at a fixed position (i.e. in the second case). The third and fourth cases have more possible positions and humans, and therefore, the agent needs more episodes. Moreover, the minimum number of steps required to pick up the object is about 30 steps, and the maximum reward is about 150.

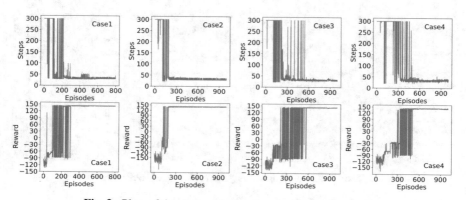

Fig. 3. Plots of the steps and rewards versus training episodes

3.2 The Optimal Pick-Up Path Generated by IRAP² Framework

The optimal pickup paths for the four cases are shown in Fig. 4, where a purple or light-blue line frame in the 3D-plots of diagonal view describes the links between the robot's joints, and each frame represents one state of the agent, as outlined in the first case in Fig. 4. The 2D-line plots under the 3D-plots are the grasping path the gripper from the top view. In viewing the 3D plots, it is seen that in the first case, the robot's joint J_1 rotates directly counterclockwise around the z-axis, the robot's two links descend almost in a straight line, and the gripper picks up the target object directly. In the second case, there is a process of keeping the robot's gripper highly parallel when it is almost close to the human (as highlighted by the green box), which implies the robot's decision to avoid the collision with the human. In the third case, the purple lines represent the robot's path when the human appears at position P1 (yellow human), while the light blue lines represent the path when human appears at position P2 (red human). Since the red human is more outward (in the positive direction of the x-axis), the path of the light blue lines is located a little more outward compared to the path of the purple lines. This is because in the fourth case, two humans are standing in the HRC workspace at the same time, and the robot tries to pick up the target object waving its arm from the outside (in the positive direction of the x-axis) around the two humans, in order to avoid a collision. In viewing the 2D-plots, it is seen that the gripper has successfully avoided the collision with the humans.

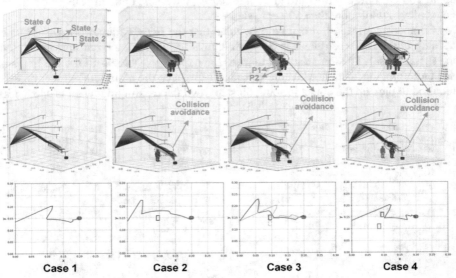

Fig. 4. The optimal pick-up paths generated by the IRAP² framework

3.3 Validation of the Robot Control

Finally, our optimal paths are successfully applied in four different scenarios. Figure 5 shows the control process in the second case as an example.

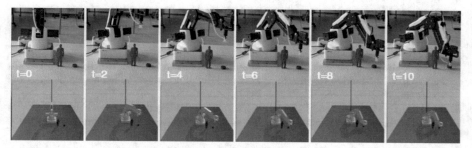

Fig. 5. Validation of the optimal path in case 'pick up the object with one human standing'

In this case, it can be seen that from 0 to 2 s, the robot arm lifts upward to avoid collision with the human. From 2 to 10 s, the robot arm moves around the human and picks up the target object. Without our algorithm, the robot would move in a straight line directly along the direction of the target object and crash into the human model. With this validation, we successfully confirmed the feasibility of the IRAP^2 framework.

4 Conclusion and Outlook

In this work, we have confirmed the feasibility of the IRAP^2 framework as well as the DDPG algorithms to generate the optimal path of robot in HRC workspaces. Moreover, in a desktop HRC workspace scenario, we studied that the case that the size of a robot is larger than humans and considered different working conditions. In terms of future work, it is suggested to upscale the implementation scenario of the IRAP^2 framework to a real industrial HRC workspace. Moreover, in this work, all humans are assumed to be standing at one position or moving between two certain positions. In the future work, a more complex moving behavior of the humans must be considered in the problem. In addition, it is to mention that big robots are relatively more dangerous than small robots, and the implementation of such HRC requires a higher level of safety measures. Furthermore, the cases in which robot and human sizes are similar should be considered in the future work because such a problem may be more complex, since the robot and human have similar velocities and workspaces dimensions, and the robot needs more accurate and fast response capabilities. Finally, another future work should be focused on the improvement of the computing efficiency of the algorithm as well as the comparison of our approach with other existing optimization approaches such as genetic algorithms or other RL approaches such as normalized advantage function.

References

1. Wang, L., et al.: Symbiotic human-robot collaborative assembly. CIRP Ann. **68**, 701–726 (2019)
2. Wang, L., Liu, S., Liu, H., Wang, X.V.: Overview of human-robot collaboration in manufacturing. In: Wang, L., Majstorovic, V.D., Mourtzis, D., Carpanzano, E., Moroni, G., Galantucci, L.M. (eds.) Proceedings of 5th International Conference on the Industry 4.0 Model for Advanced Manufacturing. LNME, pp. 15–58. Springer, Cham (2020). https://doi.org/10.1007/978-3-030-46212-3_2
3. Robla-Gomez, S., Becerra, V.M., Llata, J.R., Gonzalez-Sarabia, E., Torre-Ferrero, C., Perez-Oria, J.: Working together: a review on safe human-robot collaboration in industrial environments. IEEE Access **5**, 26754–26773 (2017)
4. Yamada, Y., Morizono, M., Umetani, U., Takahashi, T.: Highly soft viscoelastic robot skin with a contact object-location-sensing capability. IEEE Trans. Ind. Electron. **52**, 960–968 (2005)
5. Zinn, M., Khatib, O., Roth, B.: A new actuation approach for human friendly robot design. In: Proceedings of the IEEE International Conference on Robotics and Automation, ICRA 2004, vol. 1, pp. 249–254. IEEE (2004)
6. Hirzinger, G., et al.: DLR's torque-controlled light weight robot III-are we reaching the technological limits now? In: Proceedings of the 2002 IEEE International Conference on Robotics and Automation (Cat. No. 02CH37292), pp. 1710–1716. IEEE (2002)
7. Yamada, Y., Hirasawa, Y., Huang, S., Umetani, Y., Suita, K.: Human-robot contact in the safeguarding space. IEEE/ASME Trans. Mechatron. **2**, 230–236 (1997)
8. Tan, J.T.C., Duan, F., Zhang, Y., Watanabe, K., Kato, R., Arai, T.: Human-robot collaboration in cellular manufacturing: design and development. In: 2009 IEEE/RSJ International Conference on Intelligent Robots and Systems, pp. 29–34. IEEE (2009)
9. Corrales, J.A., Gómez, G.J.G., Torres, F., Perdereau, V.: Cooperative tasks between humans and robots in industrial environments. Int. J. Adv. Robotic Syst. **9**, 94 (2012)
10. Ceriani, N.M, Buizza Avanzini, G., Zanchettin, A.M, Bascetta, L., Rocco, P.: Optimal placement of spots in distributed proximity sensors for safe human-robot interaction. In: 2013 IEEE International Conference on Robotics and Automation, pp. 5858–5863 (2013)
11. Bascetta, L., et al.: Towards safe human-robot interaction in robotic cells: An approach based on visual tracking and intention estimation. In: 2011 IEEE/RSJ International Conference on Intelligent Robots and Systems, pp. 2971–2978. IEEE (2011)
12. Schiavi, R., Bicchi, A., Flacco, F.: Integration of active and passive compliance control for safe human-robot coexistence. In: 2009 IEEE International Conference on Robotics and Automation, pp. 259–264. IEEE (2009)
13. Flacco, F., Kroger, T., Luca, A. de, Khatib, O.: A depth space approach to human-robot collision avoidance. In: 2012 IEEE International Conference on Robotics and Automation, pp. 338–345. IEEE (2012)
14. Joshi, A.V.: Machine Learning and Artificial Intelligence. Springer, Cham (2020). https://doi.org/10.1007/978-3-030-26622-6
15. Sutton, R.S., Barto, A.: Reinforcement Learning: An Introduction, 2nd edn. Adaptive Computation and Machine Learning. The MIT Press, Cambridge (2018)
16. Silver, D., et al.: A general reinforcement learning algorithm that masters chess, shogi, and Go through self-play. Science **362**, 1140–1144 (2018)
17. Aydin, M., Öztemel, E.: Dynamic job-shop scheduling using reinforcement learning agents. Robot. Auton. Syst. **33**, 169–178 (2000)
18. Klar, M., Glatt, M., Aurich, J.C.: An implementation of a reinforcement learning based algorithm for factory layout planning. Manuf. Lett. **30**, 1–4 (2021)

19. El-Shamouty, M., Wu, X., Yang, S., Albus, M., Huber, M.F: Towards safe human-robot collaboration using deep reinforcement learning. In: 2020 IEEE International Conference on Robotics and Automation (ICRA), Piscataway, NJ, pp. 4899–4905. IEEE (2020)
20. Prakash, B., Khatwani, M., Waytowich, N., Mohsenin, T.: Improving safety in reinforcement learning using model-based architectures and human intervention (2019)
21. Liu, Q., Liu, Z., Xiong, B., Xu, W., Liu, Y.: Deep reinforcement learning-based safe interaction for industrial human-robot collaboration using intrinsic reward function. Adv. Eng. Inform. **49**, 101360 (2021)

A Conceptual Framework of a Digital-Twin for a Circular Meat Supply Chain

M. R. Valero[✉], B. J. Hicks, and A. Nassehi

University of Bristol, Bristol, UK
Maria.Valero@bristol.ac.uk

Abstract. Every year more than 900 million tonnes of food is wasted, contributing to almost 10% of total greenhouse gas emissions. Reducing food waste has been identified as essential to tackle the current climate crisis, and links to several UN's sustainable development goals. This is especially critical for energy and resource-intensive food products like meat, whose consumption is predicted to reach an historical maximum by 2030. Whilst wastage occurs at all stages of the supply chain, tractable data about the journey of food from production to consumer remains largely hidden or unrecorded. Powered by the latest advances in sensing like smart food packaging and digital technologies such as Big Data and IoT, Digital Twins offer a valuable opportunity to monitor and control meat products and processes across the whole supply chain, enabling food waste to be reduced and by-products reintegrated into the supply chain. This paper proposes a new framework for a Digital Twin that integrates key technological enablers across different areas of the meat supply chain towards with the goal of a "zero-waste", circular meat supply chain.

Keywords: Digital twins · Supply chain · Food

1 Introduction

The food consumed by individuals and organisations is a daily choice that has far reaching effects. Food consumption and waste can be linked directly to the 2nd and 3rd and indirectly to the 12th Sustainable Development Goals (SDGs) [1]. It is estimated that roughly one third of food produced for human consumption is either lost or wasted [2]. Wastage of food amounts to significant environmental damage, from pollution to the loss of the commodity and the energy required to create it [3]. The environmental impacts from reducing food loss and waste are far-reaching and profound, affecting SDG 6 (sustainable water management), SDG 13 (climate change), SDG 14 (marine resources), SDG 15 (terrestrial ecosystems, forestry, biodiversity), and many other SDGs, such as the Zero Hunger goal (SDG 2), calling for an end to hunger, the achievement of food security, improved nutrition and sustainable agriculture [4].

Meats are considered safety-critical foods; meaning they are susceptible to spoilage and pathogenic microorganisms which affect both food quality and safety [5]. Meat production like beef is over 30 times more carbon intensive than tofu [6], with green

K.-Y. Kim et al. (Eds.): FAIM 2022, LNME, pp. 188–196, 2023.
https://doi.org/10.1007/978-3-031-18326-3_19

and blue water requirements exceeding that of most other foods [7]. With global meat consumption set to increase by 12% by the end of the decade [8], sustainable, safe and efficient meat production will be key for the attainment of SDGs.

Technological advancements in the storage, packaging, transportation, and sale of food products have contributed to increased safety and efficiencies in the food supply chain. Emerging digital technologies such as the Internet of Things (IoT), Big Data, etc. provide a growing range of opportunities to transform food systems. Of the various digital technologies in development, Digital Twin (DT) is well poised to underpin the solution of many of the food chain's problems like loss and waste [9], increase yields and efficiencies and global meat consumption [8]. Furthermore, with supply chains moving to "value webs" characterized by complex, connected and interdependent relationships, DTs enhanced capabilities for learning and collaboration will be critical to enable real-time supply chain optimisation and resilience [10].

Whilst DT technologies are currently being trialled to enhance productivity, especially by large enterprises [11], their applications are usually limited to one or only a few areas of the supply chain, missing the more holistic opportunities and benefits DTs can offer for a "zero-waste" circular meat supply chain.

To address this gap, this paper proposes a conceptual framework of DT for a circular meat supply chain. This paper is organised with background information relating to the key areas in a circular meat supply chain in Sect. 2. Section 3 proposes a conceptual framework of DT for sustainable meat production within a circular meat supply chain, with a description of its major areas. The final section draws some conclusions and suggestions for future work.

2 Background

The circular economy is a model of production born from the idea of "zero waste" and extends this idea to reduce overconsumption and restore and regenerate ecosystems and natural capital [12]. A circular economy aims to maximise resourcefulness; reducing material resources taken, improving the efficiency of what is made and recycling what is disposed to feed back into the cycle [13].

Contrary to a "take–make–dispose", linear food supply chain, a circular supply food supply chain (CFSC) does not assume infinite resources. A CFSC seeks to reduce the loss and waste produced by all stages, from production, storage, transport to end users. With the inevitable waste that does occur, the resource is reused to extract energy (e.g., use of anaerobic digestors) and the nutrients cycled back into the food production system, thus closing the loop.

In this paper, 8 key areas have been identified for a circular meat supply chain and are depicted in Fig. 1. These are: land management, animal management, food processing, food products and packaging, transportation, retail, household and hospitality, and waste management. Land management refers to any form of husbandry that concerns soil or plants and animal management that of livestock. Food processing is differentiated from products and packaging as the management of processes rather than the output of processed and packaged food products. Transportation encompasses the movement at any stage, of the food products. Retail concerns the sale of the commodity.

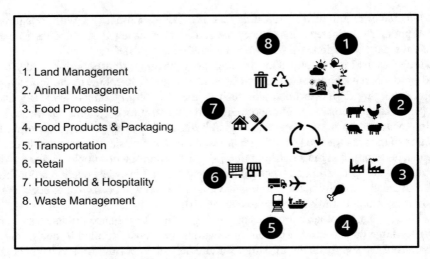

1. Land Management
2. Animal Management
3. Food Processing
4. Food Products & Packaging
5. Transportation
6. Retail
7. Household & Hospitality
8. Waste Management

Fig. 1. Key areas in a circular meat supply chain.

This is different from household and hospitality, which primarily concerns the consumption of the foodstuff. Finally, waste management covers any form of recycling of lost and wasted food offering the resource back to the cycle.

3 A Conceptual Framework of Digital-Twin for a Circular Meat Supply Chain

Digital Twins (DTs) are typically described as consisting of physical objects, their virtual counterparts, and the data connections in between [14]. DTs close the data feedback loop between the digital and the physical objects, so the data flows between the physical and a digital object in both directions. All these DT characteristics and relationships are made explicit in the proposed conceptual framework of Digital-Twin for a circular meat supply chain (Fig. 2).

As shown in Fig. 2, the *Data* between physical and digital assets has a two-way flow. In the *Sensing* stream, measured state data of *Physical Objects* flows from to the digital realm to build their corresponding *Digital Objects*. The latter generate the actionable data corresponding to strategy, decisions and actions in the physical realm, which flow back from the *Digital Objects* to the *Physical Objects* (*Actuation*).

In the physical realm, real-time state data of the *Physical Objects* in each area is collected by sensing *Edge Devices*. An array of sensors/transducers in a wireless sensor network (WSN) measure relevant physical parameters, which are transformed into electric voltages or currents. If necessary, interfacing circuits (e.g., Analogue-to-Digital Converter) convert analogue electrical signals into digital format. In each WSN node, a microcontroller (either a microprocessor or a single-board computer) collects and sends the digitalised sensor data to a transceiver. Wireless communication technologies like Bluetooth, Wi-Fi, GPRS and NFC provide connectivity capabilities for diverse edge devices.

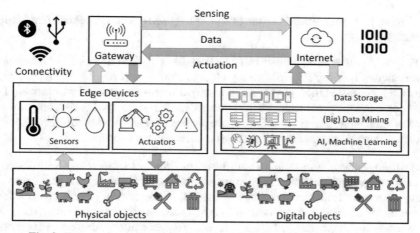

Fig. 2. Proposed conceptual framework of digital-twin and implementation.

A *Gateway* is usually located in the vicinity of the connected devices. Sometimes, a proxy server may collect and process data to send it the *Internet* by using MQ Telemetry Transport (MQTT) standards, or HTML or Extensible Messaging and Presence Protocols (XMPP). Nowadays, the use of Android smart devices and equivalent operating systems is increasing in popularity, as they can be employed as a gateway for 3G and 4G networks [15].

Given the large number of sensors, amount and variety of data collected across the different areas of the food/meat supply chain, Big-Data-type solutions are required for such large and complex datasets. Whilst structured data (e.g., temperature, location, etc.) would favour SQL databases (e.g., Oracle, MySQL, etc.) for *Data Storage*, the use of No-SQL databases (e.g., MongoDB, Cassandra, etc.) allows the inclusion of unstructured data (e.g., digital photos and video files of crops, animals, stocks, etc.).

Raw data can be stored in a data lake for subsequent *Data Mining*. Starting from descriptive analytics, data is sorted and cleaned to assess the status of the physical objects, which will inform the model(s) for their corresponding *Digital Objects*. Through *AI, Machine Learning* insights from collected, relevant data can be teased out to understand why any particular events or changes in the state of the physical objects happened. Combined with historical data, predictive analytics enable to forecast the likelihood of future events. Using prescriptive analytics, insights about what to do to achieve a particular outcome (e.g., maximise yield, etc.) can be obtained.

Informed by the real-time data collected from the physical objects, the master model underlying the DT's *Digital Objects* is used to generate actionable data regarding the decisions and actions to be performed on the physical objects (*Actuation*), namely through actuators. Given the finite resources, (e.g., water, fuel, etc.), reinforcement learning could help with resource allocation for specific goals (e.g., crop yields, animal growth, in-time delivery to warehouses, retailers, etc.). A data warehouse could be used as a repository for this filtered, actionable data (e.g., digital objects' data).

3.1 Proposed Conceptual Framework of DT Applied to a Circular Meat Supply Chain

The proposed conceptual framework of Digital-Twin applied to a circular meat supply chain is shown in Fig. 3. By integrating all areas, processes, objects and technologies, traceability, controllability and safety in the food supply chain can be improved: from feed crops, animals, processing, distribution, sale, use and disposal of the food products. Data and knowledge across the supply chain are linked (Fig. 4), so production, planning, resource use and logistics are optimised to reduce waste and costs.

Land Management. The state and attributes of *Physical Objects* like soil pH, moisture content and nutrient levels (e.g., nitrogen (N), phosphorus (P) and potassium (K)) can be monitored at the *Sensing* stage of the DT, in addition to plant/crop growth and environmental conditions (solar radiation/light levels, temperature, humidity, etc.). With scarce resources like water, DTs can help with resource allocation (e.g., irrigation) to meet specific conditions (e.g., soil moisture) that can maximise yields (*Actuation*).

Animal Management. *Physical Objects* in this area range from individual animals (e.g., level of exercise & rest and grazing patterns through GPS/location data) to environmental/housing conditions (e.g., temperature, humidity, light levels, etc.). DTs can provide insights into animal health and welfare by monitoring, predicting and influencing animal behaviour and environment conditions (*Sensing*). Combined with animal feed/crop data, DTs can help farmers and farm vets to maintain the optimum conditions for animal health and growth (*Actuation*).

Food Processing. DTs can enable the optimisation of the processes by which raw food materials are turned into final products. Namely, key environment (e.g., temperature) and operation conditions (e.g., machine blades, etc.) from processes like cutting and boning, chilling, rendering, etc.) are monitored (*Sensing*) and controlled (*Actuation*) to enhance quality and control (e.g. avoid over trimming, etc.). Combined with relevant operations and market data (e.g., raw material availability, customer demand, etc.), Big-Data, AI-powered DTs can add extra efficiencies in operations planning and scheduling via improved demand and supply forecasting.

Food Products and Packaging. Enabled by innovations in sensing and IoT, DTs can exploit technological advances like smart packaging [16] to monitor and control the safety and quality of processed/finalised products. Dotted with integrated sensors and intelligent labels, smart packaging can measure (*Sensing*) markers of freshness and/or identify the presence of harmful components in food. For *Actuation*, smart active packaging is showing promising results in providing augmented functions, from warning users when spoilage occurs to preserve the product. Examples of the latter include CO_2-emitting pads and antimicrobial preservative releasers that inhibiting microbial growth in meat and antioxidant releasers to reduce fat oxidation [16].

Transportation. The route food takes from farm to fork is complex, unique and unpredictable. DTs can capture the unique history of a product as it travels through the supply chain, thus offering improved traceability and authentication. With regards *Sensing*,

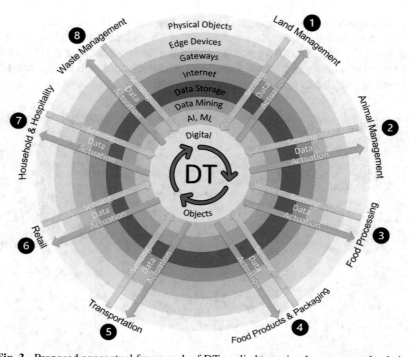

Fig. 3. Proposed conceptual framework of DT applied to a circular meat supply chain.

there are significant opportunities in the use of IoT-powered sensors in vehicles for food distribution and storage [17]. Combined with location (e.g., GPS) and smart packaging technologies, DTs could monitor of environmental conditions and quality evolution of food products during transport and storage, informing distribution and warehouse operations and planning (*Actuation*).

Retail. Reasons for food waste in the retail stage typically include inappropriate storage facilities and conditions (e.g., fridge/freezer errors), controls and quality checks at shelving (e.g., products passing beyond 'best before' dates) and inaccurate stock forecasting, like overstocking. Powered technologies like IoT and smart(er) packaging (*Sensing*), DTs can offer a valuable opportunity for the retail sector to reduce waste by providing accurate use by dates to inform stock management and planning, with their consequent impact on pricing (e.g., lower/more competitive pricing) and operational efficiencies (*Actuation*).

Household and Hospitality. Only in the UK, near 7.6Mt of food waste is generated every year by households and the hospitality sectors, up to 75% of which is avoidable [18]. The leading cause of waste generation in UK households is not using food in time, with 4.5Mt of still edible food products being thrown away and an associated loss of almost £14 billion [19]. With accurate, "real-time" use by dates, DTs can offer significant value to consumers and professionals for better meal and stock planning and ultimately less domestic and HFSS food waste.

Waste Management. Food waste and by-products can be re-integrated in the supply chain in several ways, such as by redistributing surplus food and diverting into animal feed [19]. Namely, DTs can help managing and coordinating surplus redistribution (between donors and beneficiaries) and waste reintroduction (e.g., controlling environmental conditions to turn waste into compost).

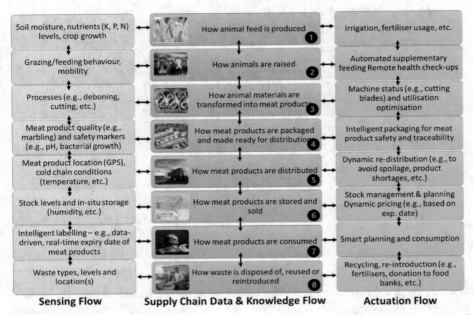

Fig. 4. Integration of data & knowledge across the circular meat supply chain with the proposed framework of DT, showing sensing and actuation flows.

4   Conclusions and Future Challenges/Work

Every year, food waste contributes to up to 10% of total greenhouse gas emissions and generates more than 900 million tonnes of residues. With a global population forecasted to reach 9 billion in the next decade, reducing food waste has never been more critical to tackle the current climate crisis and achieve UN's SDGs: Whilst wastage occurring at all areas of the supply chain, most of the proposed technology-based solutions still treat each area as silos. To overcome this limitation, this paper proposes the first framework of Digital Twin that integrates key technological enablers across all the key areas in a "zero-waste" circular meat supply chain. Meat has been purposedly chosen, as it is one of the most energy and resource-intensive food products. With an ever-increasing global demand posed to reach an historical maximum in the next decade, maximising efficiencies across all the areas of the meat supply chain are critical.

This framework represents the starting point for new conceptual thinking in food supply chain that uses digital twinning to integrate all areas, processes, objects and technologies for a sustainable, "zero-waste" food supply chain. Whilst applied to the meat sector, the proposed conceptual framework is comprehensive and versatile, looking at key food supply chain areas and thus could be used and/or adapted to other food sectors with little or no loss of relevance and/or applicability.

Future work will focus on the implementation of this framework, with the generation of a simulation model as the first step towards the creation and application of the DT in the meat/food supply chain.

References

1. UN Sustainable Development Goals. https://sdgs.un.org/goals. Accessed 25 Jan 2022
2. FAO Food Loss and Food Waste. https://www.fao.org/food-loss-and-food-waste/flw-data. Accessed 27 Jan 2022
3. Tonini, D., Albizzati, P.F., Astrup, T.F.: Environmental impacts of food waste: learnings and challenges from a case study on UK. Waste Manag. **76**, 744–766 (2018)
4. FAO: The State of Food and Agriculture 2019. Moving Forward on Food Loss and Waste Reduction, Rome (2019)
5. Mohebi, E., Marquez, L.: Intelligent packaging in meat industry: an overview of existing solutions. J. Food Sci. Technol. **52**(7), 3947–3964 (2014). https://doi.org/10.1007/s13197-014-1588-z
6. Ritchie, H., Roser, M.: Environmental Impacts of Food Production. OurWorldInData.org. https://ourworldindata.org/environmental-impacts-of-food. Accessed 27 Jan 2022
7. Kim, B.F., et al.: Country-specific dietary shifts to mitigate climate and water crises, Glob. Environ. Change **62**, 101926 (2020)
8. OECD/FAO: OECD-FAO Agricultural Outlook 2020–2029. OECD Publishing, Rome (2020)
9. Defraeye, T., et al.: Digital twins probe into food cooling and biochemical quality changes for reducing losses in refrigerated supply chains. Resour. Conserv. Recycl. **149**, 778–794 (2019)
10. Kalaboukas, K., et al.: Implementation of cognitive digital twins in connected and agile supply networks-an operational model. Appl. Sci. **11**(9), 4103 (2021)
11. Marr, B.: How Are Digital Twins Used In Practice: 5 Real-World Examples Beyond Manufacturing. Forbes. https://www.forbes.com/sites/bernardmarr/2020/08/28/how-are-digital-twins-used-in-practice-5-real-world-examples-beyond-manufacturing/. Accessed 26 Jan 2022
12. Chatham House. https://www.chathamhouse.org/2021/07/financing-inclusive-circular-economy/02-sdgs-and-how-circular-economy-finance-can-0. Accessed 27 Jan 2022
13. Sariatli, F.: Linear economy versus circular economy: a comparative and analyzer study for optimization of economy for sustainability. Visegrad J. Bioecon. Sustain. Dev. **6**(1), 31–34 (2017)
14. Jones, D., Snider, C., Nassehi, A., Yon, J., Hicks, B.: Characterising the digital twin: a systematic literature review. CIRP J. Manuf. Sci. Technol. **29**(Part A), 36–52 (2020)
15. Villa-Henriksen, A., et al.: Internet of Things in arable farming: implementation, applications, challenges and potential. Biosyst. Eng. **191**, 60–84 (2020)
16. Mustafa, F., Andreescu, S.: Chemical and biological sensors for food-quality monitoring and smart packaging. Foods **7**(10), 168 (2018)
17. Shih, C., Wang, C.: Integrating wireless sensor networks with statistical quality control to develop a cold chain system in food industries. Comput. Stand. Interfaces **45**, 62–78 (2016)

18. WRAP   UK.   https://wrap.org.uk/resources/report/courtauld-commitment-milestone-pro
    gress-report. Accessed 25 Jan 2022
19. Patel, S., Dora, M., Hahladakis, J.N., Iacovidou, E.: Opportunities, challenges and trade-offs
    with decreasing avoidable food waste in the UK. Waste Manag. Res. **39**(3), 473–488 (2021)

A Mathematical Model for Cloud-Based Scheduling Using Heavy Traffic Limit Theorem in Queuing Process

Rasoul Rashidifar[✉], F. Frank Chen, Hamed Bouzary, and Mohammad Shahin

The University of Texas at San Antonio, San Antonio, TX 78249, USA
rasoul.rashidifar@my.utsa.edu

Abstract. Cloud manufacturing (CMfg) is a service-oriented manufacturing paradigm that distributes resources in an on-demand business model. In the cloud manufacturing environment, scheduling is considered as an effective tool for satisfying customer requirements which has attracted attention from researchers. In this case, quality of service (QoS) in the scheduling plays a vital role in assessing the impacts of the distributed resources in operation on the performance of scheduling functions. In this paper, a queuing system is employed to model the scheduling problem with multiple servers and then scheduling in cloud manufacturing is classified based on various QoS requirements. Moreover, a set of heavy traffic limit theorems is introduced as a new approach to solving this scheduling problem in which different heavy traffic limits are provided for each of QoS-based scheduling classes. Finally, the number of operational resources in the scheduling is determined by considering the results obtained in the numerical analysis of the heavy traffic limit with different queue disciplines. The results show that different numbers of active machines in various QoS requirements classes play a vital role in that the required QoS metrics such as the expected waiting time and the expected completion time which are critical performance indicators of the cloud's service are intimately related.

Keywords: Scheduling · Cloud manufacturing · QoS (Quality of Service) · Heavy traffic limit · Queuing system

1 Introduction

In the cloud manufacturing (CMfg) paradigm which is inspired from the cloud computing, the connected enterprises/customers can offer their core competencies as services or acquire services on demand and the service providers share their resources in this environment [1, 2]. In terms of a central cloud management system, it is necessary to have a mechanism such as computer simulation [3] or optimization. The task scheduling in cloud manufacturing is an arduous process due to different and heterogeneous tasks (Jobs, Orders) with various functional and non-functional requirements that should be scheduled and accomplished together [4]. CMfg is customer-oriented manufacturing

© The Author(s) 2023
K.-Y. Kim et al. (Eds.): FAIM 2022, LNME, pp. 197–206, 2023.
https://doi.org/10.1007/978-3-031-18326-3_20

and in order to satisfy customer, the research papers focused on the influences of time-liness [5]. Wu et al. [6] develop a mixed linear programming model of 3D printing based on additive manufacturing cloud platform to minimize average cost per volume of material. According to [7], a cloud-based system is capable of facilitating adaptive shop floor scheduling and condition-based maintenance. When client submits all tasks into the cloud platform, then they would be decomposed into a number of sub-tasks and a scheduling plan would be built [8]. Scheduling in cloud manufacturing is fundamentally different from traditional scheduling scenarios as it deals with a "many-resources/services-to-many orders/tasks" scenario in that cloud-based manufacturing (CMfg) platforms are large-scale complex manufacturing tasks [9, 10]. In the cloud-based scheduling, Quality of Service (QoS) and low cost are requested by cloud client side while using optimal resources and making profit are brought under one roof by cloud providers [11].

In view of all mentioned so far, it may be supposed that the CMfg system with many resources in line up is a queuing system. In this paper, we identify four different types of QoS requirements that characterize the performance of the cloud-based scheduling under heavy traffic limits. Taking a design point of view, this paper attempts to show how many machines a cloud needs to support a specified system load and provide a level of quality of service. A mathematical model is developed based on heavy traffic limits in order to realize the relationship between traffic intensity and the number of machines in cloud-based environment. This model allows us to determine the number of operational machines so that they meet the QoS requirements while maintaining a cost-effective and stable operation.

A new method of formulating queueing problems is in terms of a "heavy traffic" that is developed in this study which is a new approach instead of using stochastic differential equations [12] and has received considerable critical attention in the research papers. Kingman [13] describes the behaviour of the queue system with different impartial queue discipline under heavy traffic regime. A queue system is in heavy traffic regime when the traffic intensity ρ is less than, but very near, one at which jobs are served and the number of machines n is considered very large (n $\rightarrow \infty$) [12]. Authors in [14] propose the linear combinations of queue lengths in heavy traffic in that the input-queued switch with correlated arrivals is considered. A heavy traffic limit for GI/G/1-type Markov chains in [15] is applied in which the heavy-traffic-limit formula for the moments of the stationary distribution is proposed. Maguluri et al. [16] provide a stochastic model for load balancing problems and scheduling problems under the heavy traffic limit and optimization algorithms to analyse the queue length. The performance of longest idle server first (LISF) rule and the random routing (RR) is analysed by theoretical support [17] in that the many server heavy traffic limit and quality driven (QD) are established. The term "heavy traffic" is an adequate tool when the load on the system is equal to capacity of the system [18]. A framework developed by [19] uses the Moment Generating Function (MGF) method to distribute the tasks in a queue system under heavy traffic limits. Scully et al. [20] provide a new variation of Gittins policy to optimize the response time in the heavy traffic M/G/1 in which the multi-server scheduling distributes a class of finite job size in the system. Atar et al. [21] propose formulas that approximate these performance measures and depend on the steady-state mean number of available servers in that a model with many servers and server vacations is considered.

The methodological approach taken in this study is a mixed methodology based on QoS requirements and heavy traffic limit theory in cloud-based scheduling. This paper has been organized in the following way. Section 2 describes cloud manufacturing systems (CMfg) focusing on aspects of scheduling in CMfg in order to address the scheduling concept using mathematical models. In Sect. 3 different QoS requirements are classified and then, the heavy traffic limits formulas based on QoS classes are presented. Section 4 provides the mathematical models for heavy traffic limits in each QoS class in which based some assumptions, the number of machines in each QoS class is obtained. The fifth section presents the results of the study and discusses the importance of them and finally, Sect. 6 concludes the research paper.

2 Cloud Manufacturing Platform

In a cloud manufacturing environment, a central cloud platform is responsible for handling the submission of tasks by service consumers, and result delivery to the consumers on one hand and resource publication/registration by the service providers and task dispatching to the providers on the other hand. Figure 1 illustrates various sections of a Cloud-based manufacturing system. In the first step, when a complex task is submitted into the cloud, it must go through the decomposition process so that it will be broken down into multiple single functional subtasks. Then, all the virtualized resources that exist in the cloud manufacturing platform will be retrieved based on their functional attributes in the service discovery and matching process. After forming the service candidate sets, it is the time for the service composition which can be defined as selecting one candidate service from each corresponding candidate set to compose the service while ensuring the overall QoS (a function of non-functional attributes) is optimal. Once a set of composed paths were found, the resource scheduling module will strive for finding the optimal schedule through solving the optimization problem associated with those paths.

The best definition of scheduling is given by Pinedo [22]: "Scheduling is a decision-making process that is used on a regular basis in many manufacturing and services industries; it deals with the allocation of resources to tasks over given time periods and its goal is to optimize one or more objectives." Various researchers have built up their models based on different objective functions, attributes, and constraints. These details will be summarized and discussed in the upcoming paragraph.

A mathematical model regarding objective functions and constraints in the scheduling model is presented. The scheduling provides the required service resources to perform some processes on the subtasks [8]. Scheduling mathematical models are advisable to define objective functions and optimize them. There are four main objective function

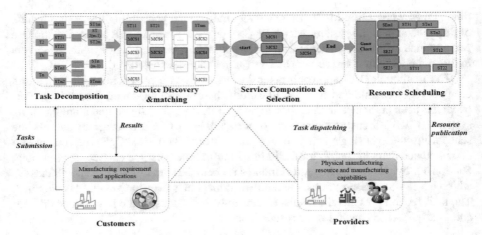

Fig. 1. Cloud-based manufacturing system [23]

Time (T), Cost (C), Quality (Q) and Utilization (U). Considering the objective function and constraints, the mathematical model is presented [24] by Eq. (1) in which a_1, a_2, a_3, a_4 are the weight values of the objective functions and x_{ij} is decision variable.

$$min \quad (a_1T + a_2C - a_3Q - a_4U)x_{ij}$$

$$S.t. \quad a_1, a_2, a_3, a_4 \geq 0, \quad T \leq T_{max}, \quad C \leq C_{max}, \quad Q \geq Q_{min}, \quad U \geq U_{min}$$

$$x_{ij}(t_x) = \begin{cases} 1, & \text{if Task } i \text{ is completed by service resource } j \text{ at time } t_x \\ 0, & \text{otherwise} \end{cases} \quad (1)$$

Quality is an important component in efficient scheduling and plays a key role in the cloud-based environment. Evidence suggests that the number of resources in the cloud manufacturing system is among the most important factors that can exert influence on the efficient scheduling since, increasing the number of machines helps the system to improve the QoS and on the other hand, it brings about the higher cost. As mentioned in the introduction, heavy traffic limit as a new approach regarding quality of service (QoS) classes is proposed to obtain the number of operational machines in a cloud-based scheduling.

3   Methodology

As mentioned above, a cloud manufacturing system with many resources is considered a queue system. To satisfy the QoS requirements and optimize the cost at the same time, the performance of the queue system based on the different QoS requirements is investigated. In order to achieve this aim, scheduling in cloud manufacturing systems has been classified based QoS requirements in which waiting time has a pivotal impact, and then each class is analysed using the heavy traffic limit theory.

3.1 Heavy Traffic Limit Theory in Different Classified QoS

Based on wating time, four different QoS is considered [12]. This system of classification is more scientific since it is based on the QoS requirements demanded by the customers.

1- Zero-Waiting-Time (ZWT): it provides the strictest of the QoS requirements in which the waiting time is zero and the tasks must be accomplished immediately when they arrive. Function P represents QoS requirements in (2).

$$\lim_{n \to \infty} P\{N \geq n\} = 0 \tag{2}$$

   where N is the total number of tasks in cloud and n is the number of operational machines.

2- Minimal-Waiting-Time (MWT): in this class, waiting time is near to zero and when the jobs arrive at the system, they are served with some probability.

$$\lim_{n \to \infty} P\{N \geq n\} = \alpha \tag{3}$$

   In Eq. 3, α represents a constant that $0 < \alpha < 1$

3- Bounded-Waiting-Time (BWT): in this case, waiting time is between 0 and t_1 and when the number of machines goes to infinity the probability is 1.

$$\lim_{n \to \infty} P\{N \geq n\} = 1, \ \lim_{n \to \infty} P\{W \geq t_1\} = \sigma_n, \ \lim_{n \to \infty} \sigma_n = 0 \tag{4}$$

   where W, t_1, and σ_n are waiting time, waiting time threshold, and decreasing rate, respectively.

4- Probabilistic-Waiting-Time (PWT): this class is the least strict QoS, waiting time is between 0 and t_2 with probability $1 - \sigma$ where σ is constant and $0 < \sigma < 1$.

$$\lim_{n \to \infty} P\{N \geq n\} = 1, \ \lim_{n \to \infty} P\{W \geq t_2\} = \sigma \tag{5}$$

The analysis of the QoS classes is based on the conceptual framework proposed by [25] in a queue system with numerous servers. As discussed above, the traffic intensity must be less than 1 so that the cloud system becomes stable, and it is close to 1 when the number of machines is infinite. In this investigation, the heavy traffic limit for the various QoS classes is developed based on proposition 1 and theorem 4 analysed in [25].

For the class 1 (ZWT) Eq. (6) is true.

$$(1 - \rho_n)\sqrt{n} \to \infty \tag{6}$$

If f(n) is defined as $(1 - \rho_n)$ in that ρ_n is traffic intensity then,

$$\lim_{n \to \infty} f(n) = 0, \ \lim_{n \to \infty} f(n)\sqrt{n} = \infty \tag{7}$$

In Class 2 (MWT) the QoS is satisfied if:

$$(1 - \rho_n)\sqrt{n} = \beta, \ \beta = \frac{(1 + c^2)f(\alpha)}{2} \tag{8}$$

β, c are constant value and coefficient of variance.

For class 3 (BWT), the QoS meets the requirements if a suitable σ_n is used in it under equations given in (9):

$$\lim_{n\to\infty} \frac{(1-\rho n)}{-\text{In}\sigma_n} = \tau \quad \lim_{n\to\infty} \sigma_n \exp(k\sqrt{n}) = \infty \tag{9}$$

where τ is $\tau = \frac{1+c^2}{2\mu t_1}$ in that μ is mean service time. k is constant and $k > 0$

In the class 4 (PWT), QoS is satisfied if

$$P\{W \geq t_2\} \approx \exp(\frac{-2n\mu(1-p)t_2}{1+c^2}) \quad \lim_{n\to\infty} (1-\rho_n)n = \gamma \tag{10}$$

γ is the probability of threshold that is represented by $\gamma = \frac{-(1+c^2)\text{In}\sigma}{2\mu t_2}$.

The results of equations provide guidelines that show how many machines should be active to meet the QoS requirements in the cloud environment. To do this, the next section develops a queuing system in which different QoS classes under heavy traffic limit terms are brought up.

4   Modelling of Queuing Process Based on Cloud Classification

The cloud-based system consists of a wide variety of machines to accomplish tasks and subtasks; therefore, it can be viewed in a queuing system. Moreover, a queuing system is in heavy traffic regime when the traffic intensity ρ is less than one in that traffic intensity is defined as the rate of job arrivals divided by the rate of served tasks [12]. In this model, it is assumed that job arrivals are independent, and the passion arrival process is considered. In addition, service time is exponentially distributed in which arrival rate and variation coefficient and mean denote λ_n, c, and $1/\mu$. Each task is completed by only one machine based on First-Come-First-Serve (FCFS) rule and resources have fully similar capability.

4.1   Number of Operational Resources in the Queuing System

In order to satisfy the QoS requirements of clients and reach the optimal performance in the cloud environment, different heavy limits are applied for different QoS classes. Based on the analysis, the minimum number of machines is computed for each QoS class in which the traffic intensity is estimated by mathematical methods. As mentioned above, an exponential service time distribution is chosen in the cloud model ($\mu = 0.3$) and variation coefficient (c = 1) is presumed [26]. The number of machines is obtained to have efficient scheduling in the cloud environment. When traffic intensity is close to one and the number of machines is large, the heavy traffic limit is an appropriate methodology. Based on the heavy traffic limit, the number of machines in each class is provided. Table 1 shows the number of machines in each QoS class with estimated traffic intensity.

Table 1. Number of machines in different QoS classes

QoS classes	The number of machines (n)	QoS classes	The number of machines (n)
Class 1 - ZWT	$f^{-1}(1-\rho)$	Class 3 - BWT	$\frac{\tau \ln \sigma_n}{\rho - 1}$
Class 2 - MWT	$(\frac{\beta}{1-\rho})^2$	Class 4 - PWT	$\frac{\gamma}{1-\rho}$

The unknown parameters in aforementioned equations are computed for each class [26]. For class 1 (ZWT), $f(n) = n^{-x}$ where x = 0.25. In class 2 (MWT), the waiting probability is $f(\alpha) = 0.01$ and c = 1, therefore form Eq. (8), we have $\beta = 0.005$. In class 3 (BWT) class, $t_1 = 1$ and $\sigma_n = \exp(-n^{0.25})$. Finally, in the class 4 (PWT), the probability of threshold (γ) is 0.2 and $t_2 = 1$ is considered. According to the assumptions mentioned above, a relationship between the number of operational machines and traffic intensity in each Cloud-based class is obtained.

5   Results and Discussion

Based on a cloud scheduling, heavy traffic limit in the queueing system is designed for formulating and solving the scheduling model based on Eqs. (2)–(11). To achieve this purpose, the aforementioned example is developed to find the unknown parameters. Based on the mathematical formulas mentioned in Table 1 for each class and assumptions indicated above, the relationship between the number of machines and traffic intensity in each QoS class for the cloud-based scheduling is illustrated in Fig. 2. It indicates that the number of machines in each specific traffic intensity ($0 < \rho < 1$) is fully dependent on the different QoS classes.

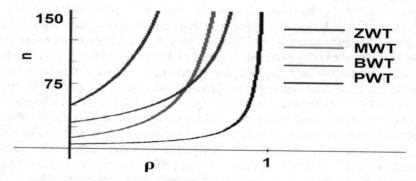

Fig. 2. Relationship between the number of machines and traffic intensity

The results illustrate that in the designed cloud model, there is an intersection between MWT and PWT classes in $\rho = 0.6$. With a given value of traffic intensity, ZWT and BWT classes need the most and least number of machines, respectively. Before $\rho = 0.6$, MWT class meets the QoS requirements in the model with less resources compared with PWT. Table 2 shows the number of machines for different traffic intensity in each QoS requirement class. It demonstrates that in the same condition, Bounded Waiting Time (BWT) class has the best performance. As mentioned above, there is a significant correlation between QoS and cost in the cloud manufacturing system, therefore, this class meets the QoS requirements with the least machine and cost in the cloud system.

Table 2. Number of machines in each QoS class for a given traffic intensity [26].

Traffic intensity (ρ)	Number of machines (n)			
	Class1- ZWT	Class 2 - MWT	Class 3 - BWT	Class 4 - PWT
0.2	75	20	5	40
0.45	220	35	7	55
0.6	300	75	15	75
0.8	1000	300	25	150
0.95	4000	2000	100	500

No significant difference between the MWT and PWT classes is evident when traffic intensity is equal and less than 0.6 however for traffic intensity more than 0.6, PWT is a more appropriate class in terms of satisfying QoS requirements and saving cost.

6 Conclusion

Cloud manufacturing is a novel paradigm of using centralized cloud for manufacturing systems to distribute resources in an on-demand business model in which the role of resource scheduling has received increased attention across a number of research in recent years. The aim of this paper is to describe the design and implementation of heavy traffic limits based on QoS requirements for the resource scheduling in the cloud-based manufacturing system in that the cloud system is considered as a queuing system. Different heavy traffic limits as new approaches are attained based on four QoS classes to find the optimal number of operational machines for each class. The results, as shown in Fig. 2, indicate that there is a significant positive correlation between the number of active machines in each class and the traffic intensity. The higher the traffic intensity becomes; the greater number of operational machines are needed to meet QoS requirements desired by customers. The analysis of each class demonstrates that the Bounded Waiting Time (BWT) class has the best performance in that both saving cost and satisfying QoS requirements arise simultaneously. As represented in Table 2, $\rho = 0.6$ is a turning point for MWT and PWT classes; however, PWT is a better class in terms of both satisfaction and cost. Results of the study provide insights for future research. Investigation of optimization algorithms and machine learning methods based on big data techniques are possible areas to analyse and optimize scheduling problems in cloud manufacturing environments.

References

1. Yang, C., Peng, T., Lan, S., Shen, W., Wang, L.: Towards IoT-enabled dynamic service optimal selection in multiple manufacturing clouds. J. Manuf. Syst. **56**, 213–226 (2020)
2. Bouzary, H., Chen, F.F., Shahin, M.: Natural language processing for comprehensive service composition in cloud manufacturing systems. Procedia Manuf. **55**, 343–349 (2021). https://doi.org/10.1016/j.promfg.2021.10.048
3. Rashidifar, R., Khataei, M., Poursina, M., Ebrahimi, N.: Comparison and evaluation of criteria ductile damage FLD & MSFLD to predict crack growth in forming processes (2014)
4. Li, F., Zhang, L., Liao, T.W., Liu, Y.: Multi-objective optimisation of multi-task scheduling in cloud manufacturing. Int. J. Prod. Res. **57**(12), 3847–3863 (2019). https://doi.org/10.1080/00207543.2018.1538579
5. Tong, H., Zhu, J.: A novel method for customer-oriented scheduling with available manufacturing time windows in cloud manufacturing. Robot. Comput. Integr. Manuf. **75**, 102303 (2022)
6. Wu, Q., Xie, N., Zheng, S., Bernard, A.: Online order scheduling of multi 3D printing tasks based on the additive manufacturing cloud platform. J. Manuf. Syst. **63**, 23–34 (2022)
7. Mourtzis, D., Vlachou, E.: A cloud-based cyber-physical system for adaptive shop-floor scheduling and condition-based maintenance. J. Manuf. Syst. **47**, 179–198 (2018)
8. Delaram, J., Valilai, O.F.: A mathematical model for task scheduling in cloud manufacturing systems focusing on global logistics. Procedia Manuf. **17**, 387–394 (2018)
9. Liu, Y., Xu, X., Zhang, L., Wang, L., Zhong, R.Y.: Workload-based multi-task scheduling in cloud manufacturing. Robot. Comput. Integr. Manuf. **45**, 3–20 (2017)
10. Jian, C., Ping, J., Zhang, M.: A cloud edge-based two-level hybrid scheduling learning model in cloud manufacturing. Int. J. Prod. Res. **59**(16), 4836–4850 (2021)
11. Bittencourt, L.F., Goldman, A., Madeira, E.R., da Fonseca, N.L., Sakellariou, R.: Scheduling in distributed systems: a cloud computing perspective. Comput. Sci. Rev. **30**, 31–54 (2018)
12. Zheng, Y., Shroff, N., Sinha, P.: Heavy traffic limits for GI/H/n queues: theory and application. ArXiv Prepr. ArXiv14093463 (2014)
13. Kingman, J.F.C.: Queue disciplines in heavy traffic. Math. Oper. Res. **7**(2), 262–271 (1982)
14. Hurtado Lange, D.A., Maguluri, S.T.: Heavy-traffic analysis of queueing systems with no complete resource pooling. Math. Oper. Res. (2022)
15. Kimura, T., Masuyama, H.: A heavy-traffic-limit formula for the moments of the stationary distribution in GI/G/1-type Markov chains. Oper. Res. Lett. **49**(6), 862–867 (2021)
16. Maguluri, S.T., Srikant, R., Ying, L.: Heavy traffic optimal resource allocation algorithms for cloud computing clusters. Perform. Eval. **81**, 20–39 (2014)
17. Sun, X.: Heavy-traffic limits for server idle times with customary server-assignment rules. Oper. Res. Lett. **46**(1), 44–50 (2018)
18. Uysal, E.: An overview of the application of heavy traffic theory and Brownian approximations to the control of multiclass queuing networks
19. Hurtado-Lange, D., Maguluri, S.T.: Transform methods for heavy-traffic analysis. Stoch. Syst. **10**(4), 275–309 (2020)
20. Scully, Z., Grosof, I., Harchol-Balter, M.: Optimal multiserver scheduling with unknown job sizes in heavy traffic. Perform. Eval. **145**, 102150 (2021)
21. Atar, R., Karmakar, P., Lipshutz, D.: Customer-server population dynamics in heavy traffic. Stoch. Syst. **12**(1), 68–91 (2022)
22. Pinedo, M., Hadavi, K.: Scheduling: theory, algorithms and systems development. In: Gaul, W., Bachem, A., Habenicht, W., Runge, W., Stahl, W.W. (eds.) Operations Research Proceedings 1991. Operations Research Proceedings 1991, vol. 1991, pp. 35–42. Springer, Heidelberg (1992). https://doi.org/10.1007/978-3-642-46773-8_5

23. Liu, Y., Wang, L., Wang, X.V., Xu, X., Zhang, L.: Scheduling in cloud manufacturing: state-of-the-art and research challenges. Int. J. Prod. Res. **57**(15–16), 4854–4879 (2019)
24. Yuan, M., Cai, X., Zhou, Z., Sun, C., Gu, W., Huang, J.: Dynamic service resources scheduling method in cloud manufacturing environment. Int. J. Prod. Res. **59**(2), 542–559 (2021). https://doi.org/10.1080/00207543.2019.1697000
25. Halfin, S., Whitt, W.: Heavy-traffic limits for queues with many exponential servers. Oper. Res. **29**(3), 567–588 (1981). https://doi.org/10.1287/opre.29.3.567
26. Zheng, Y., Shroff, N., Sinha, P., Tan, J.: Design of a power efficient cloud computing environment: heavy traffic limits and QoS, preprint (2011)

Approach for Evaluating Changeable Production Systems in a Battery Module Production Use Case

Christian Fries[1,2(✉)], Patricia Hölscher[3], Oliver Brützel[3], Gisela Lanza[3], and Thomas Bauernhansl[1,2]

[1] Fraunhofer Institute for Manufacturing Engineering and Automation IPA, Stuttgart, Germany
cfr@ipa.fraunhofer.de

[2] Institute of Industrial Manufacturing and Management IFF, University of Stuttgart, Stuttgart, Germany

[3] Wbk Institute of Production Science, Karlsruhe Institute of Technology, Karlsruhe, Germany

Abstract. Volatile markets continue to complicate manufacturing companies' production system design, leading to efficiency losses due to imperfect system setups. In such a market environment, a perfect system setup cannot be achieved. Therefore, changeable production systems that cope with immanent uncertainty gain interest in research and industry. For several decades, changeable production systems have been in the research and development stage. The advantages and disadvantages are well investigated. So far, however, they have gained only limited acceptance in industry. One of the reasons is the difficult evaluation of the benefits. Existing investment calculation methods either neglect many effects of changeability, such as easier adaptation to unpredictable events, or are too complex and therefore too time-consuming to become standard. Thus, a practical evaluation method is needed that considers these changeability aspects. This paper deviates the industry requirements regarding an evaluation method based on an industry survey and develops a practical approach for an evaluation method for a changeable production system considering monetary and non-monetary aspects. The approach is characterized by a calculation that is as accurate as possible considering the existing input factors. The method shows that changeable production systems excel in environments with frequent need for adaptation. The approach is applied to a battery module assembly in the ARENA2036 research campus.

Keywords: Changeability · Production system · Evaluation method

1 Introduction

Volatile markets continue to complicate manufacturing companies' production systems design, leading to efficiency losses due to imperfect system setups. In such a market environment, a perfect system setup cannot be achieved. Thus, changeability is required [1–4]. A wide variety of production systems with different levels of changeability has been developed over the last decades [5]. Even though the consideration of changeability in the context of production system design has been sufficiently discussed, assessing

K.-Y. Kim et al. (Eds.): FAIM 2022, LNME, pp. 207–216, 2023.
https://doi.org/10.1007/978-3-031-18326-3_21

the optimal level of adaptability is difficult for companies. Different authors analyze the aspects to be considered when deciding on the optimal degree of changeability [6]. Furthermore, they place the aspects in a structural context and describe their inter-dependencies [1]. Based on these general concepts, different evaluation methods are developed. See Sect. 4. However, based on the cooperation with industrial companies, it can be summarized that these evaluation methods are rarely applied in industrial companies. So far, classical evaluation methods such as the net present value method have been predominantly used, but these systematically disadvantage changeable production systems. Therefore, in the first step an industry survey was conducted in order to detect current obstacles and to elaborate on the requirements for future evaluation methods. Based on this, existing evaluation methods were examined and evaluated with respect to these requirements. Subsequently, a new approach was developed and exemplarily applied in the context of a battery module production.

2 Industry Survey

2.1 Survey Structure

This expert survey is based on the procedures of DIEKMANN [7] and REINECKE [8] and is structured into four parts:

Formulation and Specification of the Research Problem. The aim of this question-naire was the examination of the current situation of evaluating changeable production systems in manufacturing companies. The research question of this survey can hence be formulated as follows: "To what extent is the optimal degree of changeability for production systems determined in practice?" The formulated hypotheses to answer this research question, including the survey results, are shown in Sect. 2.2.

Planning and Preparation of the Survey. The survey's target group were profession-als in the production planning environment in industrial companies and research insti-tutions. Due to the ongoing corona pandemic, this survey was performed via an online tool.

Data Collection. The questionnaire was distributed via e-mail to the target group. The survey participants consist of selected contacts of the research team from the production environment of industrial companies. The questionnaire consists of 15 questions, the most important questions are presented within this paper.

Data Analysis. The goal of the data evaluation is the construction of an analysis-capable data file. The data collection was automated by the survey tool Lime Survey, a man-ual intervention was not necessary. The "mandatory data" function prevented incom-plete questionnaires. An error correction of the data set was therefore not necessary. Twenty-one manufacturing experts from the fields of electrical engineering, automotive engineering, mechanical engineering and IT completed the survey.

2.2 Survey Results

Hypothesis 1: "Changeability is Not Considered When Deciding on New Production Systems".

The survey results show that 79% of the survey participants stated that efforts are made to integrate changeable components into production systems, while 5% even plan new production systems exclusively with changeable components. Therefore, hypothesis 1 cannot be confirmed.

Hypothesis 2: "Evaluation Methods are Used for Investment Decisions Regarding Changeable Production Systems".

To investigate the relevance of evaluation methods in the planning process for changeable production systems, the survey participants were asked about the relevance of different factors on the required changeability. In this context, 57.1% of the survey participants stated "requirements from management" as the decisive factor. This is followed by the orientation on experience reports ("lessons learned") with 47.6%. Other decision factors include orientation towards industry trends with 28.6%. Only one-third of the survey participants said, they use evaluation methods. Consequently, decisions are based on less rational criteria and hypothesis 2 is false.

Hypothesis 3: "Evaluation Methods Do Not Sufficiently Consider All Influencing Variables and Therefore Give Incorrect Recommendations".

To identify deficiencies in general assessment methods for changeable production systems, survey participants were asked about the weaknesses of existing assessment methods. 52.4% of the survey participants think that assessment methods require too many unknown input values. Another 33.3% stated that time-consuming procedures are another deficiency of formal methods, while 14.3% even said that formal evaluation methods do not improve the decision-making process and are not needed. An additional 38.1% of the survey participants stated that evaluation methods inadequately represent the reality or give false investment recommendations. Hypothesis 3 can thus be confirmed.

The survey showed that most participants consider changeability in the planning of new production systems. Evaluation methods exceeding classical approaches are only used sporadically. In addition, most existing evaluation methods are not suitable in practice due to unknown inputs and high complexity. Beyond, survey participants complain that evaluation methods give incorrect recommendations and simultaneously require too many input variables. This shows that there is a need for a method that balances accuracy and usability.

3 State of the Art

Based on the survey and the literature research, five requirements are derived. The possibility of ex ante evaluation is as important as the consideration of all planning phases and the life cycle of the production line. A holistic view of production systems is necessary to consider all influencing factors. To counter uncertainty with changeability, planning uncertainty must also be considered in the evaluation method. To determine the best investment alternative, a consideration of monetary and non-monetary values is necessary.

In the following, the existing approaches from the literature are examined regarding their fulfillment of aforementioned requirements. Möller developed a method for determining the economic efficiency of changeable production systems using the real options theory (see [9]). The real options theory is part of the investment theory and suitable for evaluating changeable production systems under uncertainty. The set of all real options represents the field of action which is available for the decider. Using the net present value method, the individual real options are calculated and compared [10]. Stähr et al. determines the necessary technical measures to achieve the optimal degree of changeability of a production system. The optimality of production systems is determined based on the expected life cycle costs of the production system. The different probabilities of occurring events are determined by Monte Carlo simulation [11]. Heger's Integrative Evaluation of Transformability is subdivided into several analyses: First, it evaluates the potential of a factory object for transfiguration, and second it performs a monetary and a non-monetary analysis. The non-monetary unit of changeability is combined with the monetary valuation, the net present value [12]. Pachow-Frauenhofer approaches the planning and optimization of changeable production systems from a quantitative perspective using control loops and the life cycle costing method. She subdivides the assessment method into goal definition, analysis, design and evaluation. To investigate the optimal degree of changeability, control loops are used to map the change process and to identify necessary dimensions of change [13]. Kluge designs a framework for the basic planning of modular production systems by considering the life cycle costs and the development of a scenario technique [14]. Lübkemann and Nyhuis [15], Schuh et al. [6], Bürgin [16] and Sesterhenn [17] present further approaches considering frameworks for the prediction of future developments or evaluation methods. Due to their different research focus, they are not further elaborated in this paper.

The analysis of existing approaches showed that the following 3 aspects are not adequately addressed: The approaches do not provide an exhaustive list of cost factors. Beyond, most approaches do not consider non-monetary values, while other analyses, such as Heger's [12], tend to result in disproportionately high effort. Furthermore, most authors do not consider all planning phases or the life cycle of the production system. Therefore, the development of an evaluation method that considers all requirements is necessary.

4   Method

Figure 1 depicts the structure of the proposed evaluation method.

1) Changeability Potential Analysis. First, the Changeability Potential Analysis derives different options for the production systems design. The derived options should differ in the investment and operating costs and should have different potentials for changeability.

2) Changeability Profitability Analysis. The Changeability Profitability Analysis considers the options from a monetary perspective. It starts with a scenario analysis and an estimation of the costs and is completed by the monetary comparison. The scenario

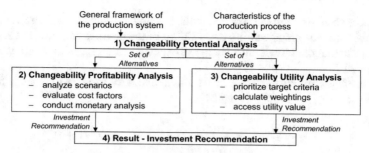

Fig. 1. Structure of the evaluation method

analysis must depict different outcomes and cover the required changeabilities. Therefore, the probability distribution over the different scenarios needs to be assessed. For the evaluation of the scenarios, the respective cost elements must be taken into account. From a cost perspective, two aspects are relevant, which are differentiated according to chronological occurrence: **System design costs** are costs that occur initially before the changeover is required. This enables the system's design to adapt. **System operation costs** describe the costs that occur due to the execution of the changeover. The two types of costs thus represent the total costs of the production system, creating a conflicting relationship. Initial higher changeability costs enable the later more favorable implementation of the changes, thus allowing overall lower total costs and vice versa (see also [1]). An overview of the costs is provided in Fig. 2.

	System design costs	**System operation costs**
tactical	– Evaluation and planning costs – Acquisition costs – Set-up and start-up costs – Disassembly and disposal costs – Area and inventory costs – Financing costs	– Disassembly and assembly costs – Logistics adjustment costs – Production downtime costs – Additional work costs
operational	– Variable product costs – Plant monitoring and operation costs	

Fig. 2. System design and system operation costs

The individual cost elements result from the combination of different approaches from the literature. The changeability costs consist of all costs related to the acquisition and disposal of a specific production system. The tactical costs are divided into costs for evaluation and planning, acquisition [12], setup and start-up [12], disassembly and disposal [12], area and inventory [15] and financing [13]. The operational costs describe variable product costs (incl. Opportunity costs due to different cycle times) and costs for plant monitoring and operation [14]. The implementation costs arise due to the conversion of the production system. They consist of disassembly and assembly costs for changeover [12], adjustment costs for logistics, production downtime costs [12], additional work costs and investments [14].

Using the scenarios and the cost elements, it is now possible to evaluate the different alternatives. First, the changeability costs $CC_{A,S}$ of the alternative A under a scenario S need to be calculated individually. Then, the expected value E_A of an alternative needs to be deduced.

The changeability costs $CC_{A,S}$ of an alternative A in a scenario S for a period T calculates as:

$$CC_{A,S} = TCC_A + \sum_{t=0}^{T}(OCC_{A,S,t} + IC_{A,S,t}) \cdot (1+i)^{-t} \tag{1}$$

with:

TCC_A tactical changeability costs of alternative A
$OCC_{A,S,t}$ operational changeability costs of alternative A and scenario S in period t
$IC_{A,S,t}$ implementation costs of alternative A and scenario S in period t
i discount rate

Based on this, the expected value E_A of an alternative A is calculated as a function of the weighted changeability costs $CC_{A,S}$ of an alternative A in a scenario S:

$$E_A = w_{S_1} \cdot CC_{A,1} + w_{S_2} \cdot CC_{A,2} + w_{S_3} \cdot CC_{A,3} \tag{2}$$

with:

w_S occurrence probability of scenario S

3) Changeability Utility Analysis. Since there are non-monetary aspects that are important to consider as well, the Changeability Utility Analysis takes those aspects into account by using a pairwise-comparison approach. After the prioritization of the target criteria and calculation of the weightings, the partial utility values and thereafter the utility values need to be assessed.

4) Result - Investment Recommendation. Finally, the results of the monetary and non-monetary analysis are combined and an investment recommendation is made.

5   Battery Module Production Use Case

The developed method was exemplarily applied and presented in a use case.

1) Changeability Potential Analysis. Within this project, two alternatives, line production and matrix production, were investigated.

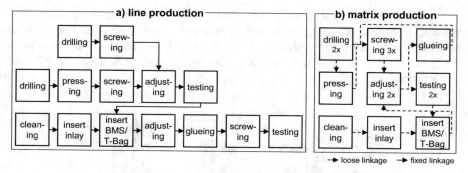

Fig. 3. Production process of line and matrix production

Figure 3 depicts the manufacturing process of a battery module use case in the a) line production and b) matrix production. The line production consists of three independent lines that are merged at different stages. Whereas the matrix production consists of independent stations without fixed linkage that can be operated in different sequences. For a detailed differentiation see also [5]. The cover is drilled and the service connector is screwed to the cover. After the housing is drilled, the four threaded inserts are pressed in, after which a circuit board is screwed to the housing together with a battery holder. The circuit board then is adjusted, and the cover is placed on the housing. Within the final assembly, the cleaned inlays are added to the battery module frame and the pre-assembled battery management system (BMS) is inserted and screwed onto the frame. The T-bag is then screwed into place and the battery module frame closed with a cover, which is glued and screwed to the battery module frame.

2) Changeability Profitability Analysis. In scenario 1 the "pressing"-station must be modified in period 3. Scenario 2 expects the changes in scenario 1 with a required modification of the battery modules in period 4. The assembly is done with an additional bond seam, for which a "bonding" station must be inserted. Scenario 3 combines scenario 1 with a modification of the battery modules in period 4, which requires the addition of two stations, while three stations must be moved. Costs that are identical to both alternatives are not considered. Figure 4. Compares the required changeability qualitatively depending on the scenarios.

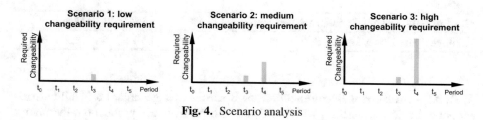

Fig. 4. Scenario analysis

The result of the Changeability Profitability Analysis shows that the matrix production is beneficial for scenarios with high-expected changeability, see Fig. 5. Using formula 2, the expected changeability costs, under consideration of the occurrence probabilities of the scenarios, were € 3.498k for the line production and € 3.229k for the matrix production.

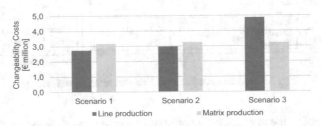

Fig. 5. Changeability costs for line and matrix production per scenario

3) Changeability Utility Analysis. In an expert workshop the two production concepts were compared based on eight criteria, which were evaluated using a scale from 1 (not applicable) to 10 (applicable). Table shows that matrix production excels particularly in process innovation, sustainability and adaption speed (Table 1).

Table 1. Changeability utility analysis

Production system	Weightings	Line production	Matrix production
1. Positive impact on corporate image and culture	12.9%	5	5
2. Increased process innovation rate	16.1%	3	7
3. Better productivity and higher transparency	12.9%	6	4
4. Improved understanding of factory and future space	12.9%	7	5
5. Positive impact on sustainability	12.9%	3	8
6. Higher employee motivation, qualification and retention	12.9%	5	5
7. Increased speed of adaptation and development	19.4%	4	8
Changeability Utility Score		4.6	6.2

4) Result - Investment Recommendation. Since the Changeability Profitability Analysis and the Changeability Utility Analysis both favor matrix production, the matrix production is the favored alternative. As the results show, higher initial investment in the changeability of production systems decreases implementation costs for new conversions. Therefore, changeable production systems do excel in volatile environments and a thorough scenario analysis is crucial to the decision for an optimal degree of changeability.

6 Summary and Outlook

Changeable production systems are a key element of future manufacturing companies. However, the benefits of these production systems are difficult to assess and often depend on the specific application. Therefore, at the beginning of this paper, industry requirements for evaluation methods of changeable production systems were collected and investigated in a structured way by means of a survey. Based on this, specific requirements were formulated and relevant methods from the literature were compared with these requirements. The existing assessment approaches have shown that a pragmatic approach for assessing the necessary changeability does not exist. They are either not accurate enough or the methods require excessive data. This paper therefore develops an evaluation method and applies it to a battery module assembly use case in the ARENA2036 research campus. The presented method combines a pragmatic and practice-oriented approach for evaluating changeable production systems by combining and extending existing approaches. The approach is characterized by a calculation that is as accurate as possible, considering the existing input factors and thus serves as a first step towards more data-driven decision-making regarding changeable production systems.

Acknowledgements. The research presented in this paper was supported by the German Federal Ministry of Education and Research (BMBF) within the research campus ARENA2036 (Active Research Environment for the Next generation of Automobiles) (funding number 02P18Q620, 02P18Q626) and implemented by the Project Management Agency Karlsruhe (PTKA). The authors are responsible for the content of this publication.

References

1. Wiendahl, H.-P., et al.: Changeable manufacturing - classification, design and operation (2007). https://doi.org/10.1016/J.CIRP.2007.10.003
2. Wiendahl, H.-P.: Wandlungsfähigkeit: Schlüsselbegriff der zukunftsfähigen Fabrik. WT. Werkstattstechnik **2002**(4), 122–127 (2002)
3. Koren, Y., et al.: Reconfigurable manufacturing systems. CIRP Ann. **48**(2), 527–540 (1999). https://doi.org/10.1016/S0007-8506(07)63232-6
4. Bauernhansl, T., Mandel, J., Diermann, S.: Evaluating changeability corridors for sustainable business resilience. Procedia CIRP **3**, 364–369 (2012). https://doi.org/10.1016/j.procir.2012.07.063
5. Fries, C., et al.: Fluid manufacturing systems (FLMS). In: Weißgraeber, P., Heieck, F., Ackermann, C. (eds.) Advances in Automotive Production Technology – Theory and Application. ARENA2036, pp. 37–44. Springer, Heidelberg (2021). https://doi.org/10.1007/978-3-662-62962-8_5

6. Schuh, G., Harre, J., Gottschalk, S., Kampker, A.: Design for changeability (DFC)-Das richtige Maß an Wandlungsfähigkeit finden. wt Werkstattstechnik online, vol. 94, no. 4, pp. 100–106 (2004)

7. Diekmann, A.: Empirische Sozialforschung: Grundlagen, Methoden, Anwendungen, 12th edn. Rowohlt Taschenbuch Verlag, Reinbek bei Hamburg (2018)

8. Reinecke, J.: Grundlagen der standardisierten Befragung. In: Baur, N., Blasius, J. (eds.) Handbuch Methoden der empirischen Sozialforschung, pp. 601–617. Springer, Wiesbaden (2014). https://doi.org/10.1007/978-3-531-18939-0_44

9. Dixit, A.K., Pindyck, R.S.: Investment UNDER UNCERTAINTY. Princeton Univ. Press, Princeton (1994)

10. Möller, N.: Bestimmung der Wirtschaftlichkeit wandlungsfähiger Produktionssysteme. Diss., TUM (2008)

11. Stähr, T., Englisch, L., Lanza, G.: Creation of configurations for an assembly system with a scalable level of automation. Procedia CIRP **76**, 7–12 (2018). https://doi.org/10.1016/j.pro cir.2018.01.024

12. Heger, C.L.: Bewertung der Wandlungsfähigkeit von Fabrikobjekten. Zugl.: Hannover, Univ., Diss., 2006. PZH Produktionstechn. Zentrum, Garbsen (2007)

13. Pachow-Frauenhofer, J.: Planung veränderungsfähiger Montagesysteme. Zugl.: Hannover, Univ., Diss., 2012. PZH Produktionstechnisches Zentrum, Garbsen (2012)

14. Kluge, S.: Methodik zur fähigkeitsbasierten Planung modularer Montagesysteme. Zugl.: Stuttgart, Univ., Diss., 2011. Jost-Jetter, Heimsheim (2011)

15. Lübkemann, J., Nyhuis, P.:"Ermittlung des Restrukturierungsbedarfs von Fabriken. Diss., Leibniz Universität Hannover

16. Buergin, J., et al.: Robust assignment of customer orders with uncertain configurations in a production network for aircraft manufacturing. Int. J. Prod. Res. **57**(3), 749–763 (2019). https://doi.org/10.1080/00207543.2018.1482018

17. Sesterhenn, M.: Bewertungssystematik zur Gestaltung struktur- und betriebsvariabler Produktionssysteme. Zugl.: Aachen, Techn. Hochsch., Diss., 2002. Shaker, Aachen (2003)

Cost-Minimal Selection of Material Supply Strategies in Matrix Production Systems

Daniel Ranke[1]([email]) and Thomas Bauernhansl[1,2]

[1] Fraunhofer Institute for Manufacturing Engineering and Automation IPA, Nobelstr. 12, 70569 Stuttgart, Germany
daniel.ranke@ipa.fraunhofer.de

[2] Institute of Industrial Manufacturing and Management IFF, Nobelstr. 12, 70569 Stuttgart, Germany

Abstract. Companies are facing changing market demands, high variance, and volatile quantities. Resilient production systems are needed to meet these challenges. The matrix production is such a system. It offers degrees of freedom in terms of operation sequence flexibility and work distribution flexibility through redundantly used resources. For the material supply this is a challenge in planning. The material must be supplied in a cost-efficient manner and without shortages.

To increase planning quality, a method for selecting the least expensive material supply strategy is developed. Depending on consumption, constraints of space, and supply framework conditions, different strategies are advantageous for each material. The developed method requires three steps.

First, required data for step 2 and step 3 is collected. In step 2, standardized process blocks combine to describe a company-specific material supply strategy. The approach is company-independent and added by cost functions to the process blocks. Through the cost functions applied to the process blocks the costs of a supply strategy is achieved. As material can be supplied in alternative ways, multiple expected costs for supplying arise. As only one supply strategy needs to be selected, step 3 is necessary. It uses the branch-and-cut algorithm on the mathematical description of the logistic selection problem to find the cost-minimal configuration of supply strategies. As the problem is in the context of matrix production, several conditions and requirements need to be included in the selection process.

The result is the assignment of a material supply strategy to each material while minimizing the costs.

Keywords: Matrix production · Material supply · Decision-making

1 Introduction

In manufacturing companies there is an ongoing transformation towards changeable and resilient production systems to encounter a high number of products, variants, and constant market fluctuation [1–4]. The matrix production system is such a system. It offers more operational flexibility and is still productive even if the environment is

K.-Y. Kim et al. (Eds.): FAIM 2022, LNME, pp. 217–226, 2023.
https://doi.org/10.1007/978-3-031-18326-3_22

turbulent [5, 6]. In addition, the matrix production consists of flexibly linked process modules and is operated by using an ad-hoc order-control, operation sequence flexibility and order distribution flexibility [6–8]. While offering flexibility to the value-adding manufacturing and assembly processes, it challenges the material supply [9].

Figure 1 shows an example and extract of a matrix production system. There are redundant stations of the types A and B. Station C is the single representation of its type. A station offers multiple processes for operation. A product/order has to choose ad-hoc, which process is operated next and which station is chosen, to do so. The decision depends on the control logic and the product's priority graph. The design, organization and controlling of these system are widely discussed [10–12]. Although, in the investigation of the material supply, there is a gap.

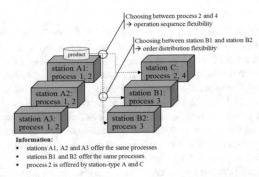

Fig. 1. Example of a matrix production system

Through the omission of a defined order sequence and pearl chain, the matching of material demand of each station with the material supply is either uncertain by continuing present methods or less cost-efficient.

This article outlines a method for selecting the cost-minimal material supply strategies by taking into regard the given uncertainty of the matrix production. Therefore, it is presented a method to standardly describe supply strategy alternatives, how to formulize the given logistic problem as a mathematical optimization problem and last, how to solve it and achieve a cost-minimal configuration.

2   State of the Art

2.1   Material Supply and the Selection Process of Supply Strategies

Material supply enables manufacturing and assembly to operate their processes by providing the right material in due time, space and quantity [13, 14]. Further requirements of the supply are providing the right quality to preferable costs [14]. There are many ways to supply material that vary e.g. in storage and handling or supply policy. An alternative and specified process to supply material is called a material supply strategy. The Kanban supply or single order commissioning are e.g. often used supply strategies and can be operated at the same time in a company and production system but for different materials [13]. For this reason, each strategy has its benefits and costs.

In practice and science, there are different approaches to select material supply strategies for all materials and thereby design the supply system [13, 15–21]. Due to the high number of materials that need an assignment of supply strategy and the different characteristics of each strategy, the selection process is a complex task. Hence, the process is widely discussed and not standardized [15]. Further approaches differ in method, selection criteria and used tools.

Some authors, e.g., Bullinger et al. or Herbert et al., design a guide with methodological support. The planner must go through this guide and make his/her own decisions. Other authors like Grünz or Cárdenas-Barrón et al. transfer the real problem into a mathematical description and submit this to a largely automated algorithm. Approaches using just methodological support offer space for discussion but require time and knowledge. Using an algorithm can be less influenced but is most likely near to optimal.

Table 1. Different evaluation criteria in designing the material supply system

	Bullinger (1994)	Köhler (1996)	Grünz (2004)	Cárdenas-Barrón (2013)	Adolph (2016)	Vojdani (2016)	Herbert (2021)	
Costs			X	X	X	X	X	
Investment		X				X		
Capital commitment		X	X		X	X		monetary criteria
Quality / Failure			X		X	X	X	
Space			X		X	X	X	
Time / Effort	X	X	X		X	X	X	temporal
Control effort	X	X			X			criteria
Lead time	X				X			
Flexibility	X	X			X	X	X	
Automation	X							qualitative criteria
Working conditions	X				X			

The criteria used in the approaches differ, as well. Some focus on costs like capital costs or costs for space, others consider the time needed for supply (Table 1).

Overall, companies require a planning method that supports them in their decision-making, manages the problem's complexity, and is application-oriented [15]. This need increases through complex production systems like the matrix production.

2.2 Challenges to the Material Supply in Matrix Production

The design of a material supply system and the selection of supply strategies is influenced by the production system. As mentioned in the introduction, the matrix production offers new degrees of freedom to the value adding processes. They challenge the material supply, since the ad-hoc order control, operation sequence flexibility and order distribution flexibility lead to uncertainty about time, space, and quantity for each order [9]. On an operational basis, without order freeze the material supply cannot act in time.

To encounter the problem and evade order freeze, the supply system must be designed flexible. In addition, the strategy selection on a mid-term perspective has to take into account the outlined challenges. In a mid-term perspective, focusing on multiple orders, the uncertainty losses on impact as shown in Fig. 2. A selection of most likely cost-minimal strategies is still possible by accepting decision-making under uncertainty and counteracting the risk of shortage.

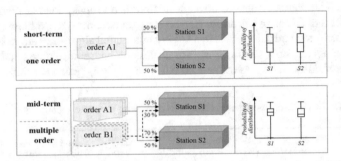

Fig. 2. Reduction of uncertainty of workload distribution over time and order sum

2.3   Cost Accounting in the Context of Logistics

Cost accounting is the method used to investigate the cost structure of processes. The use of cost accounting varies depending on the context and goal. In general, cost accounting comprises the scope, the time reference, and the execution.

Logistics costs are diverse and occur along the entire supply chain. A company's goal is the optimization of processes and resources, and the minimizing of logistics costs. The following logistics costs are often cited in literature: transport, storage, commissioning, supply, inventory, handling, space, system and control costs [22–24]. Depending on the use case and perspective, different costs have to be investigated.

2.4   Algorithms

Algorithms are defined methods to solve problems. There is a distinction between exact and inexact algorithms. The latter are called heuristics. The deployment of an algorithm depends on the application that e.g., defines the input, the goal, the required accuracy, and required output. In some cases, only specialized algorithms are applicable, other applications can be solved by using several algorithms. In practice, the simplex, the branch-and-bound, the branch-and-cut, or a genetic algorithm are often used algorithms [25, 26].

3 Method for Cost-Minimal Material Supply Strategy Selection

3.1 Structure of Method

With regard to Table 1, the developed method is based on temporal and monetary criteria for selection. These are incorporated into relevant logistics costs that are needed to differentiate between material supply strategies.

Three steps are necessary to apply the method. First, the inspected material supply system and all required data are collected. In a second step, standardized process blocks combine in a different manner and describe the alternative supply strategies. Their company-specific combination considers each company's different strategies' characteristics in the material supply. This approach is applied to all inspected materials and stations. As a result, each material can be supplied to each station by multiple alternatives. Furthermore, the expected costs for a set of strategies, material, and station can be derived. In step three of the method, the strategy selection problem is solved by describing a mathematical optimization problem and by using an automated algorithm.

The collection of the required data in step 1 results from the requirements in step 2 and 3. These two steps will be explained in more detail in the following. Step 2 in Sect. 3.2 and step 3 in Sect. 3.3.

Following the methods in Table 1, the method presented here focuses on the use of an optimal algorithm. Due to the complexity of the matrix production, the relief of the employee in the planning task is important.

3.2 Cost Side Description of Alternative Supply Strategies

Standardized process blocks are used to describe an alternative material supply strategy. The blocks describe the following activities and states: handling, transport, storing, inventory, and controlling of material supply. Through a case-specific combination of the blocks, a process chain is built representing a supply strategy. Figure 3 outlines an example. In the example, three strategy alternatives are observed. Each strategy starts at the goods receipt and ends at the provision point. The strategies pass different storage levels. This leads to a different composition of the process blocks. The first blocks of strategy 1 describe that the goods are first collected at the goods receipt, then transported to the central storage area and finally handed over and stored there. The storage process leads to inventory and storage expenses. Goods extraction takes place at regular intervals. This leads to the further process sequence. At the same time, a reorder point is identified at the central storage facility. This leads to the reordering of material and is thus a controlling process for material replenishment. In the individual characteristics of the processes, there can be differences, such as the used transport vehicles. For example, a forklift can be used in the transport process to the central storage facility, while a tugger train is used to connect the intermediate storage facility. These characteristics are represented by different cost rates in the cost functions. Additionally, the costs depend on the chosen container and the number of goods per container. A larger container requires less transports but needs more inventory space. The selection of the container per material is seen as a given input. It depends e.g. on the container policy in the company and the agreements with the supplier. If different container policies need to be displayed, alternative strategies (same sequence, but different values in the cost functions) need to be set upped. The sum of the costs of the process blocks, depending on the volume flow per material and strategy, results in the expected total costs of supplying a certain material for each strategy.

The cost functions focus on marginal costing since only those costs are relevant that differentiate the strategies. General costs like office costs or manager salaries are not important. The taken time reference differs in relation to the time of use of the method. If the logistics system is already running, the actual costs are used. On the opposite, if the logistic system is in design, normal or planned costs are used. The first, if one can take similar costs of another system into consideration, the second, if there are no actual references.

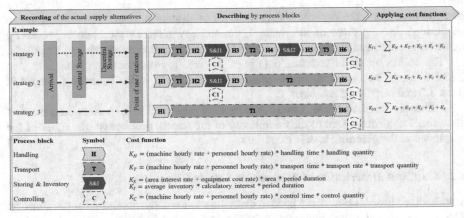

Fig. 3. Cost-side description of alternative supply strategies

3.3 Selection of the Most Cost-Effective Material Supply Strategy in Matrix Production System

The objective when selecting a material supply strategy is to identify the lowest cost alternative. The target function across all alternatives and materials considered is derived from this objective (formula 1). In addition to the target function, different constraints must be considered in the optimization problem. The constraints reflect the framework and requirements from the matrix production. These constraints are described below:

The decision variable is binary with 0 or 1. A strategy is selected or ignored (formula 2). At the same time, formula 3 applies, which allows only one strategy per station and material. In addition, the same strategy is selected for a material that is used at the same station in different processes (formula 4). This ensures transparency in provisioning and reflects the logistical practice.

Each material is supplied in or on a load carrier. This load carrier requires space. The space is either required at the station, e.g. in a Kanban shelf, or is to be provided in a system-side load carrier such as a staging trolley for shopping cart supply. The existing space at the stations and the system-side load carriers must be complied through formula 5 and 6.

In matrix production, there are redundant stations where the same processes are carried out. The individual station per order is selected on an ad-hoc basis during the production process. In order to exclude the danger of a faulty stock, the same strategies need to be selected for stations operating the same process (formula 7). This serves to prevent the following: There are two identical stations A and B, when station A is supplied via Kanban and station B is supplied via shopping cart. If operative logistics assumes that an order is manufactured at station A and therefore does not pick a shopping cart, but the order approaches station B on an ad hoc basis, a stock shortage occurs. This is prevented with formula 7, as both stations need to be supplied by Kanban or shopping cart.

$$\sum_{i \in I} \sum_{d \in D} x_{i,d} C_{i,d} \to \min \quad \forall i \in I, \forall d \in D \tag{1}$$

$$x_{i,d} \in \{0, 1\} \quad \forall i \in I, \forall d \in D \tag{2}$$

$$\sum_{i \in I} x_{i,d} = 1 \quad \forall i \in I, \forall d \in D \tag{3}$$

$$x_{i,s,m}^{p_1} = x_{i,s,m}^{p_2} \forall i \in I, \quad \forall s \in S, \forall m \in M, p_1, p_2 \in P \tag{4}$$

$$\sum_{i \in I} \sum_{d \in D} A_{i,s,d}^{station} \cdot x_{i,s,d} \le A_s^{max,station} \quad \forall i \in I, \forall s \in S, \forall d \in D \tag{5}$$

$$\sum_{i \in I} \sum_{d \in D} A_{i,d}^{system} \cdot x_{i,d} \le A_i^{max,system} \quad \forall i \in I, \forall d \in D \tag{6}$$

$$x_{i,p,m}^{s_1} = x_{i,p,m}^{s_2} \forall i \in I, \quad \forall p \in P, \forall m \in M, s_1, s_2 \in S \tag{7}$$

i: strategy index	$i \in I, I = \{1, \ldots, n \mid n \in \mathbb{N}\}$
s: station index	$s \in S, S = \{1, \ldots, q \mid q \in \mathbb{N}\}$
p: process index	$p \in P, P = \{1, \ldots, r \mid r \in \mathbb{N}\}$
m: material index	$m \in M, M = \{1, \ldots, t \mid t \in \mathbb{N}\}$
d: data set index(consisting of s, p, m)	$d \in D, D = \{1, \ldots, u \mid u \in \mathbb{N}\}$
A: Area, Space	
C: Costs	

The target function and the constraints represent the mathematical formulation of the logistic problem. The problem is solved by the branch-and-cut algorithm. This algorithm promises an exact solution in a reasonable time. Furthermore, it is applicable to this kind of an integer linear problem [27]. Problems with 100.000 variables are still solvable [28]. The logistics problem under consideration is seen within these limits.

4 Summary and Outlook

The method can be implemented using common environments. It was tested in Microsoft Excel and Python. The test set consists of 200 data sets (combination of station, process and material) and three strategy alternatives. To apply the algorithm an add-on is required for Microsoft Excel and a library for Python. In both cases, the free open source solutions by the Coin-OR Foundation (www.coin-or.org) were used and tested. The run time is negligible. The solution is by the nature of the algorithm an optimal solution. The manual search for the solution by trial and error or complete enumeration requires a larger amount of time.

The developed method supports the logistics planner by choosing material supply strategies in matrix production. The selection process is automated and focuses on minimizing the logistics costs. With the focus on the costs as a decisive criterion and the planner's relief, the acceptance of result and method increases.

Critical, however, is the extensive description of alternative strategies. The recording of all data and the derivation of the costs initially causes high manual effort. For future adjustments of the material supply strategies allocation, however, the data is reused in the optimization problem and must be adapted only slightly. The initial effort is thus relativized. Still, the automated selection of the data through tools is preferable.

Further, in additional research it has to be investigated how this method can be integrated into a superordinate assembly planning, as the selection of a material supply strategy depends on the assembly planning, as well. Also, setting up a larger data set, and investigating other strategies, like a cross-docking strategy, is of interest.

Acknowledgements. This article was written within the framework of "SE.MA.KI" (Self-learning control of a cross-technology matrix production by simulation-based AI). The research and development project SE.MA.KI is funded by the German Federal Ministry of Education and Research (BMBF). Funding code: L1FHG42421. The authors are responsible for the content of this publication.

References

1. Kampker, A., et al.: Agile low-cost montage. In: Schuh, G. (eds.) AWK Aachener Werkzeugmaschinen-Kolloquium, Internet of Production für agile Unternehmen, 1st edn., pp. 231–259. Apprimus Verlag, Aachen (2017)
2. Koren, Y.: The Global Manufacturing Revolution: Product-Process-Business Integration and Reconfigurable Systems. Wiley, Hoboken. 1 online resource (2013)
3. Bauer, D., Böhm, M., Bauernhansl, T., Sauer, A.: Increased resilience for manufacturing systems in supply networks through data-based turbulence mitigation. Prod. Eng. Res. Dev. **15**(3–4), 385–395 (2021). https://doi.org/10.1007/s11740-021-01036-4
4. Puchkova, A., Rengarajan, S., McFarlane, D., Thorne, A.: Towards lean and resilient production. IFAC-PapersOnLine **48**(3), 2387–2392 (2015)
5. Fries, C., et al.: Fluid manufacturing systems (FLMS). In: Weißgraeber, P., Heieck, F., Ackermann, C. (eds.) Advances in Automotive Production Technology – Theory and Application. ARENA2036, pp. 37–44. Springer, Heidelberg (2021). https://doi.org/10.1007/978-3-662-62962-8_5

6. Kern, W., Rusitschka, F., Kopytynski, W., Keckl, S., Bauernhansl, T.: Alternatives to assembly line production in the automotive industry. In: Proceedings of the 23rd International Conference on Production Research (IFPR), Manila (2015)
7. Fries, C., Wiendahl, H.H., Foith-Förster, P.: Planung zukünftiger Automobilproduktionen. In: Bauernhansl, T., Fechter, M., Dietz, T. (eds.) Entwicklung, Aufbau und Demonstration einer wandlungsfähigen (Fahrzeug-) Forschungsproduktion. ARENA2036, vol. 98, pp. 19–43. Springer, Heidelberg (2020). https://doi.org/10.1007/978-3-662-60491-5_4
8. Göppert, A., Hüttemann, G., Jung, S., Grunert, D., Schmitt, R.: Frei verkettete Montagesysteme. ZWF **113**(3), 151–155 (2018)
9. Ranke, D., Bauernhansl, T.: Evaluation of material supply strategies in matrix manufacturing systems. In: Weißgraeber, P., Heieck, F., Ackermann, C. (eds.) Advances in Automotive Production Technology – Theory and Application. ARENA2036, pp. 80–88. Springer, Heidelberg (2021). https://doi.org/10.1007/978-3-662-62962-8_10
10. Foith-Förster, P., Bauernhansl, T.: Changeable and reconfigurable assembly systems – a structure planning approach in automotive manufacturing. In: Bargende, M., Reuss, HC., Wiedemann, J. (eds.) 15. Internationales Stuttgarter Symposium. Proceedings, pp. 1173–1192. Springer, Wiesbaden (2015). https://doi.org/10.1007/978-3-658-08844-6_81
11. Kern, W., Rusitschka, F., Bauernhansl, T.: Planning of workstations in a modular automotive assembly system. Procedia CIRP **57**, 327–332 (2016)
12. Greschke, P., Schönemann, M., Thiede, S., Herrmann, C.: Matrix structures for high volumes and flexibility in production systems. Procedia CIRP **17**, 160–165 (2014)
13. Bullinger, H.-J., Lung, M.M.: Planung der Materialbereitstellung in der Montage. Teubner, Stuttgart (1994)
14. Wannenwetsch, H.: Integrierte Materialwirtschaft, Logistik und Beschaffung. Springer, Heidelberg (2014). https://doi.org/10.1007/978-3-642-45023-5
15. Adolph, S., Metternich, J.: Materialbereitstellung in der Montage. ZWF **111**(1–2), 15–18 (2016)
16. Grünz, L.: Ein Modell zur Bewertung und Optimierung der Materialbereitstellung. Dortmund, Univ., Diss., Shaker, Aachen (2004)
17. Köhler, R.: Disposition und Materialbereitstellung bei komplexen variantenreichen Kleinserienprodukten. München, Techn. Univ., Diss. Springer, Berlin (1997)
18. Vojdani, N., Knop, M.: Leistungsorientierte bewertung und auswahl von materialbereitstellungsstrategien mittels fuzzy axiomatic design. Logist. J. Proc. **2016** (2016)
19. Wildraut, L., Stache, U., Mauksch, T.: Planung von Produktionsversorgungssystemen. ZWF **114**(4), 167–172 (2019)
20. Cárdenas-Barrón, L.E., Porter, J.D.: Supply chain models for an assembly system with pre-processing of raw materials: a simple and better algorithm. Appl. Math. Model. **37**(14–15), 7883–7887 (2013)
21. Herbert, M., Heinlein, P., Fürst, J., Franke, J.: A systematic approach for planning, analyzing and evaluating internal material provision. Procedia Manuf. **55**, 447–454 (2021)
22. Gudehus, T., Kotzab, H.: Comprehensive Logistics. Springer, Heidelberg (2009). https://doi.org/10.1007/978-3-540-68652-1
23. Koether, R. (ed.): Taschenbuch der Logistik, 5th edn. Hanser, München (2018)
24. Schulte, C.: Logistik: Wege zur Optimierung der Supply Chain, 6th edn. Franz Vahlen, München (2013)
25. Morrison, D.R., Jacobson, S.H., Sauppe, J.J., Sewell, E.C.: Branch-and-bound algorithms: a survey of recent advances in searching, branching, and pruning. Discrete Optim. **19**, 79–102 (2016)
26. Werners, B.: Grundlagen des Operations Research. Springer, Heidelberg (2013). https://doi.org/10.1007/978-3-642-40102-2

27. Mitchell, J.E.: Branch-and-Cut Algorithms for Combinatorial Optimization Problems (2019). http://eaton.math.rpi.edu/faculty/Mitchell/papers/bc_hao.pdf. Accessed 29 Jan 2022
28. Domschke, W., Drexl, A., Klein, R., Scholl, A.: Einführung in Operations Research. Springer, Heidelberg (2015). https://doi.org/10.1007/978-3-662-48216-2

Assessment of Ergonomics Risk Experienced by Welding Workers in a Rail Component Manufacturing Organization

Khumbuzile Nedohe[1]([envelope]) [iD], Khumbulani Mpofu[2] [iD], and Olasumbo Makinde[2] [iD]

[1] University of Johannesburg, Johannesburg 2006, South Africa
Khumbuzilen@uj.ac.za
[2] Tshwane University of Technology, Pretoria 0183, South Africa

Abstract. The various types of welding workstation designs used in a rail component manufacturing system environment have drawn the attention of industrial engineers to the safety and efficiency of the workers during welding operations. Welding operations are carried out using several posture configurations, which have a negative physical ergonomic impact on the workers, especially in manual welding processes. This empirical research investigates the ergonomics conditions of welding workplaces with the aim of ascertaining the disorders that may be associated with working posture during welding operations among the South African population. Twenty-seven (27) welders were randomly selected, and data was collected using a structured questionnaire. The majority (67 percent of the welders) stated that they experience discomfort and pain whilst they carry out their task, which contradicts ergonomic guidelines for working posture. Forty-eight percent of the welding workers were frequently physically tired. Sixty-three (63) percent agree that they perform repetitive tasks, and a majority of 78% of welding workers reported neck discomfort as a result of tilting their neck posture for a longer period during welding operations. It was deduced that the correlation among risk factors associated with workstation design, repetitive tasks, contribute to the awkward posture adopted whilst welding, that, if retained for a long duration, could lead to musculoskeletal injuries, poor quality of work, and reduced productivity. Based on these results, in order to increase productivity, it was proposed to redesign the welding workstations and to prioritize interventional ergonomic programme to minimize the MSDs problems.

Keywords: Ergonomics · Welding · Workstation design

1 Introduction

The productivity of workers, their delivery time, and the quality of their work are all important factors for the success of an organization. Traditional manufacturing concepts hold the view that at the core of the production system, more focus is placed on improving the productivity of manufacturing organizations regardless of the work posture conditions of the workers [1]. Productivity improvement cannot be achieved in a

© The Author(s) 2023
K.-Y. Kim et al. (Eds.): FAIM 2022, LNME, pp. 227–236, 2023.
https://doi.org/10.1007/978-3-031-18326-3_23

welding organization without resolution of workers awkward postures during welding operations. For overall growth and sustainability of products in the existing economy, many manufacturing businesses rely heavily on human work activities/factors. As a result, humans will undoubtedly continue to be an important and crucial element of manufacturing for a long time. Henceforth, people require more comfortable and safe working environments [2]. According to [3], it is possible to design safe, effective, and productive work systems by assessing people's strengths and limitations, their occupations, equipment, and working environment, and the interactions between these variables [4]. Work-related musculoskeletal disorders (WMSDs) are common among employees who perform manual handling (MH) activities, particularly in difficult industries like structural steel welding [5]. Back discomfort and joint injuries are prevalent WMSDs that are linked to muscular tension from MH tasks. Fatigue experienced by workers are as a result of job repetitions, and uncomfortable postural conditions [6, 7] further reiterates that sprains and strains of joints/abdominal muscles account for 43% of manual handling injuries in the workplace, whereas muscular tension accounts for 33% of injuries. According to [8] legislation requires companies to measure ergonomic risks on a regular basis and document actions taken to mitigate them. [9] stated that improving workplace ergonomics, on the other hand, frequently translates into increased economic and social performance indicators for the company. Many studies have frequently used observational methods [10] such as Rapid Entire Body Assessment (REBA), Rapid Upper Limb Assessment (RULA), Ovako Working Posture Analysis System (OWAS), Job Strain Index (JSI), Ergonomic Workplace Analysis (EWA), [11–13] for assessing MSD-related risk factors; but the studies often do not consider the self-reports MSDs assessments that are subjectively reported by workers via questionnaires.

2   Methods and Materials

Twenty-seven (27) welding workers in Company XYZ within 5th, 50th and 95th percentile (See Fig. 1) of the South African population completed a questionnaire. Workers with background diseases or accidents affecting musculoskeletal system were excluded from the study. Because every participant was asked the same questions, it is a quick and easy technique to collect responses from a large group of people before conducting quantitative analysis. Refer to Fig. 1, which shows the postural configuration of the welders with body physique dimensions within the 5th, 50th and 95th percentile.

A questionnaire focused on probing the status of the body postures of the welding workers and the effect of those postures was considered in this study from company XYZ. The questionnaire consists of four sections; gender, age, years of job experience, height, and weight were among the biographical details given in section A. The information in segment A was gathered in order to determine the welding workers' social demographics. Section B includes questions that analysed the conduciveness of the work environment in order to determine how respondents felt about their work environment. Section C of the questionnaire-included questions aimed at determining how welding workers perceive their work activity constraints that contribute to musculoskeletal disorders (MSDs). Section D consists of questions aimed at determining the health state of welders while

Fig. 1. Postural configuration of the welders with body physique dimensions within the 5th, 50th and 95th percentile

performing welding operations. The questionnaire was constructed in the context of this research project to include a combination of classification questions, allowing the welding workers' responses to be classified and patterns established for further analysis.

3   Results and Interpretation

Age of Welding Workers Respondents in a Rail Component Manufacturing Environment.
The age distribution of the welding workers that responded to the questionnaire is depicted from Fig. 2.

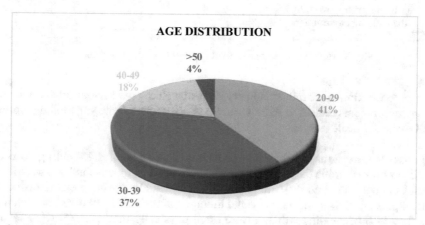

Fig. 2. Age distribution of the welding workers in the rail component manufacturing environment

As illustrated in Fig. 2, the 20 to 29 year-old age group (41%) and the 30 to 39 year-old age group (37%) represented the majority of the welding workers that responded to the questionnaire. Eighteen percent (18%) of welding workers were between the ages of 40 and 49, and four percent (4%) were older than 50. Age is a function of ergonomic risk in the sense that age is the most significant factor that influences the productivity rate and the operation efficiency as indicated from the study of [14, 15]. When working on repetitive jobs for extended periods, older welding workers are more (prone) to musculoskeletal pain as opposed to the younger welding workers [16].

Responses to Questions on Health Status Prior and During Employment.
The effect of ergonomics on welding workers is visible in terms of their health status outcomes from prior to their employment to their current employment in the welding sector. Figure 3 illustrates the health status of the respondents prior to working in the welding environment.

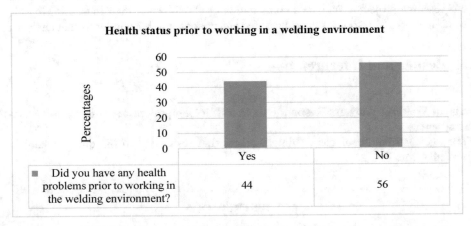

Fig. 3. Health status prior to working in the welding environment

Prior to working in the railcar industry, 44% of welding workers reported no signs of any health problems related to their work. Sixty-six percent (66%) of welding workers reported having health issues.

Responses to Questions on Physical and Mental Tiredness of the Welding Workers
In order to further, probe the response the welding workers, the respondents were asked questions about how tired they are physically and mentally during welding operation. Figure 4 presents the physical and mental tiredness of the welding workers, which indicates that 48% of welding workers were often physically tired and 3% were seldom physically tired. Fifty-nine percent (59%) of welding workers were occasionally mentally tired, and 8% were always mentally tired. The age of the welders may play a role in their health status. As stipulated in the study by [17–19] age is a factor in the relation to physical and mental health among employees.

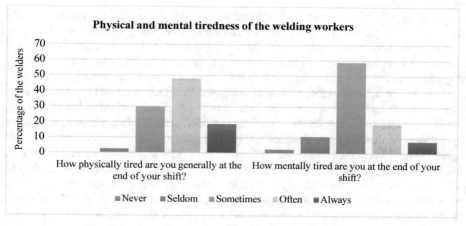

Fig. 4. Physical and mental tiredness of the welding workers

RESPondent's Response Regarding Work Activity Constraints that Contribute to MSD.

Figure 5 presents the work activity constraints that have been indicated by the respondents to be the most likely causes of MSDs during welding operations.

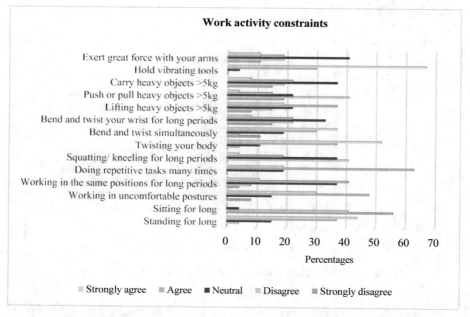

Fig. 5. Welding work body activity constraints that contribute to MSD

As shown from Fig. 5, 41% of welding workers have a neutral feeling about using their arms with great force. Sixty-six percent (66%) of welding workers strongly agree that they use a vibrating tool while welding. Eight percent (8%) strongly disagree that they carry heavy objects, four percent strongly disagree that they push or pull heavy objects, and 30% of welders agreed that they bend and twist their wrists for extended periods. While 37% of welding workers disagreed that, they bend and twist at the same time, 52% strongly agreed that they twist their body. Four percent (4%) of welding workers disagree that they squat or kneel for extended periods while performing welding operations, while 63% agree that they perform repetitive tasks. Forty-one percent (41%) agreed to working in the same positions for long periods; whereas 48% agreed to working in uncomfortable postures, and 56% strongly disagreed to sitting for long periods. Finally, 44% of welding workers strongly agreed that they must stand for long periods while performing their welding tasks. In light of this, it could be inferred that the majority of the respondents are not working in a neutral posture, which contradicts ergonomic guidelines for working posture.

Respondent's Response to Occurrence or Non-occurrence of Discomfort During and After Welding Operations.

In response to the question about the occurrence or non-occurrence of discomfort during welding operations. It was revealed in Fig. 6 that 33% of these participants stated that their welding tasks do not cause them chronic discomfort. Sixty-seven percent (67%) of welding workers reported chronic discomfort during welding operations, which was a concerning finding.

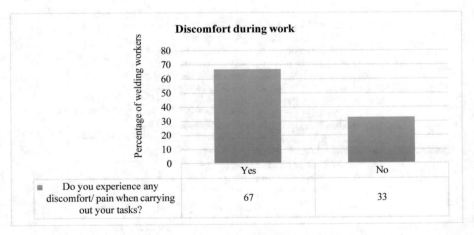

Fig. 6. Welding workers response to discomfort during and after work

The welding workers that responded to experience chronic health discomfort problems indicated that the areas of their body which experienced this discomfort included: the neck, right shoulder, left shoulder, upper back, lower back, right elbow, left elbow, right wrist, left wrist, fingers, right hip, left hip, right knee, left, knee, right ankle and on the left ankle. Therefore, as stipulated in a study by [20] measures such as workstation adjustment reduce awkward postures and allow employees to work in more neutral positions, the use of different tools or rotational clamps need to be implemented in a rail-component manufacturing environment to alleviate chronic discomfort, which is prevalent among the majority of welding workers in this environment.

Respondent's Response Regarding the Areas of Their Body Pain Are Felt During and After Welding Operations

Figure 7 depicted the various types of the body pains where respondents indicated the experience pain during and after welding operations.

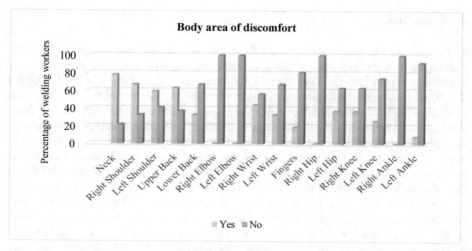

Fig. 7. Welding workers response from which pain occurs

As depicted in Fig. 7, 78% of welding workers reported neck discomfort as a result of tilting their neck posture for a longer period of time during welding operations. Sixty-seven percent (67%) of the welders reported that they experienced right shoulder discomfort. Fifty-nine percent (59%) of welding workers who participated in the study reported that they experienced left shoulder discomfort while welding. Sixty-three percent (63%) of the welding workers reported discomfort in their upper backs. The forward bending of the body trunk may cause upper back pain [21]. Thirty-three percent (33%) of welding workers reported lower back discomfort from welding activities. During welding operations, none of the welding workers in the sample reported discomfort in their right or left elbows. Welding workers may experience wrist pain as a result of working with the wrist bent from the midline. During the welding operations, 44% of the sampled participants complained of discomfort on their right wrist. Thirty-three percent (33%) of welding workers reported discomfort on their left wrist. Nineteen percent (19%) of welders reported discomfort in their fingers. All the sampled welding workers indicated that they do not experience right hip pain discomfort during welding operations. According to the data presented above, 37% of welding worker participants reported discomfort on their left hip. Welding activities performed below the knee level, which implies that the work is done while the body is bent, this has a significant impact on the torso and, in particular, the legs. Thirty-seven percent (37%) of the workers reported right knee discomfort. Twenty-six percent (26%) of welders confirm experiencing left knee discomfort during welding operations. All of the sampled welding workers stated that they do not experience right ankle discomfort during welding operations. Eight percent (8%) of the welding participants polled reported left ankle discomfort. The variation in ergonomic postural behaviour experienced by the sampled welding workers is due to variations in the welding workers' anthropometric characteristics, resilience, and adaptability rate to the welding operating conditions of a rail component manufacturing environment.

4 Conclusion

The results of the study advocate for the immediate implementation of ergonomics interventions at the rail component manufacturing organization, with proper worker knowledge and intentional awareness of common postural change. It is recommended that ergonomics laws be implemented and monitored in order to reduce morbidity due to musculoskeletal disorders. Further to this, it was observed by the authors that there is a need for improvement and redesign of welding workstation that will conducively accommodate the various body sizes and shapes. Poorly designed workstation environments indicate physical accommodation problems resulting in health effects of the welding workers, productivity losses and mediocre quality of work. Ergonomically fit and adjustable workstations, that ensure workers to adjust their workstations according to their comfort and posture to relieve physical stress are more advantageous and could be used posture to relieve physical stress, improve posture, and performance. As future work, this study must be replicated in rail component manufacturing organizations in other provinces across South Africa. The selection of the participants should be premised on anatomical justification and geometric representation of the participants.

References

1. Papetti, A., Gregori, F., Pandolfi, M., Peruzzini, M., Germani, M.: A method to improve workers' well-being toward human-centered connected factories. J. Comput. Des. Eng. **7**, 630–643 (2020)
2. Calzavara, M., Battini, D., Bogataj, D., Sgarbossa, F., Zennaro, I.: Ageing workforce management in manufacturing systems: state of the art and future research agenda. Int. J. Prod. Res. **58**(3), 729–747 (2020)
3. Mahendra, K.C., Virupaksha, G.H., Gouda, A.T.: Ergonomic analysis of welding operator postures. Int. J. Mechanical Prod. Eng. **4**(6), 922 (2016)
4. Abdousa, M.A., Delorme, X., Battini, D., Sgarbossa, F., Berger-Douce, S.: Assembly line balancing problem with ergonomics: a new fatigue and recovery model. Int. J. Prod. Res. Taylor & Francis (2022)
5. Suman, D., Debamalya, B., Shankarashis, M., Sabarni, C.: Postural stress analysis with MSD symptoms of welders and solution for workstation design. Int. J. Forensic Eng. Manage. **1**(1), 423 (2020)

6. Singh, B., Singhal, P.: Work related musculoskeletal disorders (wmsds) risk assessment for different welding positions and processes. In: 14th International Conference on Humanizing Work and Work Environment HWWE, pp. 264267 (2016)

7. Rahmana, A., Palaneeswaranb, E., Kulkarnic, A.: Virtual reality based ergonomic risk evaluation of welding tasks. In: Proceedings 19th Triennial Congress of the IEA, Melbourne **9**, 14(2015)

8. Department Of Labour. Occupational Health and Safety Act, 1993 Ergonomics Regulations, 2019, SOUTH AFRICA, Government Gazette. No. 42894 (2019)

9. Otto, A., Battaïa, O.: Reducing physical ergonomic risks at assembly lines by line balancing and job rotation: a survey. Int. J. Computers Industrial Eng. **111**, 467–480 (2017)

10. Takala, E.P., et al.: Systematic evaluation of observational methods assessing biomechanical exposures at work. Scandinavian journal of work. Environment Health, pp. 3–24 (2010)

11. Kee, D.: Comparison of OWAS, RULA and REBA for assessing potential work-related musculoskeletal disorders. Int. J. Ind. Ergon. **83**, 103140 (2021)

12. Colim, A., et al.: Towards an ergonomic assessment framework for industrial assembly workstations—a case study. Appl. Sci. **10**(9), 3048 (2020)

13. Takala, E.P., et al.: Systematic evaluation of observational methods assessing biomechanical exposures at work. Scand J. Work Environ. Health **36**(1), 3–24 (2010)

14. Calvo-Sotomayor, I., Laka, J.P., Aguado, R.: Workforce ageing and labour productivity in Europe. Sustainability **11**(20), 5851 (2019)

15. Göbel, C., Zwick, T.: Age and productivity: sector differences. De Economist **160**, 35–57 (2012)

16. Park, J., Kim, Y.: Association of exposure to a combination of ergonomic risk factors with musculoskeletal symptoms in korean workers. Int. J. Environ. Res. Public Health **17**, 9456 (2020)

17. Sousa-Ribeiro, M., Bernhard-Oettel, C., Sverke, M., Westerlund, H.: Health- and age-related workplace factors as predictors of preferred, expected, and actual retirement timing: findings from a swedish cohort study. Int. J. Environ. Res. Public Health **18**(5), 2746 (2021)

18. Scheibe, S., Yeung, D.Y., Doerwald, F.: Age-related differences in levels and dynamics of workplace affect. Psychol. Aging **34**(1), 106–123 (2019)

19. de Lange, A.H., Taris, T.W., Jansen, P.G.W., Smulders, P., Houtman, I.L.D., Kompier, M.A.J.: Age as factor in the relation between work and mental health: results of the longitudinal TAS survey. Occupational Health Psychology: European Perspectives on Res., Education Practice, ISMAI **1**, 21–45 (2006)

20. Kester, J., Monk, A.: Proper ergonomics improve welding productivity, protect welders. https://www.tregaskiss.com/proper-ergonomics-improve-welding-productivity-protect-welders/Published: 2018 Assessed 01 Feb 2022

21. Vad, V.: Is Poor Posture Causing Your Back Pain?. https://www.spine-health.com/blog/poor-posture-causing-your-back-pain] Published: 2020 Assessed 28 Jan 2022

A Survey of Smart Manufacturing for High-Mix Low-Volume Production in Defense and Aerospace Industries

Tanjida Tahmina[1], Mauro Garcia[1], Zhaohui Geng[1(✉)], and Bopaya Bidanda[2]

[1] The University of Texas Rio Grande Valley, Edinburg, TX 78539, USA
zhaohui.geng@utrgv.edu
[2] University of Pittsburgh, Pittsburgh, PA 15261, USA

Abstract. Defense and aerospace industries usually possess unique high-mix low-volume production characteristics. This uniqueness generally calls for prohibitive production costs and long production lead-time. One of the major trends in advanced, smart manufacturing is to be more responsive and better readiness while ensuring the same or higher production quality and lower cost. This study reviews the state-of-the-art manufacturing technologies to solve these issues and previews two levels of flexibility, i.e., system and process, that could potentially reduce the costs while increasing the production volume in such a scenario. The main contribution of the work includes an assessment of the current solutions for HMLV scenarios, especially within the defense of aerospace sectors, and a survey of the current and potential future practices focusing on smart production process planning and flexible assembly plan driven by emerging techniques.

Keywords: High mix low volume manufacturing · Production planning · Lean manufacturing · Robotics · Group technology

1 Introduction

Modern manufacturing is rooted in the Ford Model T production line favoring the mass production strategy to reduce production costs based on the Economies of Scale principle [1]. Its essence is to maximize the production volume based on the foreseeable demand. In this way, the prohibitive manufacturing setup cost and overhead costs can be evenly distributed among many fabricated parts, resulting in the reduction of production costs per unit. Another major effect is the training standardization and work division. The workers are only responsible for working in a certain portion of the manufacturing plant or assembly line. Even though this standardization can reduce the need for high-skill working labor, which could further reduce the production costs, the jobs become repetitive and less creative.

However, mass production has its major weakness when facing the requests for high-mix low-volume (HMLV) products, which becomes more critical with the advent of mass customization during the past few decades, especially in defense sectors and the

© The Author(s) 2023
K.-Y. Kim et al. (Eds.): FAIM 2022, LNME, pp. 237–245, 2023.
https://doi.org/10.1007/978-3-031-18326-3_24

aerospace industry [2]. HMLV manufacturing is a production scenario when producing a wide variety of parts with small demand [3, 4]. Generally speaking, the products in HMLV cases require complex fabrication processes to meet the harsh condition of usage, especially in defense and aerospace applications. This urges the industry to re-evaluate the production system and processes for a better solution to meet the HMLV demand and, more importantly, lower the production costs and energy consumption [5].

Mass customization allows the original equipment manufacturers (OEMs) or customers, in a general setting, to order the products with specified designs to fulfill their unique functional specifications, which is known as a make-to-order manufacturing strategy [6]. Suppliers also need to provide high-quality products while delivering the desired diversified families of parts in a timely manner to remain competitive in the market. Suppliers have incorporated the flexible manufacturing strategy as a remedy solution to meet these strict requirements. One way to tackle the issue is the just-in-time (JIT) approach, which can assist in increasing manufacturing effectiveness [7, 8]. JIT is an integrated scheduling system that synchronizes suppliers, schedulers, and production teams to reduce inventory and produce parts according to the specifications.

In this paper, we provide a thorough treatment of the previously proposed methods that could potentially remedy the issues and restrictions of mass customization. State-of-the-art solutions are briefly reviewed, together with their pros and cons. The solutions are summarized as the increase of two levels of flexibility: system-level and process-level. Potential solutions that are built upon current technological advancements are proposed. Specifically, mass customization can directly benefit from the major progress in Industrial 4.0/5.0, including robotics, additive manufacturing, big data, etc., based on which data-driven decision making and process planning can provide a more robust and resilient solution to defense and aerospace applications.

The remainder of this paper is organized as follows. In Sect. 2, we present the major characteristics and solutions of the modern manufacturing systems for HMLV scenarios. Major research topics are surveyed with discussions of their advantages and disadvantages. In Sect. 3, several ongoing or potential trends that can potentially further increase the readiness and lower the costs of resources for mass customization are reviewed. This paper concludes with a discussion of the current status and trend in smart manufacturing.

2 Manufacturing Strategies for HMLV in Defense and Aerospace Industries

2.1 Major Issues

In a general manufacturing setting, the widely used way of reducing cost is to connect and outsource manufacturing tasks to the regions closer to the production materials or cheap labor for cost reduction. However, since the designs and products in the defense and aerospace industries are considered sensitive and critical to a nation, inshore manufacturing is generally preferred. As a result, it is critical to provide a robust solution to strengthen preexisting manufacturing capability with smart planning and control while reshoring the high value-added manufacturing industries to keep a resilient supply chain system.

The major issues when applying traditional manufacturing strategies in inshore manufacturing, especially those involving small and medium enterprises (SMEs), with the HMLV scenario are the limited order scales and restricted manufacturing capability. The order volume is generally small for the defense and aerospace industries. In this way, the SMEs are not as profitable as the OEMs, which makes them quite vulnerable to even small demand changes. On the other hand, the small demand for each part design significantly increases the setup cost of the production process and, more importantly, prolongs the production lead time.

2.2 Advanced Processing Techniques

Several proposals are available to remedy the issues. The two major tracks of consideration are to (1) increase the production volume by grouping some parts or processes; and (2) utilize a "smart", automated process to increase the process flexibility.

Robotic-aided machining processes have been considered by researchers and practitioners, specifically in the HMLV scenario, to process workpieces in a faster and more reliable fashion than manual operations. Hu et al. [9] proposed an automated robotic deburring and chamfering system, to select features in the piece's computer-aided design (CAD) model with a human-machine interface, then target accurate gear registration and fast collision-free trajectory planning through another human-machine interface. The system also consists of a robotic manipulator, force/torque sensors, pneumatic controls, pneumatic control components, and tool changes with the aim of generating an efficient part processing plan. Similarly, another study shows the development of a flexible and automatic deburring system to process large cast parts, such as aerospace and aircraft components [10], which consists of using robotic manipulators, 3D vision sensors, lasers, and force control sensors.

Part assembly, especially a robot-guided assembly, still shows many gaps in agility, flexibility, automation technologies, and human-robot interaction, as robots require crucial data to operate in the desired manner [11]. With the goal of reducing assembly time for HMLV parts, 3D-vision systems have been developed to provide flexible robotic assembly and precise workpiece positioning. Such is the case where using vision-guided assembly robots provides subtle variations in geometries and workpiece poses rather than only using limiting expensive fixtures [12]. Using matching algorithms, part CAD models and 3D point cloud data of a part position are matched by corresponding points and features to achieve object-accurate pose estimation. Although there is some rotational and translations error, these systems show the possibility of improvement, with more vision guidance and force/torque control, to be applicable for real-life industrial applications.

Another critical advancement is the development of additive manufacturing technologies (AM) [13]. AM processes fabricate parts in a layer-by-layer fashion, which reduces the restrictions on the design geometry and intricate features. Therefore, instead of decomposing the parts into multiple subassemblies, AM provides the capability of fabricating the final product with reduced weight, higher strength, and lower costs [14, 15].

2.3 Make-to-X Strategy

The manufacturing system is conventionally considered to input raw materials and output final products or parts. However, depending on the demand behaviors, many manufacturing and production strategies are available.

The most classic manufacturing strategy is make-to-order. Job-based production system produces customized products according to particular customer demands and functional specifications. The specialized nature of the product requires varied machine setup and skilled, versatile workers [16]. Generally, one set of workers produces a particular type of product as a whole. Material and demand of the products are inconsistent, which compels the planning, and cost-effectiveness of the production to be a challenge. It is also similar to Make to Order (MTO) manufacturing which essentially involves the manufacturer producing a product based on demand placed by the customer. Wastage is low in the particular type of manufacturing; however, utilization of labor and equipment, and setup cost of the machine are challenging.

Make-to-stock is a strategy to deal with the scenario with highly variable demand. This type of production system follows mass production, whose products are generally highly demanded or consumed [17]. Standardized materials, labor, and a process flow are integral parts of this type of manufacturing. The desired product is produced as a direct output of a system, and the product is assembled by production from a different line of the system [18]. The system has significant effectiveness improvement and cost reduction potential due to the generally reduced variability [19]. Production volume demand is highly stable for the product type. In this particular kind of manufacturing, the product is produced and preserved in inventory held in stores which involves the risk of wastage in the case of goods that are not sold according to the demand forecast.

Another widely adopted manufacturing strategy is make-to-assembly. Batch production can be distinguished from job production based on the order quantity. Based on the demand, a higher number of products is produced for a certain period of time in a production facility involving a particular machine group. Batch production is also utilized to manufacture a certain portion of the product or assembly to improve cost-effectiveness and utilization. The type of production involves the make-to-assemble type of manufacturing. The order delivery lead time is relatively lower compared to job-based manufacturing, and the product parts are manufactured to assemble and customer demand relatively faster.

One of the major issues with the above sole make-to-X strategies is that they generally produce a "finished" product, which can be either the final parts or subassemblies that generally do not need further fabrication steps. A mixture of make-to-stock and make-to-order strategies is proposed to solve the unforeseen demand, mass-customization case, as presented in Fig. 1. The full product system is decomposed into two major steps: (1) semi-finished parts are produced in a make-to-stock fashion, and (2) final products are fabricated from the semi-finished parts for make-to-order purposes. The proposed strategy can potentially increase the robustness of the suppliers to the variable demand while providing the capability to produce multiple part families, which share similar manufacturing steps.

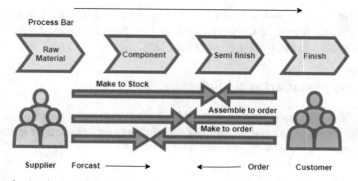

Fig. 1. A mixture of make-to-stock and make-to-order manufacturing strategies.

2.4 Group Technology

Group technology is a method that decomposes the manufacturing system into multiple subsystems by organizing different products into different clusters based on the similarities of their design and required processes. Conventionally, a cluster of products can be produced via a cellular manufacturing system, also known as a manufacturing cell [20]. This manufacturing cell can produce the final products from bulk material while remaining reconfigurable and flexible enough for this cluster of products. A more decentralized planning and control production architecture has been recently proposed to allow for better control in case of uncertainties or changes within product mix and volume via vertical integration of different production modules within the corporation [21].

2.5 Lean Practices

Mass production has been improved through lean manufacturing practices since the 1980s. The particular manufacturing type thrived with high volume, however, low mix. The dramatic changes in the market, availability of a product, and competitiveness warrant product manufacturers to increase customization suitable to customer demand. The transformation in consumer demand has resulted in low volume higher volume manufacturing practices which have brought about the need for revaluation of the manufacturing and supply chain system as a whole. Companies are renovating existing equipment with the capability to produce complex products in optimum amounts of time and reduce unnecessary waste [22, 23]. Value stream mapping is developed to improve different system components of the processes by detailed categorization of the system and finding areas for improvement [24]. Kaizen to improve existing equipment and reduction of change over time through process improvement have emerged.

Lean thinking has changed the world of manufacturing for the better and is characterized as one of the most successful approaches [25]. A fundamental part of Toyota's production system is the Kanban production system which means "to look closely." Pull and push systems are always considered superior as less inventory is involved in the system [3]. Kanban is a type of pull system where material and flow are synchronized to produce exactly what is necessary and to stock only the amount required during replenishment. Even though Kanban is a primitive process, however, the ideology is effective for continuous production and flow control. Powell showed the application of the Kanban system in the HMLV environment [25].

3 A Smart Manufacturing Solution

With the advent of new manufacturing technologies, especially smart manufacturing and artificial intelligence, new solutions are emerging to solve the major issues in mass customization.

3.1 Automation in Manufacturing

One major trend in smart manufacturing is to utilize collaborative robot technology and skilled labor to manufacture diverse parts in the same machine to cater to demand. The improvement effectiveness in HMLV is accomplished by careful inspection of current practices and through improvement initiatives driven at each step. High mix environment results are higher setup and change over time in a production line involving various repetitive tasks. The repetitive tasks and assistance for the pre-planned work process can be accomplished by introducing collaborative robots into the system. There are several techniques to enable and match the skill of the robot's task for the particular task with workers. The Cobots have the ability to learn through hand-guided teaching and programming [26]. The ease of programming and highly accurate task completion of the collaborative robots enables them to a competitive choice.

Aligning with the dynamic shift in manufacturing technological advancement to tackle the challenges has also ensued [27]. Flexible manufacturing equipment capable of unique customized product manufacturing has been gracing the factory environment. The introduction of the HMLV production scenario has ushered in the need for the inclusion of robots in manufacturing assembly. The repetitive tasks executed with the help of automation can increase accuracy, reduce time and increase the effectiveness of the system. The collaboration of lean principles with robotic automation is called lean automation. The ease of reprogramming the software to accommodate the change in product specification brings in the required flexibility to cope with the market. 3D vision and AM are often considered to be the ultimate solution to the customization landscape. The 3D printing software offers a unique customer experience, and 3D printing adapted businesses will be ahead of the competitors due to the sheer flexibility achievable. The technology allows the whole assembly or product to be manufactured together, increasing accuracy and reducing lead time. Regardless of the variations, the product will be manufactured as a whole without slowing down product flow. The paper explores the effectiveness and scrutinizes the method as an effective solution.

3.2 Smart Production Planning and Control

The dynamic art of sharing the flow of types of machinery among different cells in a factory is also an effective solution to address the HMLV scenario. Flow path design manages the flow of a product among different cells and derives the optimized flow path to share the suitable outcome. Peng et al. [28] derived a mathematical programming model from designing the flow of material in a small and mid-sized company with a higher variation. Simulation techniques, such as discrete-event simulation [29] and agent-based modeling [30], also provide a numerical solution for path, policy, and facility layout planning.

Theory of constraint (TOC) is the accumulation of managing and deriving solution based on different constraints. The physical constraint may be equipment and space; non-physical constraint can be management style, process, and demand. The limitation of the theory of constraint problem is that it is suitable for small scenarios. However, to tackle the problem genetic algorithm (GA) is formulated to work with a sample of the problem and increase throughput.

Similar to the genetic algorithm (GA) method, the evolutionary algorithm-based layered encoding cascade optimization approach is also applied to tackle the problem. The evolutionary algorithm essentially follows the survival of the fittest methodology, and the iterative method is executed with the closest to the goal batch of solution. Particle swarm optimization is a stochastic-based approach looking for a solution by searching through the objective. Neoh et al. [31] investigated and addressed the problem by combining the different methods discussed above. In the research, it was observed that the integrated GA-PSO model could efficiently solve the problem faster compared to any other model.

4 Concluding Remarks

A broad range of technologies, especially smart manufacturing techniques, can be utilized to remedy the unique characteristics in high-mix low-volume manufacturing scenarios, especially in defense and aerospace applications. This paper reviews state-of-the-art solutions to this problem, ranging from process-level to system-level. Both are critical to reducing the costs and lead time while maintaining the capability of producing a variety of products in a low volume setting. Further research studies are needed to incorporate both to further improve the performance and the capability of smart manufacturing systems in the HMLV scenario.

Another trend lies in the development of artificial intelligence and machine learning-driven methodologies, equipped with advanced metrology to further improve the smart decision making in process selection, path planning, and flexible assembly, i.e., to further reduce the restrictions on rigid tolerance design via part matching and efficient material handling.

Acknowledgment. The work of Tanjida Tahmina, Mauro Garcia, and Zhaohui Geng was financially supported by the Department of Defense (DoD) MEEP Program under Award N00014–19-1-2728.

References

1. Alizon, F., Shooter, S.B., Simpson, T.W.: Henry Ford and the Model T: lessons for product platforming and mass customization. In: International Design Engineering Technical Conferences and Computers and Information in Engineering Conference, pp. 59–66. Elsevier, Brooklyn, NY (2008)
2. Lyons, B.: Additive manufacturing in aerospace: examples and research outlook. The Bridge **44**(3), 13–19 (2014)
3. Zhang, Q., Tseng, M.M.: Modelling and integration of customer flexibility in the order commitment process for high mix low volume production. Int. J. Prod. Res. **47**(22), 6397–6416 (2009)
4. Rajgopal, J., Bidanda, B.: On scheduling parallel machines with two setup classes. Int. J. Prod. Res. **29**(12), 2443–2458 (1991)
5. Hu, G., Wang, L., Chen, Y., Bidanda, B.: An oligopoly model to analyze the market and social welfare for green manufacturing industry. J. Clean. Prod. **85**, 94–103 (2014)
6. Thürer, M., Stevenson, M., Silva, C., Land, M.J., Fredendall, L.D.: Workload control and order release: a lean solution for make-to-order companies. Prod. Oper. Manag. **21**(5), 939–953 (2012)
7. Swanson, C.A., Lankford, W.M.: Just-in-time manufacturing. Bus. Process. Manag. J. **4**(4), 333–341 (1998)
8. Cleland, D.I., Bidanda, B.: Project management circa 2025. Project Management Institute, Newtown Square, PA (2009)
9. Hu, J., Kabir, A.M., Hartford, S.M., Gupta, S.K., Pagilla, P.R.: Robotic deburring and chamfering of complex geometries in high-mix/low-volume production applications. In: 2020 IEEE 16th International Conference on Automation Science and Engineering (CASE), pp. 1155–1160. IEEE, Hong Kong, China (2020)
10. Onstein, I.F., Semeniuta, O., Bjerkeng, M.: Deburring using robot manipulators: a review. In 2020 3rd International Symposium on Small-scale Intelligent Manufacturing Systems (SIMS), pp. 1–7. IEEE, Gjovik, Norway (2020)
11. Johansen, K., Rao, S., Ashourpour, M.: The role of automation in complexities of high-mix in low-volume production–a literature review. Procedia CIRP **104**, 1452–1457 (2021)
12. Ogun, P.S., Usman, Z., Dharmaraj, K., Jackson, M.R.: 3D vision assisted flexible robotic assembly of machine components. In: Eighth international conference on machine vision (ICMV 2015), pp. 98751O. International Society for Optics and Photonics. SPIE, Bellingham, WA (2015)
13. Bidanda, B., Narayanan, V., Billo, R.: Reverse engineering and rapid prototyping. In: Dorf, R.C., Kusiak, A. (eds.) Handbook of Design, Manufacturing and Automation, pp. 977–990. John Wiley & Sons Inc, Hoboken, NJ (1994)
14. Najmon, J.C., Raeisi, S., Tovar, A.: Review of additive manufacturing technologies and applications in the aerospace industry. In: Frobes, F., Boyer, R. (eds.) Additive manufacturing for the aerospace industry, pp. 7–31. Elsevier, Cambridge, MA (2019)
15. Geng, Z., Bidanda, B.: Medical applications of additive manufacturing. In: Bártolo, P.J., Bidanda, B. (eds.) Bio-Materials and Prototyping Applications in Medicine, pp. 97–110. Springer, Cham, Switzerland (2021)
16. Chua, T.J., Cai, T.X., Low, J.M.: Dynamic operations and manpower scheduling for high-mix, low-volume manufacturing. In: 2008 IEEE International Conference on Emerging Technologies and Factory Automation, pp. 54–57. IEEE, Hamburg, Germany (2008)
17. Tien, J.M.: Manufacturing and services: from mass production to mass customization. J. Syst. Sci. Syst. Eng. **20**(2), 129–154 (2011)

18. Iravani, S.M., Liu, T., Simchi-Levi, D.: Optimal production and admission policies in make-to-stock/make-to-order manufacturing systems. Prod. Oper. Manag. **21**(2), 224–235 (2012)
19. Quante, R.: Management of Stochastic Demand in Make-to-Stock Manufacturing. Peter Lang International Academic Publishers, Bern, Germany (2018)
20. Billo, R.E., Bidanda, B., Tate, D.: A genetic cluster algorithm for the machine-component grouping problem. J. Intell. Manuf. **7**(3), 229–241 (1996)
21. Didden, J.B., Dang, Q.V., Adan, I.J.: A semi-decentralized control architecture for high-mix-low-volume factories in Industry 4.0. Manufacturing Letters **30**, 11–14 (2021)
22. Bidanda, B., Kadidal, M., Billo, R.E.: Development of an intelligent castability and cost estimation system. Int. J. Prod. Res. **36**(2), 547–568 (1998)
23. Martinez, S.E., Smith, A.E., Bidanda, B.: Reducing waste in casting with a predictive neural model. J. Intell. Manuf. **5**(4), 277–286 (1994)
24. Seth, D., Seth, N., Dhariwal, P.: Application of value stream mapping (VSM) for lean and cycle time reduction in complex production environments: a case study. Prod. Planning Control **28**(5), 398–419 (2017)
25. Powell, D.J.: Kanban for lean production in high mix, low volume environments. IFAC-PapersOnLine **51**(11), 140–143 (2018)
26. El Zaatari, S., Marei, M., Li, W., Usman, Z.: Cobot programming for collaborative industrial tasks: an overview. Robot. Auton. Syst. **116**, 162–180 (2019)
27. Duray, R.: Mass customization origins: mass or custom manufacturing? Int. J. Oper. Prod. Manag. **22**(3), 314–328 (2002)
28. Peng, Y., Guan, Z., Ma, L., Zhang, C., Li, P.: A mathematical programming method for flow path design in high-mix and low-volume flow manufacturing. In: 2008 IEEE International Conference on Industrial Engineering and Engineering Management, pp. 1169–1173. IEEE, Singapore, Singapore (2008)
29. Balaban, C., Sochats, K.M., Kelley, M.H., Bidanda, B., Shuman, L.J., Wu, S.: Dynamic discrete decision simulation system. U.S. Patent No. 8,204,836. U.S. Patent and Trademark Office, Washington, DC (2012)
30. Wu, S., Shuman, L., Bidanda, B., Kelley, M., Sochats, K., Balaban, C.: Agent-based discrete event simulation modeling for disaster responses. In: Fowler, J., Mason, S. (eds.): Proceedings of the 2008 Industrial Engineering Research Conference, pp. 1908–1913. Institute of Industrial and Systems Engineers (IISE), Atlanta, GA (2008)
31. Neoh, S.C., Morad, N., Lim, C.P., Aziz, Z.A.: A layered-encoding cascade optimization approach to product-mix planning in high-mix–low-volume manufacturing. IEEE Trans. Syst., Man, Cybernetics-Part A: Syst. Humans **40**(1), 133–146 (2009)

Feasibility Analysis of Safety Training in Human-Robot Collaboration Scenario: Virtual Reality Use Case

Morteza Dianatfar[✉] ⓘ, Saeid Heshmatisafa ⓘ, Jyrki Latokartano ⓘ, and Minna Lanz ⓘ

Tampere University, Tampere, Finland
morteza.dianatfar@tuni.fi

Abstract. Design and modification of human-robot collaboration workspace requires analysis of the safety of systems. Generally, the safety analysis process of a system commences with conducting a risk assessment. There exists a number international standards for design robotics work cells and collaborative shared workspaces. These guidelines expound on principles and measures to identify hazards and reduce risks. Measures of risk reductions include eliminating hazards by design, safeguarding, and providing supplementary protective measures such as user training. This study analyzed the technical feasibility and industrial readiness of Virtual Reality (VR) technology for safety training in manufacturing sector. The test case of a VR-based safety training application is defined in the human-robot collaboration pilot-line of diesel engines. The Analytic Hierarchy Process method was utilized for conducting a quantitative analysis of the survey with ten experts. The participants performed the importance rating with respect to two hierarchy level criteria. Regarding the evaluation of safety training methods in a human-robot collaboration environment, two alternatives of traditional and Virtual Reality -based training are compared. The results indicates that the VR-based training is valued over the traditional method, with a scored proportion of approximately 65 percent over 35 percent.

Keywords: Human-robot collaboration · Virtual reality · Safety training · Manufacturing assembly · Learning transfer

1 Introduction

Today, the industry is tackling with changing cutomer needs that leads to the increased need of mass-customisation and personification, and to the reconfiguration of production lines. In order to implement product customization and reconfigurability, there is a need to incorporate human dexterity and intelligence with machine accuracy and repeatability by developing human-robot collaboration (HRC) concepts. In order to keep the operator safe and overall system productive, there is a need to tackle safety risks for the operator. However, the continuous change, risks mitigation and training is expensive and time consuming.

© The Author(s) 2023
K.-Y. Kim et al. (Eds.): FAIM 2022, LNME, pp. 246–256, 2023.
https://doi.org/10.1007/978-3-031-18326-3_25

Among the emerging technologies, Virtual Reality (VR) became prevalent in the gaming industry and the social community (E.g., the advent of metaverse). Principally, Reality-Virtuality Continuum consists of two environments category, whereas one defined as pure real environment and another as virtual environment. VR provides solely of virtual objects for user where it utilizes computer graphic simulation for creating objects where it could be seen with monitor-based displays, partially or fully immersive head-mounted display [1]. As these categories is expanding and more hardware and implementation are introduced, the term Extended Reality(XR) is defined to consist of all these hardware in Reality-Virtuality continuum. While the ICT sector is advancing in the use of XR, the manufacturing sector is hesitant due to technical maturity and cost of use.

This paper proceeds to study the technical feasibility of VR technology as a tool in the safety training process and assess industry readiness level with a HRC use case.

2   Related Work

The safety of an operator in close proximity of robot system is a challenge in industry, regarding interfaces, interaction, trust, and technical validation of these systems, several comprehensive studies were explored in [2–5]. Virtual reality has been used in assembly applications in manufacturing and consists of different categories such as simulation design, as discussed in the literature [6, 7]. To examine different possible interactions, a study was conducted by [12] to compare mouse-based, phantom haptic, two configurations with the markless motion capture method, and video-based training. The result demonstrated no notable differences between these five groups' experiments. Another finding was that haptic feedback in case of the collision did not improve learning transfer. In parallel, Augmented Reality has been largely investigated in the same categories. Multiple researchers studied applications where the goal was to provide augmented information on 2D surfaces with projector-based and camera systems. These systems monitor the operator's movements and dynamically project working areas on a table [8–10]. Various studies were conducted to contribute to training applications for providing information, instruction, and visual presentation. In the following sub-sections, literature on VR in training, and safety training is presented.

2.1   Virtual Reality Training

Due to the demand of the manufacturing industry, integrating digital manufacturing and technology is growing rapidly. The users are needed to be trained for changing environments while providing relevant information for the tasks at hand. As virtual reality can provide a safe environment for training purposes various studies have been carried out. Gorecky [11] targeted the two main challenges: training system design,,and training content generation from data management that is required for virtual training. They proposed a methodology consisting of VISTRA's architect, knowledge platform, training simulator, and knowledge sharing center for the content generation of the virtual training system based on semantic technologies.

Training users in online mode (factory floor) is time-consuming and difficult. Therefore, game engines such as UNITY are playing a big role in the advancement of the training system. Pérez [12] implemented a training system, where lab environments such as 3D scanners were mapped into the real environment and simulated the kinematic and dynamic of the robot and VR could provide a promising solution for companies regarding costs, standardization, usability, and training time. In [13] study is conducted where the operator worked in close proximity of an industrial robot where VR was used to increase the immersiveness. From the first study, the survey result depicted that assembly tasks with their hands (VR controller) without the presence mass of components improved the feeling of presence and immersion. Later on Mastas et al. [14] conducted the experiment consisting of two techniques: passive, which aimed to prepare proactive human behavior by utilizing cognitive help via VR, and active approach where the intention was to examine robot adaptive behavior in case of deceleration of trajectories and modification of them.

2.2 Virtual Reality Safety Training

Virtual reality safety training has been studied in fewer industry sectors; however, they had common reasoning behind, reducing injuries concerning user safety while confronting hazardous events. The same concept of training, whereas operators are trained to act when facing hazardous event, and how the operator could react on such circumstances. Buttussi [15] conducted a study in the aviation safety domain for the door opening process. The result demonstrated the significant usability of VR headsets over printed materials. Additionally, the study displayed that VR headset applications led to more engagement and satisfaction compared to other methods. Moreover, a similar result was carried out in the study of effectiveness compared between traditional and VR training approaches in a generic environment of power plants regarding theoretical and practical learning purposes [16].

Among research communities in collaboration with companies, safety training was examined in the construction and mining industry, where operator's safety could have been violated more frequently. Zheng [17] developed a use case that demonstrates the effectiveness of safety training with respect to memorizing critical points when facing hazardous situations in a construction site of urban cities. In addition, improving training contents of safety for construction professionals and workers are studied in [18] and performed empirically experiment that result in user's identification of hazardous events. Additionally, recent research was carried out on developing a framework to study immersive learning in the safety training process of mine rescuers [19]. Their model proposed multiple criteria such as trainee experience with gaming and technology, in addition to the learning experience with real case scenarios, utilizing VR features, and understandability of VR usability. The study demonstrated the positive impact of such criteria in immersive VR-based training. Tichon in [20] performed comprehensive literature based on the effectiveness of virtual reality safety training in the mining industry regarding different skills such as problem-solving, decision making, cognition.

3 Research Methods, Material and Approach

The research methods in this study consist of the definition of a case study, and assessment of the approach with AHP method. The aim of the study is to include the acknowledgement of safety risks and guidelines for operator, and how to avoid or react in case of hazardous events. The VR training will guide and explain to the user what events can trigger the safety system and lead to stopping a robot. The overall goal is to reduce the production stops that origin from the user errors.

The test case is defined in this study relates to the HRC pilot line for assembly of a diesel engine with the industrial robot ABB IRB 4600. In this regard, a VR-based safety training is presented in our research laboratories to investigate the possibility of reducing user mistakes and hazardous events while facing robots in close proximity. The concept of this training was introduced in [21], and 2D snapshot of environment is depicted in Fig. 1.

Fig. 1. VR based safety training environment

In this study, a VR HMD with high resolution selected as user required to experience fully immersed environment with realistic detail of each component of laboratory. The 3D models were rendered carefully to represent replica of real components. For an instance, the engine's CAD models were 3D scanned, and robot models were used from simulation software library with detail rendering. The users can interact with VR HMD controllers to interact with virtual components for movement and assembly, where virtual components contain game engine Physics features. In addition, it is required to walk in the full-scale simulation area to perceive safety distances and working areas.

The hardware selected for this application are HTC VIVE Pro headset with resolution of 2880 x 1600 pixels, refresh rate of 90 Hz and field of view of 110 degrees, and gaming desktop with 32 GB of RAM, NVIDIA Geforce GTX 1060 graphic card. The reason for selecting wireless HMD is that the virtual environment representation is on 1:1 scale of laboratory to improve feel of presence of an operator. HTC VIVE pro controllers have haptic feedback feature, but this feature has not integrated to the system as psychological effect of that in training system requires investigations. The UNITY game engine v 2020.3 is utilized to create visual elements such as working area borders and User Interface such as instruction for acknowledging of different safety measures. For creating more realistic laboratory presentation, manufacturing simulation's software Visual Components v4.4 and Blender adds-on for animation rendering of robot's movements were used.

4 Assessment Approach

Investigating technology feasibility and integration of technical development for manufacturing follows the path between the research community and SMEs and finally adaptation of it in the industry sector. Regarding this, technical development should go through a process to get mature enough to be accepted in the industry. Tirmizi in [22] provided an assessment method to analyze use cases in the development process and steps they have taken toward maturity. Pöysäri [23] proposed a feasibility analysis concept of reconfigurable pilot lines. In this study, KPIs in the aforementioned studies were employed, which are important and relatable for the VR training concept and conduct a technology feasibility assessment. An assessment of this multi-criteria was carried out through the AHP method introduced by [24]. Performing this analysis could be beneficial in the decision-making of adaptation VR-based safety training based on these criterions.

At first, the definition of the overall goal is to evaluate safety training in a HRC environment. Two alternatives of training was chosen: traditional training (i.e., paper-based, video instruction, and instructor-based training) and VR based training using fully immersive VR HMD devices. Following the hierarchy Fig. 2, at the second level of criteria, twelve KPI's are considered, which is respect to first level criteria, use case, and technical criteria. First, use case criteria focus on managerial level, and there are common factors that companies consider when using different solutions. These criteria are as follows: hardware cost, development cost, implementation cost, ergonomic, and lifecycle. Second, technical criteria focus on technical aspects consist of seven KPI's where there are studied in [22, 23], and the definition originated in the context of reconfigurability in [25]. The definition of KPIs is included in Table 1.

A questionnaire with 10 participants from university researchers who are experts in industrial robotics and manufacturing has been conducted. Researchers were informed about the definition of VR application through the text and later by a demo video of application. This step taken toward clarification of application opportunities and demonstrated possible level of realistic representation of laboratory in VR HMD.

Table 1. Multi-criteria definition

KPI	Definition
Hardware cost	A price paid to a third party for safety training equipment
Development cost	An expense incurred for researching, growing, and introducing the safety training equipment product
Implementation cost	Costs to install and/or implement safety training content and equipment measures
Ergonomic	The efficiency and safety concerns regarding using a safety training program for the operator
Lifecycle	The length of time a safety training content and equipment is first introduced until it is outdated or loses official support
Availability	The accessibility and readiness of safety training content and equipment in the market
Integrability	The capability to unite or blend the safety training content and equipment into different use case scenarios
Modularity	Ability to construct standardization for flexibility and variety in use
Customization	The capability to build, fit, or alter the safety training content and equipment according to individual specifications
Scalability	The capability of being easily expanded or upgraded on demand
Diagnosability	To recognize any issues and/or condition associated with the operator involved in the safety training program
Support	The ease of access to receive help in case of malfunction

5   Results

The local weights of safety training represented in Table 2 including two criteria levels. The table provides the interpretation of the critical features that contributed to the final decision. The results indicate that technical criteria, with a rate of approximately 60 percent, are significantly influential. Whereas use case criteria account for the remaining proportion with approximately 40 percent. The second level criteria are devoted to the fundamental components constructing each use case and technical criteria. Regarding use case criteria, the development cost is the most crucial KPI at first rank, and Ergonomic takes the second rank, with 21.36% of the local wight. In the case of technical criteria, customization alongside modularity was at the highest rate. It is worth mentioning that customization, modularity, and availability account for more than 55% of the importance of local weight.

Figure 3 provides an overview of the participants' response distribution to the various KPIs. The hardware cost appears to be right-skewed, with three participants dedicating a weight of more than 0.2. This signifies the weight of hardware cost is relatively spread out with a median of about 0.1. However, 75% of the hardware cost survey results fall under 0.2, and the standard deviation is 0.1. As Table 2 depicted, the respondents allocated greater importance to development costs. This is also evident in the boxplot,

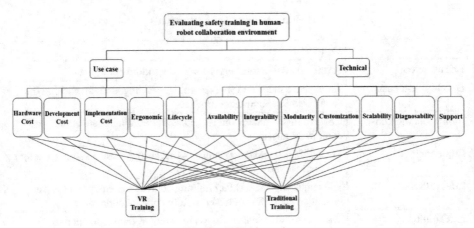

Fig. 2. AHP hierarchy

the highest median of the use case criteria is associated with development cost, where 75% of responses are in the range of 0.18 to 0.42. Moreover, the implementation cost and lifecycle weights seem to experience less variability. Ultimately, the ergonomic criteria, the second local influential weight, has the median of 0.25 and upper quartile of 0.27.

Table 2. Estimated local weights of 1st and 2nd level criteria

1st level criteria	Local weights	2nd level criteria	Local weights
Use case	39.83%	Hardware Cost	14.53%
		Development Cost	**34.17%**
		Implementation Cost	13.76%
		Ergonomic	21.36%
		Lifecycle	16.19%
Technical	**60.17%**	Integrability	11.42%
		Customization	**22.85%**
		Support	8.66%
		Availability	16.82%
		Diagnosability	9.17%
		Modularity	19.57%
		Scalability	11.51%

In technical criteria, the given weights of integrability, customization, support, and diagnosability are to some degree concentrated. The median of integrability, support, and diagnosability is comparably close to one another, 0.1, 0.8, 0.8, respectively. Customization criteria appear left-skewed, representing that the majority of participants have a consensus of great importance towards this measure. Such trend can be seen by the standard deviation of 0.05 and close proximately of the median of 0.22 and maximum value of 0.29. Additionally, the median of modularity and availability is about 0.18. This highlights that the experts have a moderately indistinguishable position concerning both criteria. However, on average, partakers tend to favor modularity compared to availability. The criteria of support follow a normal distribution. Finally, the specialists in the field of factory layout design yield outliers of diagnosability and integrability.

During the survey, participants were asked to provide an importance scale from 0 to 9 in pairwise comparison for both training methods regarding second level criterion. Afterward, the average local weight compared to each second level criteria, and the proportional ranks are depicted in Fig. 4. It can be seen that development cost and ergonomics are instrumental in the VR. Correspondingly, the proportion of hardware cost and lifecycle in the VR outweighs the traditional method. Nevertheless, partakers believe that implementation cost is more important in the traditional method, and the valuation of implementation costs for both training methods is insignificant. Furthermore, scalability and modularity are critical in the VR. Experts asserted that diagnosability and support are comparatively vital in the case of VR. Moreover, customization and integrability in the VR method are third and fourth. It is worth mentioning that eleven out of twelve local priorities were valued more in the VR-based training method.

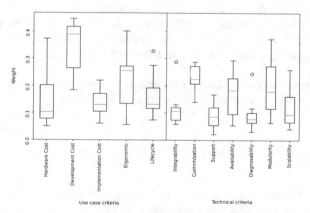

Fig. 3. Average local criteria weights in upper and lower levels of hierarchy among participants

Ultimately, we conclude our assessment results by presenting global priority sensitivity. Generally, the VR training is appreciated over the traditional method with 65.22%. Seven out of ten experts tend to stress their interest in the VR-based method. One participant ranked both traditional and VR training methods nearly equivalent.

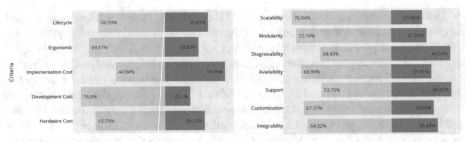

Fig. 4. A comparison between traditional (red) and VR-based training methods (blue) based on average local proportion in respect to second level criteria (colour figure online)

6 Conclusions

This study introduced a novel application of VR for safety training in human-robot collaboration, assessment method and indicative results of its technical feasibility, and user acceptance. This study is founded on the notion that in HRC, VR are a potentially innovative type of safety training. The results suggest that VR is a viable solution to design and train operators regarding required safety concerns in collaborative workcells. Technical criteria plays a vital role in investigating solutions with potential adaptation for industrial use. However, the survey results indicate the importance of availability, modularity, and customization of VR-based safety training solutions.

Similar to any study, this research has several limitations. The assessment is done by experts in human-robot collaboration involved in small-scale use case projects. Due to the Covid-19 pandemic, receiving survey results from the respondents was challenging, which limited our assessment to 10 experts. The larger studies are planned to be launched with different users such as managers, machinery safety experts, and shop-floor technicians can synergies different perspectives on the question.

Acknowledgements. This project has received funding from the European Union's Horizon 2020 research and innovation program under grant agreement No. 825196.

References

1. Milgram, P., Takemura, H., Utsumi, A., Kishino, F.: Augmented reality: a class of displays on the reality-virtuality continuum. **2351**, 282–292 (1995)
2. Villani, V., Pini, F., Leali, F., Secchi, C.: Survey on human-robot collaboration in industrial settings: Safety, intuitive interfaces and applications. Mechatronics **55**, 248–266 (2018). https://doi.org/10.1016/j.mechatronics.2018.02.009
3. Maurtua, I., Ibarguren, A., Kildal, J., Susperregi, L., Sierra, B.: Human-robot collaboration in industrial applications: safety, interaction and trust. Int. J. Adv. Robot. Syst. **14**, 1 (2017). https://doi.org/10.1177/1729881417716010
4. Arents, J., Abolins, V., Judvaitis, J., Vismanis, O., Oraby, A., Ozols, K.: Human–robot collaboration trends and safety aspects: a systematic review. J. Sens. Actuator Networks **10**(3), 48 (2021). https://doi.org/10.3390/JSAN10030048

5. Valori, M., et al.: Validating Safety in Human–Robot Collaboration: Standards and New Perspectives. https://doi.org/10.3390/ROBOTICS10020065
6. Berg, L.P., Vance, J.M.: Industry use of virtual reality in product design and manufacturing: a survey. Virtual Reality **21**(1), 1–17 (2016). https://doi.org/10.1007/s10055-016-0293-9
7. Seth, A., Vance, J.M., Oliver, J.H.: Virtual reality for assembly methods prototyping: a review. Virtual Real. **15**(1), 520 (2011). https://doi.org/10.1007/S10055-009-0153-Y
8. Vogel, C., Walter, C., Elkmann, N.: Safeguarding and supporting future human-robot cooperative manufacturing processes by a projection- and camera-based technology. Procedia Manuf. **11**, 39–46 (2017). https://doi.org/10.1016/j.promfg.2017.07.127
9. Hietanen, A., Halme, R.-J., Latokartano, J., Pieters, R., Lanz, M., Kämäräinen, J.-K.: Depth-sensor-projector safety model for human-robot collaboration
10. Vogel, C., Poggendorf, M., Walter, C., Elkmann, N.: Towards safe physical human-robot collaboration: a projection-based safety system. In: IEEE International Conference Intelligent. Robots System, pp. 3355–3360 (2011). https://doi.org/10.1109/IROS.2011.6048158
11. Gorecky, D., Khamis, M., Mura, K.: Introduction and Establishment of Virtual Training in the Factory of the Future. **30**, 182–190 (2015)
12. Pérez, L., Diez, E., Usamentiaga, R., García, D.F.: Industrial robot control and operator training using virtual reality interfaces. Comput. Ind. **109**, 114–120 (2019). https://doi.org/10.1016/J.COMPIND.2019.05.001
13. Matsas, E., Batras, D., Vosniakos, G.C.: Beware of the robot: a highly interactive and immersive virtual reality training application in robotic manufacturing systems. IFIP Adv. Inf. Commun. Technol. **397**, 606–613 (2012)
14. Matsas, E., Vosniakos, G.C., Batras, D.: Prototyping proactive and adaptive techniques for human-robot collaboration in manufacturing using virtual reality. Robot. Comput. Integr. Manuf. **50**, 168–180 (2018). https://doi.org/10.1016/j.rcim.2017.09.005
15. Buttussi, F., Chittaro, L.: A comparison of procedural safety training in three conditions: virtual reality headset, smartphone, and printed materials. IEEE Trans. Learn. Technol. **14**, 1–15 (2021). https://doi.org/10.1109/TLT.2020.3033766
16. Avveduto, G., Tanca, C., Lorenzini, C., Tecchia, F., Carrozzino, M., Bergamasco, M.: Safety training using virtual reality: a comparative approach. In: De Paolis, L.T., Bourdot, P., Mongelli, A. (eds.) AVR 2017. LNCS, vol. 10324, pp. 148–163. Springer, Cham (2017). https://doi.org/10.1007/978-3-319-60922-5_11
17. Xu, Z., Zheng, N.: Incorporating Virtual Reality Technology in Safety Training Solution for Construction Site of Urban Cities. Sustain **13**(1), 243 (2020)
18. Jeelani, I., Han, K., Albert, A.: Development of virtual reality and stereo-panoramic environments for construction safety training. Eng. Constr. Archit. Manag. **27**, 1853–1876 (2020). https://doi.org/10.1108/ECAM-07-2019-0391
19. Pedram, S., Palmisano, S., Skarbez, R., Perez, P., Farrelly, M.: Investigating the process of mine rescuers' safety training with immersive virtual reality: a structural equation modelling approach. Comput. Educ. (2020). https://doi.org/10.1016/J.COMPEDU.2020.103891
20. Tichon, J., Burgess-Limerick, R.: A review of Virtual Reality as a medium for safety related training in Mining. J. Heal. Saf. Res. Pract. **3**, 33–40 (2011)
21. Dianatfar, M., Latokartano, J., Lanz, M.: Concept for virtual safety training system for human-robot collaboration. In: Procedia Manufacturing (2020)
22. Tirmizi, A., et al.: Technical Maturity for Industrial Deployment of Robot Demonstrators. pp. 310–317 (2022). https://doi.org/10.1109/ICAR53236.2021.9659436
23. Pöysäri, S., Siivonen, J., Lanz, M.: Dimensions for reconfiguration decision-making and concept for feasibility analysis of reconfigurable pilot lines in industry, research and education
24. Saaty, T.L.: How to make a decision: the analytic hierarchy process. Eur. J. Oper. Res. **48**, 9–26 (1990). https://doi.org/10.1016/0377-2217(90)90057-I

25. Wang, G.X., Huang, S.H., Yan, Y., Du, J.J.: Reconfiguration schemes evaluation based on preference ranking of key characteristics of reconfigurable manufacturing systems. Int. J. Advanced Manufacturing Technol. **89**(5–8), 2231–2249 (2016). https://doi.org/10.1007/s00 170-016-9243-7

Enabling Technologies

Energy Efficiency for Manufacturing Using PV, FSC, and Battery-Super Capacitor Design to Enhance Sustainable Clean Energy Load Demand

Olukorede Tijani Adenuga$^{(\boxtimes)}$ ⓘ, Khumbulani Mpofu ⓘ, and Thobelani Mathenjwa

Tshwane University of Technology, Staartsartillirie Rd, Pretoria West, Pretoria 0183, South Africa
olukorede.adenuga@gmail.com, mpofuk@tut.ac.za

Abstract. Energy efficiency (EE) are recognized globally as a critical solution towards reduction of energy consumption, while the management of global carbon dioxide emission complement climate change. EE initiatives drive is a key factor towards climate change mitigation with variable renewable technologies. The paper aimed to design and simulate photovoltaic (PV), fuel cell stack (FCS) systems, and battery-super capacitor energy storage to enhance sustainable clean energy load demand and provide significant decarbonization potentials. An integration of high volume of data in real-time was obtained and energy mix fraction towards low carbon emission mitigation pathway strategy for grid linked renewables electricity generation was proposed as a solution for the future transport manufacturing energy supplement in South Africa. The interrelationship between energy efficiency and energy intensity variables are envisaged to result in approximately 87.6% of global electricity grid production; electricity energy demand under analysis can reduce the CO_2 emissions by 0.098 metric tons and CO_2 savings by 99.587 per metric tons. The scope serves as a fundamental guideline for future studies in the future transport manufacturing with provision of clean energy and sufficient capacity to supply the demand for customers within the manufacturing.

Keywords: Decarbonization · Microgrid linked system · Sustainable clean energy · Energy-efficient technology

1 Introduction

The increasing need for clean and sustainable technologies for energy sources in the power generation sector has motivated researchers to develop power generation sys-tems using renewable energy sources [1, 2, 3]. The quest for low carbon emission mitigation requires process technology diffusion. Energy efficiency for manufacturing plays a critical role in the economic growth, cost advantage against global competitors and provide a competitive advantage on a national scale. Additionally, interconnection of renewable

© The Author(s) 2023
K.-Y. Kim et al. (Eds.): FAIM 2022, LNME, pp. 259–270, 2023.
https://doi.org/10.1007/978-3-031-18326-3_26

energy sources into the distribution substation grid networks is constrained by the limitation of the network's capacity and power flow behaviour [4, 5]. Microgrid captive energy supply and demand balances help to coordinate stable service within defined grid resources boundaries [6]. The provision for significant decarbonization potentials has motivated the quest for a conceptual design as a solution to enhance the integration of renewable energy sources into the grid linked networks. Researcher's responses to greenhouse effect using solar PV, fuel cell stack and battery super capacitor energy storage with microgrid operation complexities increases when on a grid-connected mode, prompting cost increase. However, modelling in Matlab/Simulink along with electronic interfacing circuits are not widely used due to its disadvantages [7, 8]. Renewables energy penetration is enhanced by microgrid development in relations to dependable power supply, reduced transmission system expansion and ensures converter-interfaced generation units based on renewable power sources [9]. Recent research revealed that 80% savings are achieve in energy losses through localized control compared to centralized control [10]. The user's need for sustainable lower energy cost, low carbon mitigation, reliable and resilient energy supply services, arising from issues regarding grid security and survivability [11]. The vast rise of energy demand necessitates actions towards grid optimization, conges-tions relief, and integration of ancillary services [12]. Energy efficient practices complement disaggregated econometric estimations of manufacturing operations. Energy demand through manufacturing operations can alter understanding of consumption from energy sources with direct implications on significant decarbonization potentials, while analysed data from microgrid are related to EE initiative, costs performance, and energy savings.

2 Methodology

2.1 Microgrid Design Configuration

The designed microgrid system uses monocrystalline PV modules is simulated with the substation in Matlab/Simulink, including interconnected solar PV cell, and energy storage rated at 100 295 kWp (kilowatts-peak). 50-kW fuel cell stack, and battery and super capacitor storage at 500 V dc voltage to provide backup for the system in an emergency [13]. The microgrid has sets of battery storage unit with individual lithium cells connected in series, with rated capacity of 21.7 kWh. Figure 1 shows the proposed microgrid system simulated configuration comprising solar PV array connection to DC bus through a dc/dc boost converter. PV array generates DC voltage by converting solar PV modules energy with variable irradiance temperature to electrical energy. And. Fuel cells stack generate a DC voltage by converting chemical energy to electrical energy with the inputs of hydrogen and oxygen. DC/DC converters step-up/step-down the voltage. A battery serves as energy storage device, connected to same dc bus via dc/dc bidirectional converter. The ac and dc buses are connected through ac/dc bidirectional converter. The controller collects the energy supply data and delivers through the communication link via control demands. The power converters coordinated the power flow signals for energy management system (EMS) through the power link. Energy storage (super capacitor) is used for storage of energy in case of insufficient power from photovoltaic array, since this source is weather dependent hence, the inputs cannot be control, therefore prompting the use of MPPT controller to extract maximum power from the PV. The output voltages are synchronized via the common DC bus, while the three-phase inverter converts direct current to alternating current for AC loads and the three-phase transformer steps up the voltage.

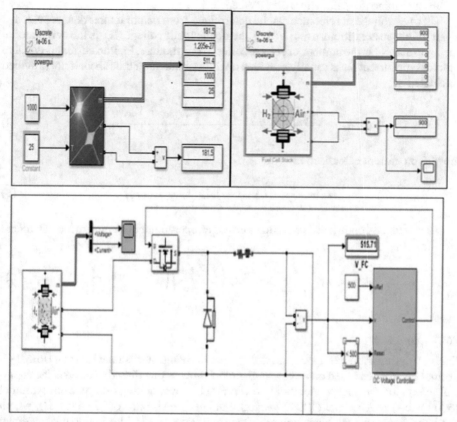

Fig. 1. Proposed design and simulated diagram solar PV, fuel stack cell and super capacitor grid linked hybrid system configuration.

2.2   Microgrid Design Modeling

2.2.1   Mathematical Model of a Solar PV Cell

The solar cell circuit shown in Fig. 1 have I_{ph} is the photon current (A), ID is the diode current (A), R_s is the series resistance (Ω), R_{sh} is the shunt resistance (Ω), and I, V are the current (A) and voltage (V) of the solar cell respectively. Applying Kirchhoff's law I_S as the saturation current (A); A is the thermal voltage (V); K_B is the Boltzmann's constant (J/K); q is the electronic charge (C) results in the following equation: the four parameters I_{ph}, R_s, A and I_s determined from the manufacturer data depicting $A = \frac{mkT_c}{q}$

$$I = I_{ph} - I_s \left\{ \exp\left(\frac{V + IR_s}{A}\right) - 1 \right\} - \frac{V + IR_s}{R_{sh}} \tag{1}$$

To determine the parameters of a PV cell:

Thermal voltage at reference can be determined from manufacturer data provided at given reference conditions using V_{OC} is the open circuit voltage, $E_{.g}$ is the cell material band gap and N_S is the number of cells in series in one module. Photon current is equal to short circuit current. Rs is usually provided on the manufacturer data sheet if not provided calculated as

$$R_s = \frac{A_{ref} \ln\left(1 - \frac{I_{mpp,ref}}{I_{ph,ref}}\right) - V_{mpp,ref} + V_{oc,ref}}{I_{mpp,ref}} \tag{2}$$

the photon current calculated as:

$$I_{ph} = \frac{irr}{irr_{ref}}\left(I_{ph,ref} + \mu I_{sc}(T_c - T_{c,ref})\right) \tag{3}$$

At reference conditions, saturation current at operating cell temperature calculated as:

$$I_s = I_{s,ref}\left(\frac{T_c}{T_{c,ref}}\right)^3 \times \exp\left\{\left(\frac{N_s E_g}{A}\right)\left(1 - \frac{T_{c,ref}}{T_c}\right)\right\} \tag{4}$$

2.2.2 Fuel Cell

Fuel cell systems modelling was set with proton exchange membrane fuel cell (PEMFC) normal parameters based on an electrochemical process that directly converts the chemical energy of the fuel to electrical power. Stack power settings range from Nominal (5000 watts) to Maximal (120400 watts). Fuel cells resistance of (0.66404 Ω), which work like batteries with Nerst voltage of one cell (En = 1.1342 V), but do not run down or need recharging. The fuel cell consists of two electrodes namely, hydrogen, which feed the anode and the air, fed to the cathode produce electricity and heat. Due to the activation losses and losses caused by diffusion and resistivity, the output voltage of series connected fuel cells expressed as:

$$V_{FC} = E_O - Aln\left(\frac{I_{fc}}{I_o}\right) \times \frac{1}{\frac{T_D + 3}{3}} - I_{fc}R \tag{5}$$

where E_O and A are empirical coefficients, I_{fc} and I_o are the fuel exchange current in ampere, T_D denotes the settling time of a cell due to a change in current, and R denotes the diffusion resistance., N is the number of cells connected in series to maximize the voltage.

2.2.3 Boost Converter

In the proposed microgrid system, the boost converter mainly used to step up the output voltage of the PV, an MPPT controller that uses perturb and observe technique optimizes the switching duty cycle. This MPPT system automatically varies the voltage across the terminals of the PV array to get the maximum possible power. The ratio of open circuit and short circuit voltage relationship for the inductor L is:

$$\frac{V_s T_s D}{L} = -\frac{(V_d - V_o)T_s(1 - D)}{L} \tag{6}$$

2.2.4 Buck Converter Components Design

The buck converter will be mainly used to step down the output voltage of a fuel cell stack, the fuel cell stack found in Simscape library with an output voltage of 625 V. the DC voltage control block will be used to control the output voltage of the buck converter and the output of the DC voltage control will regulate the MOSFET ensuring that the output voltage does not exceed the reference voltage. Inductor size can be express as:

$$L_{crit} = \frac{(V_s - V_o)DT_s}{2I_o} \tag{7}$$

where f_s is the switching frequency and ΔI_L is the inductor current ripple, and the inductor ripple current is usually smaller than output current. Hybridization of a supercapacitor connected with a battery to a buck-boost converter is limited by a rate limiter block.

2.2.5 Low Carbon Emission Conversion from Energy Mix

Low carbon emission mitigation evaluation is relative to energy generation and consumption in manufacturing sector using applicable emission factor $(7.0555 * 10^{(-4)})$ for energy mix carbon emissions estimation is model as.

$$\text{Carbon emission}[\text{kg}tCO_2)] = \pounds_j * \text{CES}^{\text{TM}}[\text{kg}tCO_2/GJ] \tag{8}$$

Non-baseload CO_2 output emission rate are used to convert reductions of kilowatts hours into avoided units of carbon dioxide emissions. Coal energy content of 25.532 MJ/kg is adopted for the equivalent of emissions reductions from energy efficiency programmes assumed to affect non-baseload generation (power plants that brought online as necessary to meet demand).

3 Results and Discussion

Power electronic converters used in renewable energy sources mainly in hybrid microgrid system. However, certain precautions are considered in the circuit operations such as frequency variations and voltage synchronization from different sources during system performance evaluation. The results displayed some frequency variations that shows large current and voltage harmonics in the proposed hybrid microgrid system done under different circumstances to ensure reliable, efficient, and cost-effective solution for the intended users. The solar PV were tested with different temperatures and irradiances to see how the system behaves when operating at fixed temperature, while varying the irradiance and vice versa. The change in irradiance with constant temperature of 25 °C. Cells per module (Ncell) is 96, number of series connected modules per string (5), using 5 series modules and 112 parallel strings. The maximum current and voltage [Solar irradiance (input)] obtained at 1000 W/m^2, when irradiance decreases the current, power, and the voltage slightly decreases. The decrease in irradiance results to insufficient power generation. Operating temperatures are varied at 0, 25 and 50 °C; Temperature coefficient of ISC is 0.061745%/°C; Light generated current IL at 6.0092 A, Maximum power (305.226 W); VOC (64.2 V), ISC (5.96 A), VMP (54.7 V), IMP (5.58 A); Diode saturation current IO (6.3014e−12 A), Diode ideality factor (0.94504), Shunt resistance Rsh (269.5934 Ω) and Series resistance Rs (0.37152 Ω).

3.1 Photovoltaic Simulated Results

The photovoltaic is simulated in different irradiances and temperatures; the signal builder with two input signals temperature and irradiance for PV array was utilized in the design. Figure 3 is the simulation of the PV array demonstrating the behavior of the PV array when temperature or irradiance changes. A drastically decrease in irradiance results to a drastically decrease in power delivered by the PV array this behavior validates the theory based on the solar energy. The mean voltage is almost constant since an MPPT controller extracting the possible maximum power from the PV array regulates the duty cycle (Fig. 2).

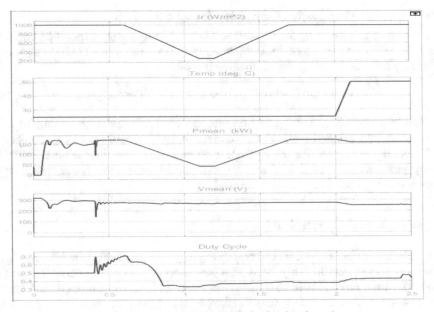

Fig. 2. PV inputs and Photovoltaic simulated results.

In a hydrogen fuel cell, a catalyst at the anode separates hydrogen molecules into protons and electrons, which take different paths to the cathode. The electrons go through an external circuit, creating a flow of electricity. The protons migrate through the electrolyte to the cathode, where they unite with oxygen and the electrons to produce water and heat. To increase the low voltage of a single fuel cell, many cells connected in series to form a fuel cell stack. The fuel cell signal variation {x_H$_2$ = 99.95%] and {y_O$_2$ = 21.0%}. Hydrogen utilization range between [Nominal 417.3 l pm and Maximum 1460 l pm]. The system temperature is set to [T = 338 K], Fuel supply pressure [P$_{fuel}$ = 1.5 bar] and Air supply pressure [Pair = 1 bar].

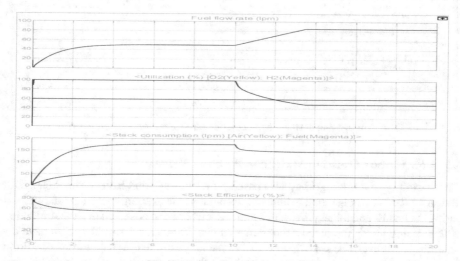

Fig. 3. Fuel cell simulation result.

The integrated system of photovoltaic assisted by fuel cell stack operating in parallel yielded the results shown in Fig. 3 after conversion from DC-AC through the universal bridge, the voltage is precisely the desired DC voltage at the DC bus. The time span of the scope was set to 0.5, the scope shows the current and voltage characteristics of the three-phase inverter. Large harmonics were observed during the operation of the system and due to such system behavior, it is not advisable to directly connect AC loads assuming that the voltage has been converted to AC this could possibly result to overheating of components connected directly to the converter, for AC loads to be properly connected these harmonics must be filtered. Figure 3 demonstrate the characteristics of current and voltage at the reduced time span (Fig. 4).

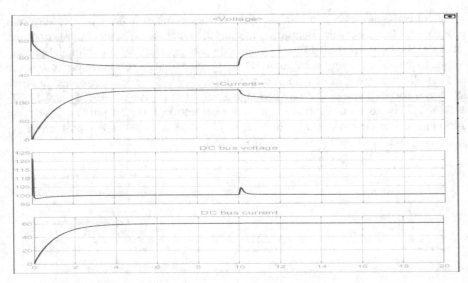

Fig. 4. Voltage and current characteristics ac/dc and measured on the grid.

3.2 Low Carbon Emission Mitigation Pathway Strategy

The feasibility of a microgrid system based on a PV, FSC, and battery-super capacitor design has been demonstrated to enhance sustainable clean energy load demands. Comparing the results and the requirements of the user it has been observed that the simulation of this design yields the precise requirements, the designed microgrid system have the capability to deliver 170 kW which is within the range of the user's requirements. From the simulated and tested results, we concluded that the requirements of the user were met. Table 1 is the simulated data from microgrid tied with PV, FSC, and battery-super capacitor in MATLAB/SIMULINK.

Table 1. Simulated data from microgrid tied with PV, FSC, and battery-super capacitor in MATLAB/SIMULINK.

Hour	Grid	Fuel stack cell	Solar	Fuel cell/grid	Solar/Grid	Consumption
1	5.5	4.2	2.4	0.763	0.436	0.2
2	5.1	4.0	2.8	0.780	0.549	0.4
3	5.0	3.8	3.0	0.760	0.600	0.6
4	4.8	3.6	2.8	0.750	0.583	1.2
5	4.8	3.0	3.5	0.625	0.729	1.5
6	5.35	2.0	2.8	0.373	0.523	1.6
7	5.1	1.8	1.0	0.353	0.196	1.8

(*continued*)

Table 1. (*continued*)

Hour	Grid	Fuel stack cell	Solar	Fuel cell/grid	Solar/Grid	Consumption
8	6.1	1.0	2.0	0.164	0.327	3.0
9	6.4	1.0	2.5	0.156	0.390	4.5
10	5.7	1.1	3.4	0.193	0.596	6.7
11	5.2	0.7	4.1	0.135	0.788	7.8
12	4.7	1.1	3.9	0.234	0.829	8.9
13	4.3	1.15	4.0	0.267	0.930	10.8
14	4.5	1.1	3.9	0.244	0.866	11.4
15	5.8	0.9	3.8	0.155	0.655	11.0
16	6.3	1.8	4.0	0.285	0.634	8.9
17	10.4	1.0	5.0	0.096	0.481	10.0
18	12.2	1.5	6.5	0.123	0.533	9.8
19	11.6	2.0	5.2	0.172	0.448	10.0
20	10.4	2.5	5.0	0.240	0.480	9.8
21	9.0	3.1	4.8	0.344	0.533	8.0
22	8.6	4.25	3.55	0.494	0.413	6.7
23	7.5	4.5	5.5	0.600	0.733	14
24	6.2	4.2	0.15	0.677	0.024	0.1
Total	160.55	53.1	85.6	8.983	15.035	137.7

The hourly-amassed energy production is 160.55 kW/h for grid tie supply, 85.6 kW/h (53.3%) from Solar PV; 53.1 kW/h (33.0%) from fuel cell; 138.7 kW/h (86.39%) Solar PV and Fuel Cell combined. The simulated energy consumption is 137.7 kW/h (85.76%) of grid supply from Table 1 shows in Fig. 5, Fig. 6 and Fig. 7 for Grid tied supply, Solar

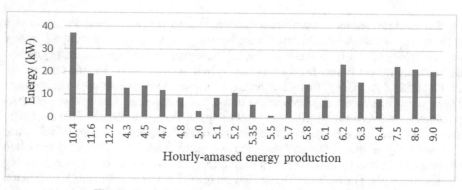

Fig. 5. Sum of simulated hourly data of Grid tied supply

PV, and Fuel Stack Cell respectively. Figure 8 is the sum of simulated hourly data of Energy Consumption.

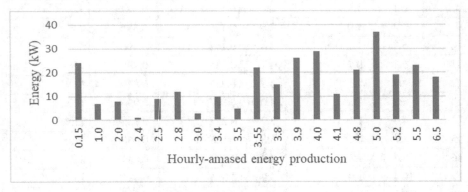

Fig. 6. Sum of simulated hourly data of Solar PV

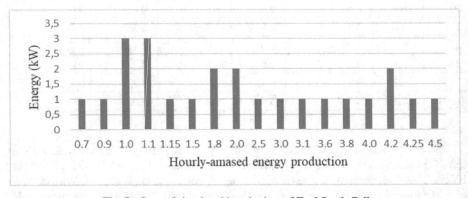

Fig. 7. Sum of simulated hourly data of Fuel Stack Cell

The need for renewable energy mix fraction towards low carbon emission mitigation pathway strategy require reduction of fossil fuel usage in energy generation by 87.76% of grid electricity supply if the study is adopted. Applying country-specific energy carbon emission equivalent factors for coal mining derived from 2019 [14, 15, 16], coal energy content of 25.532 MJ/kg. This study brought forward the argument of the importance of renewable energy use appropriateness design and simulation in comparison of inter-relationship between energy efficiency and energy intensity variables f of the electricity energy demand under analysis, which can reduce the CO_2 emissions by 0.098 metric tons and CO_2 savings by 99.587 emission per metric tons. The substantial reduction of carbon emissions by the design and simulated systems could assist in decarbonizing the environments. The microgrid integrated energy system could mitigate CO_2 emission from use of coal as fuel source to make the environment cleaner and ecofriendly. More-over, the outcome of this paper can provide sustainable design strategy and guide for

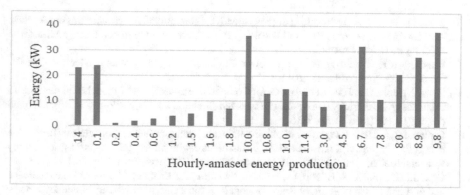

Fig. 8. Sum of simulated hourly data of Energy Consumption

policy makers and investors in alternative energy mix (fuel source) towards a low-carbon carbon emission pathway for more profitable transition to enhanced energy microgrids.

4   Conclusion

The hybrid microgrid system proposed in this study has been successfully simulated and system performance analysis has been tested and evaluated when the system operating in off-grid mode with an AC load of 120 kW. It has been observed that the system yields stable output voltage and current. Approximately 170 kW real power added to the grid when operating in grid-connected mode. Hence, the proposed hybrid microgrid system considered as power generation for rural areas for the ongoing power crisis.

References

1. Renewable Energy Policy Network for the 21st Century. REN21. Renewable 2017 Global Status Report (Paris: REN21 Secretariat). 2017 ISBN 978–3–9818107–6–9
2. Biel, K., Glock, C.H.: Systematic literature review of decision support models for energy-efficient production planning. Comput. Ind. Eng. **101**, 243–259 (2016)
3. Adenuga, O.T.: Mpofu.: Ramatsetse, B.I.: Energy efficiency analysis modelling system for manufacturing in the context of industry 4.0. Procedia CIRP: 2019 80:735–740
4. Adenuga, O.T., Modise, R.K.M.: An approach for enhancing optimal resource recovery from different classes of waste in South Africa: selection of appropriate waste to energy technology. Sustain. Fut. 2, 100033 (2020). ISSN 2666-1888, https://doi.org/10.1016/j.sftr.2020.100033
5. Adenuga, O.T., Mpofu, K., Ramatsetse, B.I.: Exploring energy efficiency prediction method for Industry 4.0: a reconfigurable vibrating screen case study. Procedia Manuf **51**, 243–250 (2020)
6. Ali, A., Li, W., Hussain, R., He, X., Williams, B.W., Memon, A.: Overview of current micro-grid policies, incentives and barriers in the European Union, United States and China. Sustainability 9, 1146 (2017). https://doi.org/10.3390/su9071146, www.mdpi.com/journal/sustainability

7. Irena. Hydrogen: A renewable energy perspective, International Renewable Energy Agency, Abu Dhabi Hydrogen: A Renewable Energy Perspective. 2nd Hydrogen Energy Ministerial Meeting in Tokyo, Japan (2019)

8. Lasseter, R.H. Micro-grids, IEEE Power Engineering Society Winter Meeting, Vol.01, pp. 305–308, New York, NY, 2015

9. Ryan, N., Johnson, J.X., Keoleian, G.A.: Comparative assessment of models and methods to calculate grid electricity emissions. Environ. Sci. Technol. **50**(17), 8937–8953 (2016). https://doi.org/10.1021/acs.est.5b05216

10. Okundamiya, M.S.: Integration of photovoltaic and hydrogen fuel cell system for sustainable energy harvesting of a university ICT infrastructure with an irregular electric grid. Energy Convers. Manag. **250**, 114928 (2021). https://doi.org/10.1016/j.enconman.2021.114928

11. Cox, S., Hotchkiss, E., Bilello, D., Watson, A., Holm, A.: Bridging climate change resilience and mitigation in the electricity sector through renewable energy and energy efficiency. emerging climate change and development topics for energy sector transformation, National Renewable Energy Laboratory Jennifer Leisch I U.S. Agency for International Development. https://www.nrel.gov/docs/fy18osti/67040.pdf

12. Motlagh, N.H., Mohammadrezaei, M., Hunt, J., Zakeri, B.: Internet of Things (IoT) and the energy sector. Energies 13, 494 (2020). https://doi.org/10.3390/en13020494. www.mdpi.com/journal/energies

13. Adenuga, O.T., Mpofu, K., Modise, K.R.: An approach for enhancing optimal resource recovery from different classes of waste in South Africa: selection of appropriate waste to energy technology. Sustainable Fut. **2** (2020)

14. IPCC, Global warming of 1.5 °C, I.P.C.C. Change, p. 616 (2019)

15. International Energy Agency (IEA). Energy efficiency: Policies and measures database (2012). http://www.iea.org/textbase/pm/?mode=pm

16. Renewable Energy Policy Network for the 21st Century. REN21. Renewable 2017 Global Status Report (Paris: REN21 Secretariat), pp. 6–9 (2017). ISBN 978-3-9818107

The Impact of Learning Factories on Teaching Lean Principles in an Assembly Environment

Fabio Marco Monetti[1]([✉]) [ID], Eleonora Boffa[1] [ID], Andrea de Giorgio[2] [ID], and Antonio Maffei[1] [ID]

[1] Department of Production Engineering, KTH Royal Institute of Technology, 11428 Stockholm, Sweden
monetti@kth.se

[2] SnT, University of Luxembourg, Esch-sur-Alzette, Luxembourg

Abstract. Learning factories are realistic manufacturing environments built for education; many universities have recently introduced learning factories in engineering programs to tackle real industrial problems; however, statistical studies on its effectiveness are still scarce. This paper presents a statistical study on the impact of learning factories on the students' learning process, when teaching the lean manufacturing concepts in an assembly environment. The analysis is carried out through the Lean Manufacturing Lab at KTH, a learning factory supporting the traditional educational activities. In the lab, the students assemble a product on an assembly line; during three rounds, they identify problems on the line, apply the appropriate lean tools to overcome the problems, and try to achieve a higher productivity. The study is based on the analysis of the times recorded during the sessions of the lab. A questionnaire submitted to the students after the course evaluates the level of knowledge of lean production principles that the students achieved. The results are twofold: the improvement of the assembly times through the implementation of the lean tools and the positive effect of a hands-on experience on the students' understanding of the lean principles, highlighted by the answers to the questionnaire. The main contributions are that applying the lean tools on an assembly line improves the productivity even with inexperienced operators, implementing a learning factory is effective in enhancing the learning process, and, lastly, that a first-hand experience applying the lean tools in a real assembly environment is an added value to the students' education.

Keywords: Learning factory · Lean manufacturing · Lean principles · Assembly · Education

1 Introduction

The term Learning Factory (LF) identifies a system that integrates elements of education in a realistic production environment [1], thus implementing the

© The Author(s) 2023
K.-Y. Kim et al. (Eds.): FAIM 2022, LNME, pp. 271–283, 2023.
https://doi.org/10.1007/978-3-031-18326-3_27

"factory-to-classroom" concept [2]. The teaching-learning process is moved closer to real manufacturing problems through emulation of a simplified shop-floor [3], so that learners can be tested in concrete industrial conditions while practicing with hands-on activities [2]. LFs support the experiential learning process [3] as defined by Kolb's theory [4]: a concrete experience helps in realizing the knowledge transfer, according to a cycle in which the students (a) actively engage in a practical task; (b) observe, reflect, and discussed the activity with their peers; (c) abstract conceptualization (analysis) and generalize the events suggesting improvements (conclusions); (d) try to put knowledge into practice, thus making sure that the information is retained. A LF represents an approach that embraces the theory of experiential learning; engineering students' learning process requires experience in laboratories to facilitate their understanding of the theoretical concepts learnt during regular classroom activity [5]. This makes LF a suitable learning activity to be embedded in engineering programs.

Lean Manufacturing (LM) is a philosophy intended to systematically reduce waste – i.e., non-value-added activities – in the entire product's value stream, while promoting continuous improvement [6,7]. Such philosophy articulates in five basic principles: (i) identify the value from customer's perspective; (ii) map the value stream to include only value adding processes; (iii) create continuous flow; (iv) establish pull system; (v) strive for perfection by continuously improve the company's processes [8]. Several lean tools have been developed to implement these principles, such as Value Stream Mapping (VSM), Kanban/Pull production, 5S, and production smoothing [8,9]. Such approach promotes active involvement of workers that iteratively question and solve problems related to their tasks, thus acquiring knowledge from being engaged first-hand in the problem solving activity – "learning by doing" [10]. In view of the above, LFs appear to be a suitable teaching and learning activity for LM education and training of engineers. Numerous LFs have been built in recent years in academic facilities, institutes, and companies [11] as experience-based methodology improves the learners' knowledge of lean principles while transferring the skills necessary for a real implementation [12].

This paper analyzes the impact of LF on the students' learning process, when teaching the LM concepts in an assembly environment. The LM Lab (LML) developed at the Royal Institute of Technology in Stockholm (henceforth KTH) is taken as a case study. The analysis shows that applying lean tools on an assembly line improves productivity even with novice operators, that using a LF effectively enhance the learning process, and that a first-hand experience applying the lean tools in a real assembly environment is an added value to the students' education.

The paper is structured as follows: Sect. 2 presents a brief overview of related literature, Sect. 3 describe the setup for the LML activity and how the analysis is performed, Sect. 4 presents the results of the analysis that are then discussed in Sect. 6.

(a) Initial setup of the assembly line for round 1 (station 1-6)

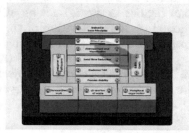

(b) The final product assembled during the LML

Fig. 1. The experimental setup for the Lean Manufacturing Lab Learning Factory

2   Background

Several studies report the potential and benefits of simulation games for LM training and education [13]. Such simulation games create a hands-on learning and training environment for better engagement of the learners, promoting a deeper understanding of the lean principles [13,14]. Simulation games environment for training LM highly engage participants in realistic industrial situations, where they are asked to solve the encountered problem [14]: as a result, the participants gain a good and practical understanding of LM principles and tools and show the interest to play and learn even further [15].

Traditional teaching methods do not provide the desired improvement in the acquisition of knowledge and development of necessary skills to apply LM principles and tools. One study shows the importance of different teaching processes by implementing 5S methodology in the preparation of a laboratory environment, emphasizing the role of applying lean principles already in higher education institutes [16].

Employing "learning by doing" methods demonstrates to facilitate the acquisition of the concepts related to LM and to be effective in developing the skills for LM successful implementation. Therefore, in recent years several academic institutes and companies have promoted the construction of LF facilities [11]. The "Lean School" at the University of Valladolid (Spain) is the result of a pedagogical project in collaboration with Renault Consulting [12], while Karlstad University (Sweden) designed and built their own lean factory on site [13]: both studies describe how the LF involves students in the emulated production environment, by being asked to improve the efficiency of the production line applying LM tools. The LF creates the suitable conditions for students to practically engage in different situations by applying the concepts and the tools presented in class. The students gain hands-on experience on the proposed experimental set-up, showing a better take in of the concepts with respect to those who have not taken the lab. From a pedagogical perspective, this means that courses and training programs that include a LF approach aim at fulfilling higher educational objectives than regular lectures. Looking at the Bloom's taxonomy model for the cognitive domain – a classification of the level of understanding based on six different types of knowledge (remembering, understanding, applying, analyzing, evaluating, and creating) [17] – the LF through the "learning by doing" approach allows to step from basic cognitive objectives, i.e., remembering or understanding, to more complex ones, i.e., applying and analyzing.

3   Methods

3.1   Lean Lab Description

The LML is a learning activity based on the concept of the LF. The laboratory simulates a realistic manufacturing environment, whose main goal is to involve the students to develop solutions applying appropriate lean tools (Table 1). The LML consists of an assembly line with 6 workstations (Fig. 1a). The assembled product has a simple temple structure composed of several blocks and labels that differ from each other (Fig. 1b), thus requiring the operators to change tools between operations.

The LML is structured in 3 rounds, each of them lasting 12 min. The rounds are designed to show a step-wise improvement of the line applying LM tools. The first round is set in a disorganized way to highlight the waste in the whole assembly process. At the end of this round, the students engage in a group discussion aimed at identifying problems and suggesting relevant lean tools to solve them. The second round starts with the students implementing the suggested LM tools and then it is run with the proposed improvements in place. Another group discussion raises the problems still occurring. Additional LM tools are evaluated and suggested for implementation in the third round. The last round consists of major improvements to the previous setup, e.g., reduced number of stations, use of conveyor belt for transportation, use of product kits and advanced tools.

The students are actively involved in the LML as the production line needs (a) operators, performing the assembly operations at the workstations, (b) logistics, handling the transportation of components and WIP, (c) observers, measuring the cycle time at each station and inspect the problems on the line.

In every round, at each station the operators assemble a specific sub assembly, part of the final product. The observers measure the time to complete one sub

Table 1. Lean tools covered in the LML

Tool	Description
5S	5S is a simple tool for organizing your workplace in a clean, efficient and safe manner to enhance productivity, visual management and to ensure the introduction of standardized working. 5S stands for: Sort, Set, Shine, Standardise and Sustain
Pull production	Pull production is based on customer demand. The production is triggered by a demand signal from a subsequent process. The production signal is sent upstream
Kanban	Kanban is a simple method to pass demand signals between processes. It is a signal to refill it with a specific number of parts or send back a card with detailed information about the part location
One piece flow	It is an ideal state of the operations where production works on one product at a time. It can be achieved by calculating a takt time, introduce pull system with one piece flow, ensure a feasible layout for one piece flow. "Make one, move one"
Line balancing	It aims at eliminating overburden to people and equipment, level out the workload of all manufacturing processes. To match production rate and takt time. Quantity of workers, work and machines assigned to each task should be re-balanced to meet optimal production rate
Built-in quality	Every operator is an inspector and works to fix problems at the station before passing them on. If defects do get passed on, they are detected quickly and problem can be immediately diagnosed and corrected
Takt time	It sets the pace of production aligning it with the customer demand. The aim of Takt is to detect deviations. If a product has not left the production flow when the takt time is out, it is a signal that there is waste in the process

assembly (cycle time) at each station, for every sub assembly they complete. The eventual waiting time between one product and the following is not considered.

After each round, the students switch stations and occupation, moving from production operators to logistics and observers, so that they cannot develop skills and memorize the sequence of required operations to complete a sub-assembly.

In the first round, the assembly operations run slower and are less likely to complete their assemblies since no LM tool has yet been applied; during round 2 the LML already comprises many improvements that facilitate the operators in their task; round 3 is the best possible set-up for the LML; however, it should be noted that – due to reasons of time and availability of the facility – less time is spent performing the activity and recording completion times in round 3 than in the other two rounds.

3.2   Analysis

Throughput Time. The time to fully assemble a single product consists of the sum of the times for said product to be assembled at each station (1 to 6) – we call this "line throughput time". The throughput time of every completed product for every iteration of the LML is recorded, even the ones that exceed the 12 min mark: we give additional time to complete the assembly, but we label these cycles as "overdue".

We divide the throughput times of each completed assembly by round (round 1-2-3) and compare them to the target time of 12 min, to see how many are "on time" and how many are "overdue". Then, we check the three rounds assembly time populations for normality by looking at the normal probability plot and performing a Shapiro-Wilk test. We also plot a box plot with the three populations to have a first insight into how far apart and different they look.

We test separately round 1 against round 2, and round 2 against round 3 to check if the differences are statistically significant, using a t-test with two samples and unequal variances; we want to check if there is evidence that applying the lean tools through the rounds of the LML helps reducing the throughput time for the product. Since strong indications of non-normality should not be ignored and the performance of the t-test can be affected, we choose to perform a non-parametric Mann-Whitney U test on the same hypothesis.

After course completion, the student show correct learning pattern and are thus considered for a second stage of investigation of their perceived impact of the LF.

Population. The course has a total of 88 enrolled students who do the lean lab activity. All the students are enrolled at KTH and are following a Master's program in production engineering: they have a background in manufacturing and learn the concepts of LM and the lean tools of Table 1 during the course before the lab takes place; the majority does not have previous hands-on experience in assembly activities and is not trained to perform such tasks, so they can be considered as novice operators for the assembly operations they perform in the lab.

Due to different availability, the students autonomously form groups of 8 to 10 participants for the sessions, where they can individually work at the various stations, collect time data or note what upgrades could be made on the line.

Since not all the students that take part in the laboratory participate in the questionnaire, we calculate the margin of error – or confidence interval – for the answers, using the inverse equation for calculating a significant sample size [18]:

$$n_0 = \frac{Z^2 \cdot p \cdot q}{e^2}, \tag{1}$$

where n_0 is the adjusted sample size, Z is the critical value of the Normal distribution at $\alpha/2$ (where $\alpha = 0.05$ for a confidence level of 95%, and the critical value is 1.96), e is the margin of error or confidence interval, p is the estimated sample proportion, and $q = (1 - p)$. We use the adjusted sample size because our population is small: a Finite Population Correction is applied to Eq. 1 [19]:

$$n = \frac{n_0}{1 + \frac{(n_0 - 1)}{N}}, \tag{2}$$

where n is the actual sample size, N is the population size and n_0 is the adjusted sample size.

Questionnaire. After the students complete all the rounds of the laboratory, we submit a questionnaire to collect their opinions about the experience as a whole and its usefulness to understand and learn the LM principles. Also, we want to understand whether the LF experience successfully engage the students more than a standard set of lectures on the topic, and if it would be valuable to add more hands-on activities in similar courses.

After collecting the responses, we conduct a $\chi^2(4, N = 25)$ test to examine the significance of the students picks in the questionnaire – i.e., if the answer that the students select more frequently significantly represents the group's general preference.

4   Results

Throughput Time. Table 2 shows how many products the students complete in each round and how many of them are "on time", along with the percentage of "on time" completion. During round 1 they complete the least number of assemblies, because of the implied difficulties in the LML set-up: this is reflected in the percentage of assemblies completed "on time", which is 0%. The number increases during round 2, as well as the percentage of product completed within the 12 min mark. In round 3, despite completing all assemblies before 12 min, the total number of products is lower because we allocate less time to this round – compared to the other two – due to restrictions in time and availability of the facility.

Table 2. Number of completed products, 'on time' and % in each round

Round	N of completed products	N of "on time" products	Percentage of "on time" products
1	12	0	0%
2	26	18	69%
3	24	24	100%

We record the throughput times for all completed assemblies separately during the three rounds, and Table 3 shows the average, standard deviation and variance values for the three populations.

A visual representation of the time distributions is presented with the box-plot of Fig. 2, which shows that the three populations are quite distant from one another, and also highlights that they have unequal variances.

Table 3. Average line throughput time, standard deviation and variance values for each round

Round	Average line throughput time [s]	Standard deviation [s]	Variance [s^2]
1	1195.67	240.16	57678.79
2	669.96	93.18	8681.88
3	393.71	54.39	2957.87

Since doubts regarding the normality of the distributions arise through visual representations such as histograms and normal probability plots (not reported here), we perform Shapiro-Wilk tests. They do not show evidence of non-normality for round 1 ($W = 0.934$, $p = .45$) and round 3 ($W = 0.93$, $p = .09$), but they highlight that round 2 is significantly non-normal ($W = 0.90$, $p < .02$).

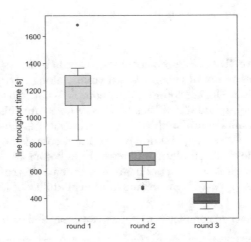

Fig. 2. Box-plot of the three rounds time distribution

Given the large difference in variance between the three rounds, we apply a t-test with two samples and unequal variances to test if such differences are significant (separately for round 1 against round 2 and for round 2 against round 3). The line throughput times recorded during the first round of the lab ($M = 1195.67$ s, $SD = 240.16$ s) compared to the times recorded during the second round ($M = 669.96$ s, $SD = 93.18$ s) demonstrate to be significantly higher, $t(13) = 7.33$, $p < .01$; likewise for the times of the second round ($M = 669.96$ s, $SD = 93.18$ s) compared to the times of the third ($M = 393.71$ s, $SD = 54.39$ s): $t(41) = 12.92$, $p < .01$.

Furthermore, given that the evidence of non-normality of round 2 might affect the performance and power of the t-test, we also perform a non-parametric test, namely the Mann-Whitney U test; it is generally less powerful than its parametric equivalent, but its value is not hindered by non-normality. Thus, a positive result might confirm the implications of the t-test. The median line throughput time in rounds 1 and 2 are 1232 and 681 s; the distributions in the two groups differ significantly (Mann-Whitney $U = 0.0$, $n_1 = 12$, $n_2 = 26$, $p < .01$ two-tailed); likewise, median line throughput time in rounds 2 and 3 are 681 and 377 s; the distributions in the two groups differ significantly (Mann-Whitney $U = 8.0$, $n_2 = 26$, $n_3 = 24$, $p < .01$ two-tailed).

The t-test and the Mann-Whitney U test results are the same, so we can conclude that they are reliable: given the box-plot information and these results, we can confirm that the time for assembling a complete product with the setup from round 1 is significantly higher than with the setup from round 2, and

Table 4. Confidence interval from sample size

N	Population size (# of students attending the lab)	88
$(1 - \alpha)$	Confidence level	95%
Z	Critical value	1.96
n	Actual sample size (# of students taking the questionnaire)	25
n_0	Adjusted sample size	34.52
p	Likely sample proportion	50%
e	Margin of error	16.68

consequently that the setup of round 3 makes the operation significantly quicker than round 2.

The time layout is in line with the expected outcome and thus a correct learning process to raise the expected learning outcomes of the course; this is the base for the second part of the analysis.

Population. Only 25 students out of the 88 enrolled in the laboratory activity participated in the submitted questionnaire, so we evaluate if the sample size significantly represents the answers from the whole group of students, as explained in Sect. 3.

Data and results of the margin of error calculation are presented in Table 4. The 95% confidence interval (CI) level is a very common choice for any standard application, the values of N ($N = 88$) and n ($n = 25$) are known, while the value of n_0 derives from Eqs. 1 and 2; the value of p is set based on what the expected answers are: in this case the result is unknown, and the most conservative value of 50% is selected.

Questionnaire. Figure 3 shows the answer to the questionnaire. The majority of them highlights that the students found the lean lab experience to be useful and satisfactory, and that it helps in learning the LM principles.

Table 5 shows the results for the $\chi^2(4, N = 25)$ test for each question: the majority pass the test with p-value lower than $\alpha = 0.05$, except two (questions 3.4 and 3.6), as highlighted in the table.

5 Discussion

The aim of our study is to verify the significance of implementing LM tools on a experimental assembly line in terms of number of products completed on time and in terms of line throughput times and to assess how much this lesson is effectively learned by our students participating to the LML. Results seem to confirm the usefulness of LM tools and that the experimental setup is appreciated and increase students' perception of lean tools.

LM Tools for Improved Productivity. The results show an increase in the number of assemblies that are completed within the 12 min mark, going from

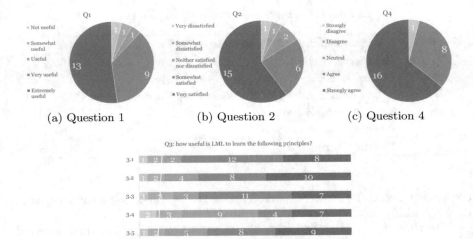

(a) Question 1   (b) Question 2   (c) Question 4

(d) Questions 3.1 to 3.6

Fig. 3. Distribution of students' answers in the questionnaire

Table 5. $\chi^2(4, N = 25)$ test results (* indicates a p-value > .05)

N	Question	χ^2 value	p-value
Q1	How much is a hands-on experience useful for your learning process?	25.6	< .001
Q2	How satisfied are you with the Lean lab activity?	28.4	< .001
	How useful is LML to learn the following principles?		
Q3.1	Create a continuous process flow to bring problems to the surface	18.4	.001
Q3.2	Use pull system to avoid overproduction	12	.017
Q3.3	Level the workload out (heijunka)	12.8	.012
Q3.4	Built in the equipment capability of detecting problems (built in quality)	6.8	.147*
Q3.5	Standardized tasks to empower employee and develop continuous improvement	10	.040
Q3.6	Become a learning organization through continuous reflection and improvement	7.6	.107*
Q4	Would you include more similar activities in other courses?	39.2	< .001

0% to 100% through the three rounds; the average throughput shows a decrease in times, with round 3 averaging results almost four times shorter than round 1. The t-test and the Mann-Whitney U test confirm the previous results, showing high significance in the differences between the time populations of the three rounds. This means that the implemented tools actually serve their purpose of easing the operators' jobs and making an assembly line as lean as possible, enhancing the productivity.

It is worth noting that the productivity increases even when inexperienced operators perform the required operations, thus showing that the improvement is not due to the ability of a particular worker, instead it depends solely on the efficacy of the lean tools. Companies deploying an assembly line should carefully consider applying those tools because they have beneficial effects that reverberates to the productivity of the line.

As mentioned, less time is spent with round 3 set-up, thus only 24 completed products; allocating the same amount of time to all rounds would mean being able to compare total numbers of completed assemblies as well as percentages, to have a more complete picture of the assembly process under different set-up combinations.

Enhanced Learning Process Through LF. The LML serves as a hands-on experience to give students at KTH Production Engineering Department a different approach to learning manufacturing concepts and specifically lean production. The knowledge transfer occurs in a practical set-up, where the learner is immersed in a recreated production environment, and can test the previously acquired concepts of lean manufacturing and the efficacy of applying those tools.

The students partake in a well designed laboratory that makes them analyze and solve the problems on the line and assess the efficacy of lean tools by applying them in a realistic manufacturing environment. Seeing the assembly times steadily decrease from one round to the other allows them to comprehend how the application of lean manufacturing principles influences production. Given this valuable experience, they increase their level of educational objectives in the cognitive hierarchy of Bloom's taxonomy: from just a knowledge and comprehension level that comes from learning in the classroom, they go to an application and even analysis level, which gives them potential to become better learners and better professionals. A practical approach to teaching is commonly believed to be a good way of keeping the students engaged in the learning process and gives them a way to get a deeper understanding of what is taught during standard lectures.

The answers from the submitted questionnaire confirm that our students appreciate the hands-on experience and that they are very satisfied with the activity, they also think it is useful to improve their learning process: including more similar activities in other courses where a lab could be implemented could be a useful teaching approach. Finally, the students believe that this lab is helpful for learning and remembering most main concepts of LM and specifically the LM tools, since they can see them applied in reality and analyze how they prove beneficial during production; however, for two of the lean tools – namely built-in quality and continuous improvement – no evidence is given from the questionnaire that the lab activity helps with the learning process.

6 Conclusions

In this paper we analyze the impact of LF, to teach LM concepts in assembly, on the students' learning process; the aim of the lab is to let the students experience the efficacy of the lean tools first-hand, how those tools reduce the assembly times, assess the good lab design and achieve specific higher level educational learning objectives. The LML at KTH is taken as a case study. This activity produces an increase in students' understanding of the lean tools and their ability to indicate, analyze and deploy actual improvements on a real assembly line. The application of such tools show how they affect assembly times and productivity, thus highlighting how they can be exploited to promote efficiency and increase the output of companies.

The work hereby presented, and the collected data come from the very first iteration of the experimental setup of the course, and all the answers contained in the questionnaire come from a sample of the total number of students that attended the LML, thus this study does not compare results with previous studies and has a wide CI when analyzing the answers. Future work includes collecting more data from future iteration of the course, to analyze a wider sample of the population, to reduce the CI and better capture the significance of the students' experience, and to compare the results over the years. Moreover, the participation to the questionnaire will be made mandatory for all students taking part in the course.

References

1. Wagner, U., AlGeddawy, T., ElMaraghy, H., Müller, E.: The state-of-the-art and prospects of learning factories. Procedia CIRP **3**, 109–114 (2012)
2. Rentzos, L., Mavrikios, D., Chryssolouris, G.: A two-way knowledge interaction in manufacturing education: the teaching factory. Procedia CIRP **32**, 31–35 (2015)
3. Müller, B.C., Menn, J.P., Seliger, G.: Procedure for experiential learning to conduct material flow simulation projects, enabled by learning factories. Procedia Manuf. **9**, 283–290 (2017)
4. Kolb, D.A.: Experience as the Source of Learning and Development. Prentice Hall, Upper Sadle River (1984)
5. Jasti, N.V., Kota, S., Venkataraman, P.: Development of a lean manufacturing simulation laboratory for continuing education students. J. Adult Contin. Educ. **27**(2), 292–323 (2021)
6. Shah, R., Ward, P.T.: Defining and developing measures of lean production. J. Oper. Manag. **25**(4), 785–805 (2007)
7. Taj, S., Berro, L.: Application of constrained management and lean manufacturing in developing best practices for productivity improvement in an auto-assembly plant. Int. J. Product. Perform. Manag. **55**(3/4), 332–345 (2006)
8. Stentoft Arlbjørn, J., Vagn Freytag, P.: Evidence of lean: a review of international peer-reviewed journal articles. Eur. Bus. Rev. **25**(2), 174–205 (2013)

9. Bhamu, J., Singh Sangwan, K.: Lean manufacturing: literature review and research issues. Int. J. Oper. Prod. Manag. **34**(7), 876–940 (2014)

10. Alves, A.C., Dinis-Carvalho, J., Sousa, R.M.: Lean production as promoter of thinkers to achieve companies' agility. Learn. Organ. **19**(3), 219–237 (2012)

11. Goerke, M., Schmidt, M., Busch, J., Nyhuis, P.: Holistic approach of lean thinking in learning factories. Procedia CIRP **32**, 138–143 (2015)

12. Ruano, J.P., Hoyuelos, I., Mateo, M., Gento, A.M.: Lean school: a learning factory for training lean manufacturing in a physical simulation environment. Manag. Prod. Eng. Rev. **10**(1), 4–13 (2019)

13. De Vin, L.J., Jacobsson, L.: Karlstad lean factory: an instructional factory for game-based lean manufacturing training. Prod. Manuf. Res. **5**(1), 268–283 (2017)

14. Badurdeen, F., Marksberry, P., Hall, A., Gregory, B.: Teaching lean manufacturing with simulations and games: a survey and future directions. Simul. Gaming **41**(4), 465–486 (2010)

15. Messaadia, M., Bufardi, A., Le Duigou, J., Szigeti, H., Eynard, B., Kiritsis, D.: Applying serious games in lean manufacturing training. In: Emmanouilidis, C., Taisch, M., Kiritsis, D. (eds.) APMS 2012. IAICT, vol. 397, pp. 558–565. Springer, Heidelberg (2013). https://doi.org/10.1007/978-3-642-40352-1_70

16. Sremcev, N., Lazarevic, M., Krainovic, B., Mandic, J., Medojevic, M.: Improving teaching and learning process by applying lean thinking. Procedia Manuf. **17**, 595–602 (2018)

17. Bloom, B.S., Englehart, M.D., Furst, E.J., Hill, W.H., Krathwohl, D.R.: Taxonomy of Educational Objectives, Handbook I: the Cognitive Domain. David McKay Co., Inc., New York (1956)

18. Cochran, W.G.: Sampling Technique, 2nd edn. Wiley, New York (1963)

19. Israel, G.D.: Determining sample size. University of Florida Cooperative Extension Service, Institute of Food and Agriculture Sciences (1992)

Generation of Synthetic AI Training Data for Robotic Grasp-Candidate Identification and Evaluation in Intralogistics Bin-Picking Scenarios

Dirk Holst$^{(\boxtimes)}$, Daniel Schoepflin , and Thorsten Schüppstuhl

Hamburg University of Technology (TUHH), Am Schwarzenberg-Campus 1, 21073 Hamburg, Germany
dirk.holst@tuhh.de

Abstract. Robotic bin picking remains a main challenge for the wide enablement of industrial robotic tasks. While AI-enabled picking approaches are encouraging they repeatedly face the problem of data availability. The scope of this paper is to present a method that combines analytical grasp research with the field of synthetic data creation to generate individual training data for use-cases in intralogistics transportation scenarios. Special attention is given to systematic grasp finding for new objects and unknown geometries in transportation bins and to match the generated data to a real two-finger parallel gripper. The presented approach includes a grasping simulation in Pybullet to investigate the general tangibility of objects under uncertainty and combines these findings with a previously reported virtual scene generator in Blender, which generates AI-images of fully packed transport boxes, including depth maps and necessary annotations. This paper, therefore, contributes a synthesizing and cross-topic approach that combines different facets of bin-picking research such as geometric analysis, determination of tangibility of objects, grasping under uncertainty, finding grasps in dynamic and restricted bin-environments, and automation of synthetic data generation. The approach is utilized to generate synthetic grasp training data and to train a grasp-generating convolutional neural network (GG-CNN) and demonstrated on real-world objects.

Keywords: Synthetic data · Grasp Simulation · Tangibility · Bin picking · Automation

1 Introduction

Grasping under uncertainties and in highly dynamic environments remains an unsolved problem and has repeatedly been a central target of state-of-the-art research. Current approaches use empirical solution strategies and try to derive rules for robotic grasping by applying machine learning. To identify a valid grasp for a given object different strategies can be applied such as the 6D pose estimation of objects in RGB images [1] or a graspability analysis based on point clouds [2] or depth images [3]. The availability of individualized training data is a constant problem in this research area and prevents

© The Author(s) 2023
K.-Y. Kim et al. (Eds.): FAIM 2022, LNME, pp. 284–292, 2023.
https://doi.org/10.1007/978-3-031-18326-3_28

a wide adoption of developed technologies. The manual generation of real-world data is considered time- and cost-intensive and is prone to errors [4], especially concerning the ground truth which is used for supervised machine learning algorithms. On the other hand, synthetic data generation is a promising approach for generating large and high-quality data sets for training neural networks. It overcomes many of the problems of real data generation, such as scalability of data synthesis, inaccurate annotations, or the introduction of unwanted effects such as dataset bias. Existing solutions for synthetic data generation often cover only a limited fraction of possible annotations, which means that the generated datasets can only be used for a subset of possible grasping research [1, 3, 4]. To support this branch of research and to ensure, that results are more comparable, it is important to cover as many different solution strategies as possible with one data set. Therefore, this work presents a holistic toolchain that creates synthetic data and annotations for object identification and segmentation, object pose localization, grasp candidate detection, and grasp candidate quality evaluation.

Following previous work [5], an existing synthetic data generator is used for the creation of training images, to identify objects on load carriers in intralogistics settings. This tool will be extended with functionalities to annotate grasps for a two-finger parallel gripper. Of particular importance is the systematic graspability analysis of single objects under uncertainties in the physics simulation software Pybullet, to integrate insights from analytical graspability research into synthetic data generation. For the determination of suitable grasp candidates, the underlying geometry of the object model will be analyzed and simplified to an object skeleton. The calculated bones are used to approach individual points with a two-finger parallel gripper and then be tested for force-fit under varying simulation parameters. Afterward, valid grips are transferred under a strict set of rules to fully loaded carriers in the 3D rendering software Blender. This method is intended to ensure that context-based intralogistics information is preserved and not altered by the tangibility analysis.

2   Related Works

Research in the field of robotic grasping can mostly be separated into analytical and empirical methods [6]. The former tries to calculate valid grasps for different types of grippers, with varying amounts of contact points [7, 8], by solving analytical problems. The complexity of this approach is strongly based on the number of equations and geometric properties, to be calculated for determining valid grasp candidates. The latter describes a set of data-driven methods, to derive rules for robotic grasping based upon examples [9]. They represent a broad topic in machine learning and the field of grasp planning has been strongly influenced by the progress of Convolutional Neuronal Networks (CNN) [3, 10]. It is possible to solve a variety of problems like object detection, image segmentation, 6D pose estimation [1], and grasp candidate generation [10] or evaluation [3]. All of them have in common, that they need training data to adapt their underlying algorithm to a given task. Since training on real robotic systems can be time-consuming, expensive, and sensitive to changes to the physical setup [11] and creating annotations for real image datasets is error-prone, another approach seems feasible: Synthetic Data. This approach is using 3D-Rendering software to create image databases and is able to create perfect ground truth to aid supervised machine learning algorithms.

To incorporate knowledge from analytical research into empirical solutions and synthetic data generation, the use of physics simulations has become popular [4, 12, 13]. They offer the possibility to rapidly recreate grasping situations and check whether a grasp candidate results in a successful grasp or fails to hold an object. Two main challenges arise from this approach: The first one is the free choice of simulation parameters. Since there is often no unique solution to the parameter space, it is necessary to determine intervals from which to choose values and evaluate simulation results under a given uncertainty [3, 14]. The second problem is based upon geometric analyses and the varying complexity of shapes. Since a grasp planning algorithm should be able to evaluate a huge variety of objects, techniques have been developed to simplify mesh structures into a couple of geometric primitives [15] or to collapse a volumetric body into lines and segments to build a skeleton out of its structure. This reduced model can then be used to perform path planning of a simulated gripper and to evaluate high-quality grasp candidates [16].

Recent work shows the use of synthetic data generation in the space of robotic grasping and bin picking. Depending on a given task, these generators create a certain amount of annotations, like 6D object pose information [4] or grasp candidates with a quality value to differentiate between several grasp candidates and their rate of success [3]. There is no one method that outperforms all others in grasping situations and creating several datasets for various applications with separate pipelines is laborious. Therefore, this paper presents a method that includes different types of annotations and rendering methods and combines techniques from different research areas such as Geometric Analysis, Physical Simulation, and Synthetic Data Generation.

3   Methodology - Synthetic Grasp Data Generation

This chapter presents the systematic generation of synthetic grasp data. As shown in Fig. 1, the process consists of six sub-steps across two programs and two separate databases.

As a foundation for the general graspability analyses in Pybullet, the pipeline extends an existing Graspit! [12] clone [17] with features such as geometric analyses, partial domain randomization of simulation parameters, and the ability to store information in a database. For the final data generation and the transfer of found grasps, the 3D rendering software Blender is. The sub-steps are described in the following:

1) Deriving **Object Skeletons**, to enable Grasp Simulation: Graspability analyses start with an understanding of the object to be grasped. To determine suitable grasp candidates, the mesh structure of an object is repeatedly collapsed until only a skeleton, consisting of individual interconnected points, emerges [16]. The simplified structure is located either on or within the mesh model and can consist of several hundred points. To further reduce the amount of data points and to identify meaningful grasp centers, the individual points of the determined skeleton are clustered, based on their

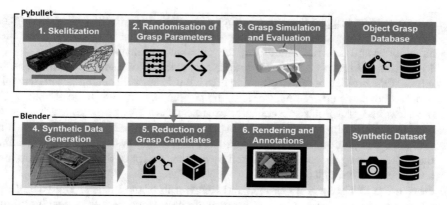

Fig. 1. Pipeline for synthetic grasp data generation, consisting of geometric and graspability analysis in Pybullet and synthetic image data generation, including annotations, in Blender. Results of the respective simulations are stored in databases and are available for analyses and machine learning algorithms.

spatial coordinates, using k-means clustering. The determining centers of the individual clusters then serve as an approaching point for the simulated gripper, which is moved to the object from different directions using spherical coordinates.

2) **Randomisation** of **Grasp Parameters** and rigid body **Simulation**: A physical simulation offers the possibility to adjust a large number of parameters and thus significantly influence the outcome. Since it is usually not possible to determine exact and correct values a priori, intervals for parameters such as mass, grasp force, and friction coefficient are determined based on physical constraints. These parameters can directly be derived from real objects and the robotic grasping systems used later on. From this value space, randomized values are chosen for each grasp candidate and are then checked for force-fit under consideration of the weight force. This process is repeated at least ten times for each grasp found so that it can be analyzed which parameter combinations are leading to successful grasps. If it turns out that an object can only be gripped under difficult conditions, the simulation should be repeated more frequently in order to generate meaningful results.

3) **Evaluation** of Grasps: A grasp is considered successful if the object to be grasped has remained in the gripper under the action of its weight. Due to the randomly chosen simulation parameters and in order to create the possibility to compare grasps with each other and to enable statistical analyses, like distributions of succesfull grasp parameters, the contact points of a these grasp including their simulation parameters are saved.

4) Transfer **to Synthetic Data Generation** Toolbox [5]: A structured set of rules is used to generate packed transport boxes, ensuring that logical consistencies are maintained within the training data domain. These include lighting conditions, camera positions, positioning of objects, and packaging material used. The aim is to generate scene compositions that are as realistic as possible and to identify intervals for partial randomization for the parameters mentioned. The program developed for Blender generates random variations of packed transport boxes and simulates the insertion

of objects to be gripped through rigid-body simulation. The tightly integrated set of rules ensures for each generated scene that visual material matches the generated annotations and does not contain any unwanted inconsistencies.

5) **Reduction of Grasp-Candidates** with respect to the boxing scenario and reachable grasps: One problem in grasp planning is to identify suitable candidates in dynamic and unpredictable environments. Here, the gripping system has to be able to reach the identified grasp and subsequently execute it. For this purpose, ray tracing is used to examine the reachability of the previously identified grasp candidates. It is evaluated whether the individual fingers of the gripping system can be placed and the object to be gripped is not covered by another object or the grip is not too low and prevented through collisions.

6) **Derivation of Grasp Annotations** and **Rendering** of the Images: The goal of the developed pipeline is to cover a wide range of possible machine learning algorithms and to enable further research on different methods, with only one data generation tool. Core tasks in grasp planning are image classification, object identification, object localization, image segmentation, 6D pose determination, depth images processing as well as the possibility to evaluate the quality of grasps. The solution presented here consists of a database, which archives all annotations from simulations parameters, in addition to a render pipeline, which creates the corresponding depth maps and segmentation masks from the RGB images generated in Blender, as shown in Fig. 2. Furthermore, information about the generated scene compositions, such as camera and light positions or packaging material used, is also archived. This offers the possibility to statistically evaluate the bias of a dataset and to generate new data if necessary to increase diversity.

Fig. 2. Showcase of the rendering pipeline, (left) realistic rendering with the Cycles engine, (middle) derived depth image, (right) corresponding segmentation mask

Besides the 6D pose determination of objects or the generation of grasp candidates on depth images, the field of grasp evaluation procedures is still available for research. For example, DEX-Net [3] uses the method of identifying grips analytically on depth images and then having them evaluated by a neural network. For learning such a method, grips have to be distinguishable in terms of their quality. This can be done using the stored simulation parameters of the graspability analysis (Step 3). An object to be grasped can be grasped and held with different ease at different points, which is why grasping data with unusual parameter combinations can be statistically identified and evaluated with a lower quality. To enable further research in this branch, each grip in the final training data is uniquely identifiable and can be subsequently modified. This provides the opportunity to observe the outcome for various metrics of a given grasp quality index.

4 Demonstration of the Developed Toolbox and Training Results

With the presented toolbox, a test data set for different geometric objects such as cubes, pyramids, and rings are created. This includes the sex step approach from general graspability analysis, the archiving of found grips, as well as the final transfer of possible grasps into packed transport boxes, as they are commonly found in intralogistics settings (Fig. 3).

Fig. 3. (left) Synthetic depth image of three objects in a transport box with packaging flips, including annotated grasp in red. (right) Real depth image, with a found grasp for a test cube by the GG-CNN Architecture, trained only on synthetic data.

The parameters for the gripping simulation from step 2 were adapted to the technical specifications of the two-finger parallel gripper of the Panda Franka Emika robot and performed with a virtual replica of the system. The randomized grasping simulation produced results to the expected degree, such as objects with lower mass being easier to hold, or that increased closing force is associated with increased grasping success rate.

To demonstrate that purely synthetic data can be used to determine grasp candidates in real images, a non-pretrained Gras Quality Evaluation network (GQ-CNN) [10] is used. This architecture was chosen for its state-of-the-art performance, but the dataset created is not limited to it, as a wide range of annotation is generated. The training dataset consists of 525 depth images, with slightly varying bird's eye camera angles, and a total of 176,000 grasp candidates for 9 different objects. Annotations for grasps are given in the format: Grip center and finger spread in pixels and rotation angle in radians. The data is split into 80% training data and 20% validation data.

To demonstrate that the domain gap between synthetic- and real-world data is successfully overcome, a RealSenseL515 lidar camera is used to capture depth images of heavily cluttered test scenes (s. Figure 4). The scaling of the grayscale was normalized to the minimum and maximum height values of the recorded scene. Test series with real depth images have shown that the used architecture is able to identify grasp candidates for the examined objects, by giving them the highest grasp probability. The network was able to distinguish between box edges, unknown objects, and objects to be grasped and did not output unwanted grasps.

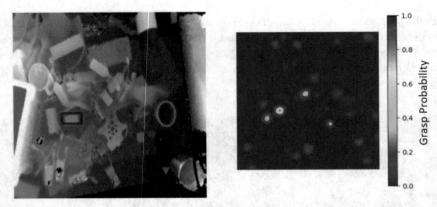

Fig. 4. (left) Successfully plotted grasp in red for a specific test object, using a wide variety of interfering sources, including objects similar to the test object. (right) The generated output of the GG-CNN [10] shows the highest probability for a possible grasp, including location and gripper width in pixel coordinates and angle of rotation.

Additionally, grasping tendencies towards objects that have parallel faces were detected, which can be attributed to the simulated two-finger parallel gripper and a significantly higher number of annotated grasps for this form of objects. This tendency can be minimized in further work by evenly distributing candidate grasps for different objects. Even in test environments with a large number of interfering sources, as shown in Fig. 4, grasps for the target object could be successfully identified.

5   Conclusion

In this work, a cross-disciplinary method for the generation of synthetic grasp data in highly dynamic environments was presented. Methods for geometric object analysis of mesh models, detailed graspability analysis under uncertainty, and techniques for transferring possible grasps into context-based scenes of packed transport boxes from intralogistics are used. The developed toolchain enables further research in grasp planning and execution by a wide range of annotations and offers the possibility to produce targeted data sets for individual problems.

Test series have shown that it is possible to identify grasp candidates on real depth images by using only synthetic data in combination with the GG-CNN architecture. The results were robust to sources of interference and offer a promising approach to enable further research in robotic grasping in a highly dynamic environment.

In further research, the developed pipeline can be used to generate grasp data for a wide variety of objects and geometric shapes, to test the dataset with different types of machine learning algorithms.

Acknowledgments. Research was founded by the German Federal Ministry for Econoic Affairs and Climate Action under the program LuFo VI-1 WLIBoro.

References

1. Kehl, W., et al.: SSD-6D: Making RGB-based 3D detection and 6D pose estimation great again. In: 2017 IEEE International Conference on Computer Vision (ICCV), pp. 1530–1538 (2017)
2. Kleeberger, K., et al.: Automatic Grasp Pose Generation for Parallel Jaw Grippers (2021)
3. Mahler, J., et al.: Dex-Net 2.0: Deep Learning to Plan Robust Grasps with Synthetic Point Clouds and Analytic Grasp Metrics (2017)
4. Periyasamy, A.S., Schwarz, M., Behnke, S.: SynPick: a dataset for dynamic bin picking scene understanding. In: 2021 IEEE 17th International Conference on Automation Science and Engineering (CASE), pp. 488–493 (2021)
5. Schoepflin, D., et al.: Synthetic training data generation for visual object identification on load carriers. Procedia CIRP **104**, 1257–1262 (2021)
6. Sahbani, A., El-Khoury, S.: Bidaud P An overview of 3D object grasp synthesis algorithms. Robot. Auton. Syst. **60**, 326–336 (2012)
7. Li, J.-W., Liu, H., Cai, H.-G.: On computing three-finger force-closure grasps of 2-D and 3-D objects. IEEE Trans. Robot. Autom. **19**, 155–161 (2003)
8. Ponce, J., et al.: On characterizing and computing three- and four-finger force-closure grasps of polyhedral objects. In: 1993 Proceedings IEEE International Conference on Robotics and Automation, vol. 2, pp. 821–827 (1993)
9. Kleeberger, K., Bormann, R., Kraus, W., Huber, M.F.: A survey on learning-based robotic grasping. Current Robot. Reports **1**(4), 239–249 (2020). https://doi.org/10.1007/s43154-020-00021-6
10. Morrison, D., Corke, P., Leitner, J.: Closing the Loop for Robotic Grasping: A Real-time, Generative Grasp Synthesis Approach (2018)
11. Pinto, L., Gupta, A.: Supersizing self-supervision: learning to grasp from 50 K tries and 700 robot hours. In: 2016 IEEE International Conference on Robotics and Automation (ICRA), pp. 3406–3413 (2016)
12. Miller, A.T., Allen, P.K.: Graspit! A versatile simulator for robotic grasping. IEEE Robot. Autom. Mag. **11**, 110–122 (2004)
13. Kleeberger, K., Landgraf, C., Huber, M.F.: Large-scale 6D Object Pose Estimation Dataset for Industrial Bin-Picking (2019)
14. Mahler, J., et al.: Dex-Net 1.0: A cloud-based network of 3D objects for robust grasp planning using a Multi-Armed Bandit model with correlated rewards. In: 2016 IEEE International Conference on Robotics and Automation (ICRA), pp. 1957–1964 (2016)
15. Miller, A.T., et al.: Automatic grasp planning using shape primitives. In: 2003 IEEE International Conference on Robotics and Automation (Cat. No. 03CH37422), vol. 2, 1824–1829 (2003)
16. Vahrenkamp, N., et al.: Planning high-quality grasps using mean curvature object skeletons. IEEE Robot. Autom. Lett. **3**, 911–918 (2018)
17. Carlyn Dougherty Robot Graspit! Project (2019). https://github.com/carcamdou/cr_grasper. Accessed 24 Jan 2022

Objectifying Machine Setup and Parameter Selection in Expert Knowledge Dependent Industries Using Invertible Neural Networks

Kai Müller[1]([⊠]), Andrés Posada-Moreno[2], Lukas Pelzer[3], and Thomas Gries[1]

[1] Institute for Textile Technology (ITA) of RWTH Aachen University, Otto-Blumenthal-Street 1, 52074 Aachen, Germany
`kai.mueller@ita.rwth-aachen.de`
[2] Institute for Data Science in Mechanical Engineering, Dennewartstr. 27, 52068 Aachen, Germany
[3] Institute for Plastics Processing (IKV) in Industry and Craft at RWTH Aachen University, Seffenter Weg 201, 52074 Aachen, Germany

Abstract. The textile industry is one of the oldest and largest industries in the world. The fields of application for textile products are diverse. Although the technologies for manufacturing textiles are extensively researched, the industry is still highly dependent on expert knowledge. To date, manual process- and machine adjustments and quality control are the norms rather than the exception. Heat setting is used in the process chain to dissolve or selectively introduce tensions from the weaving or knitting process and to prepare the products for digital printing. The correct setting of the machine depends on a large number of different materials-, processes- & environmental parameters. For each product, the machine has to be set up again by an experienced textile engineer. To ease the training for new workers and shorten the machine setting process, this study aims to use machine learning to facilitate and objectify the setting of the heat-setting process. Machine parameters are generated using an invertible neural network (INN) based on predefined target parameters. The results can be used to identify trends in machine settings and respond accordingly. Thus, a reduction of machine setting time could be realized.

Keywords: Neural network · Process parameters · Production management · Textile production

1 Introduction

The textile industry is one of the oldest industries in the world with a wide variety of product applications. These range from medical applications (e.g. stents) and technical textiles (e.g. protective equipment) to clothing and building materials (e.g. insulation) [1]. Similar to all manufacturing companies, the industry is exposed to new challenges due to global change. For example, batch sizes are becoming increasingly smaller due to individual customer requirements [2, 3], experts are retiring [4] and qualified employees

© The Author(s) 2023
K.-Y. Kim et al. (Eds.): FAIM 2022, LNME, pp. 293–300, 2023.
https://doi.org/10.1007/978-3-031-18326-3_29

are becoming increasingly difficult to find [5]. As the textile industry is characterized by manual labor and expert knowledge dependency, these challenges pose a threat to a large number of companies.

The advancement of production technologies as well as the introduction of digital sensing and networking technologies are opening up new opportunities for companies. Machine learning (ML), for example, can be used to record and analyze product/process data, as well as derive recommendations for action [6–8]. Expert knowledge can even be objectified to a certain extent using ML [9, 10]. This objectification could help support decisions in production environments. In the textile industry, finding appropriate process parameters is often done by varying the parameters in an empirically defined range. This results in a significant workload since theoretically, a full factorial experimental design needs to be conducted for every new material. Traditionally, a process can be modelled using ML techniques, then, said model can be used as a part of an optimization process (e.g. evolutionary algorithms [11]) to estimate the required parameters to produce a result of certain characteristics [11–16]. Yet, these approaches can not only be computationally expensive, but the used models are applied in non-bijective settings and are often not trivially invertible [12–14]. This setting limits severely the amount of knowledge that can be represented in the models. In contrast, the usage of deep invertible models allows an accurate modelling of a process, efficiency in its usage as well as a correct objectification of the parameter prediction problem, in a non-bijective setting.

In this paper, the use of invertible neural networks (INN) is suggested to generate the machine setting of a textile thermosetting machine to shorten setup time and reduce the complexity of the process. Implementing expert knowledge in the training and setup of the INN, machine settings are generated for specific product quality. This partial objectification of expert knowledge supports inexperienced personnel in process handling. For this purpose, experimental data is gathered, an INN is trained, an artificial dataset is generated, its consistency is reviewed for model evaluation, and the accuracy is validated in a set of experiments.

2   Background and State of the Art

2.1   Thermosetting

Thermosetting is a finishing process for textiles. It is used to relieve residual stresses resulting both from the spinning process and from subsequent deformation (knitting, warp knitting, etc.) [17, 18]. Mechanical properties and the crystallinity of the fibers are determined in the spinning process and vary for each production batch. Therefore, the thermosetting process needs to be adapted for every product.

During heat setting, the textile is pre-tensioned and heat is applied. As soon as the textile cools down, new intermolecular bonds are formed which hold the textile in the "deformed" structure [19]. Directional shrinkage occurs in the process [20]. In our application, a textile armband is clamped in a machine, fed to a padder via several deflection rollers, then dried with infrared, and finally fixed via two heated deflection rollers (Fig. 1).

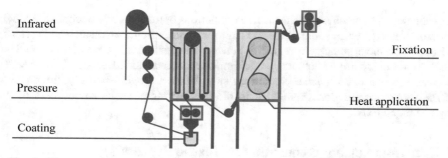

Fig. 1. Thermosetting process for an elastic wrist band at ITA, RWTH Aachen University.

The settings that an employee without expert knowledge can adjust on the machine are listed in Table 1. In addition, empirically recorded minimum and maximum values are given, which must not be exceeded to ensure an error-free process. In particular, the intensity of the heat input and the dwell time of the textile in a heating zone has a major influence on the test result.

Table 1. Identified parameters for machine setting prediction

Parameter	Unit	Minimum	Maximum
Feed	m/min	2.5	6
Infrared intensity	%	60	85
Temperature	°C	160	190
Pressure	10^5 Pa	2.2	4

2.2 Invertible Neural Networks

The task of process parameter prediction has traditionally been modeled as an optimization process [11, 15, 16]. This approach requires the approximation of the process dynamics through a function, which later is wrapped in an optimization process, changing the input parameters to minimize the difference against the desired quality and the prediction of the process model. Invertible neural networks (INN) allow an approximation of a process that can be evaluated as a forward or inverse function [21]. This replaces the optimization loop of the traditional parameter prediction with an inverse evaluation of the estimated function describing the process.

INNs base layers were introduced by Dinh et al. [22] as 'inverse coupling layers', which in contrast with other neural networks can be inverted trivially. Inverse coupling layers require an equal dimensionality of input and output vectors, which is obtained by adding the input and output, as well as adding an extra latent space (bootstrapped to a normal distribution) to the output of the network [21]. During training, the minimized loss is composed of four terms. First is the maximum mean discrepancy (MMD) norm between the input batches and the inverse predictions of the network. Second, the MMD norm between the extra latent space and a normal distribution with mean 0 and variance 1. Finally, the mean square errors (MSE) between the predictions of the network, and the labeled dataset inputs and outputs. In addition, one term was added as a conditional MSE, to ensure forward and inverse predictions stay within a range. Similarly, the gradient of the forward prediction was added to the loss, to increase the smoothness of the approximated INN.

3   Machine Setting Generation in Textile Processing

3.1   Data Acquisition

The data for this approach was acquired in two steps. First, data was gathered in a production environment at the Digital Capability Center (DCC), Aachen, Germany over four days. For that, probes of a woven armband were prepared. Each probe had an initial length of 20 cm and a mean weight of 3.5 g. After thermosetting, the deviance in length, weight, and residual moisture was measured for each probe. An experimental plan with a total of 43 different machine settings was prepared and conducted. For each setting 5 probes were produced, resulting in 215 probes (i.e., data points) in total. Figure 2 shows the deviation of the length of the armband before and after heating.

Fig. 2. Deviation of armband length before and after heating.

Second, an INN was trained with the data and created an enlarged artificial dataset. This set was used to validate the INN performance. Textile experts reviewed the dataset to ensure the reflection of physical phenomenons within the model (see Sect. 3.2 for detailed information).

All experiments were conducted on a coating machine MFR2 2C from Jakob Mueller AG, Frick, Switzerland. Pressure was monitored using an SV-Advanced-Box-Mobile, WAGO Kontakttechnik GmbH & Co. KG, Minden, Germany. The residual moisture after processing was measured using an HE53 Moisture Analyzer Scale, Mettler-Toledo GmbH, Greifensee, Switzerland. The length before after processing was measured with a steel rule by Bayerische Maßindustrie A. Keller GmbH, Hersbruck, Germany. All data were collected in the middleware Kepware provided by PTC Inc., Boston, USA.

3.2 Parameter Generation

Since the acquired data for training the neural network was limited, the model had to be improved in another way. Therefore, different measures were taken to train the model to fit the limited acquired data.

First, extra thresholds and the derivative terms were added to the loss for training the model. By limiting the range of values, only values that empirically allow a correct process flow can be specified.

After that, a hypercube with 5 values per variable was created. The values ranged between -2.5 and 2.5 times the previous variable limits. This is to show not only how the predictions behaved for the known regions of the variables, but also how they behaved for extreme values outside of training.

Finally, the trained model was used to compute forward predictions of every data point on the created hypercube, obtaining a representative dataset of the behavior of the model with 16,000 data points. In the first run, the range of the data was not constrained. To enhance the performance, the range of values was narrowed, excluding unrealistic high or low values.

4 Results and Discussion

The generated data were analyzed, revealing several values that could not be produced by the production process under consideration. Furthermore, values were identified, which did not exactly reflect the physical relationships between the influencing parameters (e.g., increasing weight despite lower moisture and higher heat input). Constraints for the data set were therefore developed together with data scientists to increase the data quality.

The real data was plotted with the artificial data to identify correlations between machine parameters and measured quality parameters (Fig. 3). It was possible to show trends between the measured data and the generated set. The green dots indicate the original data, the blue dots represent the artificial data. The generated data traces the relationship between feed rate and weight well, for example. With increasing speed, a higher weight was measured. At the same time, a higher infrared intensity reduces the weight. Based on these findings, it is reasonable to assume that the model reflects a large part of the physiological phenomena in thermosetting. In some experiments, armbands were stretched rather than shrunk (bottom left, green dots over the yellow and red line). This is not displayed by the artificial data, indicating that our model is still to be improved. The deviance in length after processing is significant between measured and generated data sets (see the deviation in the "Length" row). There are two theories for that problem.

The first one is, that there is no data available regarding the material used nor from previous processing. Therefore, errors already worked into the material (e.g., from fiber production or weaving) are not considered.

The second thought is missing information on at least one significant variable in the process. It can be assumed, that the gathering of environmental data such as room temperature and humidity, enhances the accuracy significantly. Despite the inaccuracies, the model at hand is still suitable to limit the range of machine settings to a certain degree.

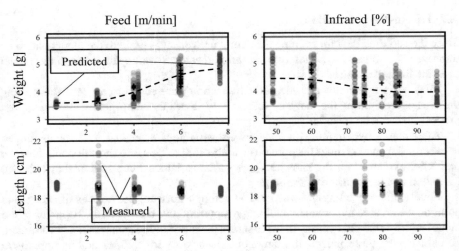

Fig. 3. Measured data (green) and generated data (blue) after constraining the parameter range.

For validation, a total of 5 different values of the quality parameters weight and length of the wristband were determined. These values were fed into the INN and the resulting machine parameters were used for the process. The accuracy for predicted and measured length and weight varied between 63.4%–88.12% (length) and 71.63%–91.8% (weight). These results show, that the model can predict a trend and limit the options for machine settings to a specific interval. At the same time, the accuracy varies significantly, which needs to be addressed in further research.

5   Summary and Outlook

In this paper, a study to generate machine settings based on desired product properties using an invertible neural network (INN) was presented. To do so, a small data set from the production environment was acquired and used to train an INN. A larger artificial dataset was generated to evaluate the model, and finally, the accuracy of the generated process parameters was validated in an experiment.

Although it was not possible to compile accurate machine settings due to the very complex process and unavailable data regarding material properties from previous production steps as well as environmental data, results are still promising. The model showed clear trends and correlations between the different parameters that were expected by textile experts. For example, the increasing weight of the wristband corresponds with a higher feed rate. Additionally, the influence of heat treatment was detectable. The accuracy for predicted quality parameters varied between 63% and 91%. Though the gap between values is considerable, the results significantly minimize the number of adjustment possibilities for the process and therefore the number of iterations needed for machine setup.

Further research lies in the possibility of transferring the approach to other industries. On top of that, further tests taking into account additional factors such as environmental data, the chemical composition of the coating material, or process data from previous process steps could increase the accuracy of the prediction.

Acknowledgement. Funded by the Deutsche Forschungsgemeinschaft (DFG, German Research Foundation) under Germany's Excellence Strategy – EXC-2023 Internet of Production – 390621612.

References

1. Gries, T., Veit, D., Wulfhorst, B.: Textile Fertigungsverfahren: Eine Einführung, Zweite überarbeitete und erweiterte Auflage, Hanser, München (2015)
2. Klocke, F., et al.: Self-optimizing production technologies. In: Brecher, C., Özdemir, D. (eds.) Integrative Production Technology, pp. 745–875. Springer, Cham (2017). https://doi.org/10. 1007/978-3-319-47452-6_9
3. Piller, F.T.: Kundenindividuelle Massenproduktion (Mass Customization). In: Piller, F.T. (Ed.), Mass Customization, pp. 200–266. Deutscher Universitätsverlag, Wiesbaden (2001). https://doi.org/10.1007/978-3-663-08187-6_7
4. Krüger, K.: Herausforderung Fachkräftemangel. Springer Fachmedien Wiesbaden, Wiesbaden (2018). https://doi.org/10.1007/978-3-658-20421-1
5. Pultarova, T., Onita, L.: A problem we can't afford to ignore [labour resources]. Eng. Technol. **10**, 30–33 (2015). https://doi.org/10.1049/et.2015.0317
6. Colosimo, B.M., Huang, Q., Dasgupta, T., Tsung, F.: Opportunities and challenges of quality engineering for additive manufacturing. J. Qual. Technol. **50**, 233–252 (2018). https://doi. org/10.1080/00224065.2018.1487726
7. Dai, H.-N., Wang, H., Xu, G., Wan, J., Imran, M.: Big data analytics for manufacturing internet of things: opportunities, challenges and enabling technologies. Enterp. Inf. Syst. **14**, 1279–1303 (2020). https://doi.org/10.1080/17517575.2019.1633689
8. Dekhtiar, J., Durupt, A., Bricogne, M., Eynard, B., Rowson, H., Kiritsis, D.: Deep learning for big data applications in CAD and PLM – Research review, opportunities and case study. Comput. Ind. **100**, 227–243 (2018). https://doi.org/10.1016/j.compind.2018.04.005
9. MacInnes, J., Santosa, S., Wright, W.: Visual classification: expert knowledge guides machine learning. IEEE Comput. Grap. Appl. **30**, 8–14 (2010). https://doi.org/10.1109/MCG.2010.18
10. Holzinger, A.: From machine learning to explainable AI. In: 2018 World Symposium on Digital Intelligence for Systems and Machines (DISA), Kosice, vol. 82018, pp. 55–66. IEEE (2018)
11. Mia, M., Królczyk, G., Maruda, R, Wojciechowski, S.: Intelligent optimization of hard-turning parameters using evolutionary algorithms for smart manufacturing. Materials **12**, 879 (2019). https://doi.org/10.3390/ma12060879
12. Nassef, A.M., Sayed, E.T., Rezk, H., Abdelkareem, M.A., Rodriguez, C., Olabi, A.G.: Fuzzy-modeling with particle swarm optimization for enhancing the production of biodiesel from microalga. Energy Sour. Part A Recov. Utiliz. Environ. Effects **41**, 2094–2103 (2019). https:// doi.org/10.1080/15567036.2018.1549171
13. Ronowicz, J., Thommes, M., Kleinebudde, P., Krysiński, J.: A data mining approach to optimize pellets manufacturing process based on a decision tree algorithm. Eur. J. Pharm. Sci. **73**, 44–48 (2015). https://doi.org/10.1016/j.ejps.2015.03.013

14. Chuang, K.-C., Lan, T.-S., Zhang, L.-P., Chen, Y.-M., Dai, X.-J.: Parameter optimization for computer numerical controlled machining using fuzzy and game theory. Symmetry **11**, 1450 (2019). https://doi.org/10.3390/sym11121450
15. Gurrala, P.K., Regalla, S.P.: Multi-objective optimisation of strength and volumetric shrinkage of FDM parts. Virtual Phys. Prototyp. **9**, 127–138 (2014). https://doi.org/10.1080/17452759.2014.898851
16. Dey, A., Yodo, N.: A systematic survey of FDM process parameter optimization and their influence on part characteristics. JMMP **3**, 64 (2019). https://doi.org/10.3390/jmmp3030064
17. Mittal, K.L., Bahners, T.: Textile Finishing. John Wiley & Sons Inc., Hoboken (2017)
18. A. Kumar, R. Choudhury, Principles of Textile Finishing, Woodhead Publishing, Cambridge (2017)
19. Choi, Y.J., Kim, I., Kim, S.H.: Effect of heat-setting on the physical properties of chemically recycled polyester nonwoven fabrics. Text. Res. J. **89**, 498–509 (2019). https://doi.org/10.1177/0040517517750643
20. Horrocks, A.R. Anand, S.C.: Handbook of Technical Textiles. Woodhead Publishing (2016)
21. Ardizzone, L., et al.: Analyzing inverse problems with invertible neural networks. In: ICLR 2019 (2019)
22. Dinh, L., Sohl-Dickstein, J., Bengio, S.: Density estimation using Real NVP (2016)

Integration of Machining Process Digital Twin in Early Design Stages of a Portable Robotic Machining Cell

Panagiotis Stavropoulos[(✉)], Dimitris Manitaras, Harry Bikas, and Thanassis Souflas

Laboratory for Manufacturing Systems and Automation (LMS), Department of Mechanical Engineering and Aeronautics, University of Patras, 26504 Rio Patras, Greece
pstavr@lms.mech.upatras.gr

Abstract. Industrial robots have been getting a more important role in manufacturing processes during the last decades, due to the flexibility they can provide in terms of reachability, size of working envelope and workfloor footprint. An especially interesting application are material removal processes and specifically machining. Use of robots in machining has opened new pathways for the development of flexible, portable robotic cells for several use cases. However, the peculiarity of such cells compared to traditional machine tools calls for novel approaches in their design and dynamic analysis. To this end, this work proposes an approach that integrates the digital twin of the machining process to set the boundary conditions for the design and dynamic analysis of the robotic cell. Physics-based modelling of milling is coupled with a Multi-Body Simulation of the robotic arm to define the inputs for the design of the cell. The design and dynamic analysis of the robotic cell is performed in a commercial FEA package, taking into account the requirements of the machining process.

Keywords: Robot machining · Digital twin · Modal analysis · Harmonic response

1 Introduction

Industrial robots have been an integral part of the manufacturing sector since several decades. Lately, they have upgraded their role and are directly involved in manufacturing processes, forming the base structure of manufacturing cells for several processes, one of which is machining [1]. Machining with robots is an interesting opportunity due to the ability to machine large scale parts with complex geometries in single operation setups, thanks to their wide working envelopes and flexible kinematics [2].

Together with the introduction of machining with robots, another application that has emerged as well is the use of robots for hybrid manufacturing. Hybrid manufacturing, which combines additive and subtractive processes on a single workstation, has drawn significant attention due to its ability to capitalize on the advantages of independent processes, while minimizing their disadvantages [3]. Reduced production times, desirable geometric accuracy and surface characteristics, multiple material parts, and the ability

K.-Y. Kim et al. (Eds.): FAIM 2022, LNME, pp. 301–315, 2023.
https://doi.org/10.1007/978-3-031-18326-3_30

to repair damaged parts are some typical advantages coming with the use of hybrid manufacturing [4]. The use of robots for hybrid manufacturing is a much more flexible alternative compared to hybrid machine tools, in terms of integration complexity and, as a result, it has been highly pursued [5].

Hybrid manufacturing also opens a pathway for in-situ repair of high-value industrial components for industries, such as aerospace, oil and gas, etc. [6]. A portable robotic cell for repair enables the solution to move to the problem, instead of the opposite that was the norm for years [7]. Through portable manufacturing cells, supply management practices, such as just-in-time manufacturing, can achieved. The waste is eliminated in terms of time, inventory, transportation and the supply chain of a production or maintenance operation could be shortened [8]. Apart from significantly shortening the supply chain, utilizing in-situ repair methods for large-scale industrial components can dramatically increase the sustainability of the repair operation, by significantly reducing the required transports [9].

Although the development of portable manufacturing cells based on hybrid manufacturing robots are a promising solution for in-situ manufacturing and repair, the design of the load bearing structure of the cell becomes significantly complicated, since it needs to provide the required stiffness to support the machining loads, while minimizing the weight to enhance portability. The load bearing structure cell is an essential functional component inside the machining system. The final accuracy of the machining system depends on the behavior of the frame structure which is loaded under static, dynamic, and thermal loads [10]. To support this design optimization effort, the digital twins of the machining process can be exploited, taking into account both the process and the robotic arm to set the boundary conditions for the design of the load bearing structure.

2   Literature Review

The use of the digital twins of the process to support the design and development of the machining system has been explored in literature, in several applications that are mainly related to high-accuracy manufacturing processes. Huo et al. [11] have proposed an integrated dynamic design and modelling approach for the development of a micro-milling machine. By integrating the modelling of the micro-milling process and the dynamic models of the sub-components of the machine, they managed to quantify the impact of their design in the accuracy of the process and take it into account in the optimization of the design. The importance of the digital twins of the process during the virtual prototyping phase of machine tools has been also highlighted by Altintas et al. [12], who introduced the virtual machine tool concept that comprised of Finite Element modelling of the machine tool components, kinematic modelling of the machine tool, Multi-Body Simulations, as well as dynamic modelling of the cutting process. Leonesio et al. [13] proposed a set of process related KPIs, targeting process quality and economic efficiency, which can be integrated in the machine tool design process to deliver the optimal product for a specific application. Regarding portable equipment, Law et al. [14] have developed a dynamic substructuring framework for in-situ machining systems to predict the dynamics of the assembled machine tool-workpiece system and facilitate the design of portable machine tools, enabling also the identification of their position-dependent dynamics [15]. Checchi et al. [16] have proposed a stiffness identification methodology for portable machine tools and offline compensation of their deflections, with a focus on machining of wind turbine components. Garnier et al. [17] have investigated the stability of the mounting structures of mobile machining robots, focusing on naval applications. In the field of robotics, Hazel et al. [18] developed a portable, track-based robot for in-situ repair of hydropower equipment through welding and grinding. By using the dynamic model of the grinding process, they were able to develop a robotic platform that would ensure the required quality of the processed surface. There have been some additional solutions proposed for portable manufacturing cells, either based on portable platforms [19] or mobile robots [20]. Apart from the research related to portable robotic cells for machining, there are also industrial efforts for such solutions that can be transferred to remote or harsh environments and operate [21]. This indicates the need for concrete developments that can support the design on such platforms.

Based on the literature review, it can be concluded that a gap exists on the development of portable machining solutions, based on robotic platforms, from the perspective of having an integrated representation of the interrelation between the machine and the process and being able to explore this during the design phase of the portable robotic cell. To this end, this paper presents an approach for the design, analysis and development of such systems, based on the integration of the digital twin of the machining process to facilitate design optimization.

3   Approach

The whole approach is based on the integrated machine-process modelling concept. In order to set accurate and realistic boundary conditions for the dynamic analysis of the load bearing structure of the robotic cell it is important to know the input loading conditions. These loading conditions are dependent on the physics of the machining process, as well as the dynamic behavior of the robot during the process. This calls for the formation of the robot machining Digital Twin that will act as an input for the robotic cell design. The formation of this Digital Twin is based on the integration of previous works of the authors. The Multi-Body Simulation (MBS) of the machining robot captures its dynamic behavior during the process [22], where the cutting forces that are exerted during the machining process are used as an input. The cutting forces are modelled with the mechanistic modelling approach. The specific force coefficients required for this model (that is explained in detail in the next section) can be constantly captured by indirectly monitoring the cutting forces during the robot machining process [2], thus closing the loop with the simulation and forming the Digital Twin. The Digital Twin architecture is depicted in Fig. 1.

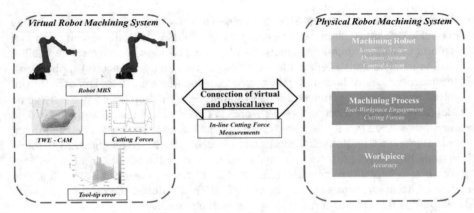

Fig. 1. Robot machining digital twin architecture

The whole approach of integrating the robot machining Digital Twin in the whole design of the cell is presented in detail in the next sections, while Fig. 2 provides an overview of the approach.

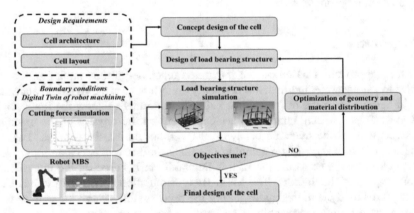

Fig. 2. Overview of the proposed approach

3.1 Simulation of the Milling Process

The first step of the whole approach is the determination of the cutting forces that will take place during the machining process. For this purpose, the well-established mechanistic cutting force model of Altintas and Lee [23] is utilized, which provides a fast and flexible modelling approach that takes into account the tool geometry and material, process parameters, workpiece material data and tool orientation. Moreover, the amplitude of specific force coefficients regarding different cutting tools and materials that are available in literature provides an opportunity for a vast design space exploration regarding different machining operations that the robot will have to serve. In short, this model considers the cutting forces to be linearly linked with the feed per tooth (f_z) and axial depth of cut (z), while being a function of the immersion angle of the j-th cutting edge of the cutting tool (φ_j). The cutting force coefficients (K_c) and edge force coefficients (K_e) are experimentally determined or they can be sourced from related literature. The formulation of the cutting force model is presented below.

$$
\begin{cases}
dF_{t,j}(\varphi_j, z) = K_{te}dS(z) + K_{tc}f_z\sin(\varphi_j)dz \\
dF_{r,j}(\varphi_j, z) = K_{re}dS(z) + K_{rc}f_z\sin(\varphi_j)dz \\
dF_{a,j}(\varphi_j, z) = K_{ae}dS(z) + K_{ac}f_z\sin(\varphi_j)dz
\end{cases}
\tag{1}
$$

The instantaneous cutting forces in the tangential ($dF_{t,j}$), radial ($dF_{r,j}$) and axial ($dF_{a,j}$) direction are integrated along the axial depth of cut to calculate the cutting forces at each cutting edge over one revolution of the tool. The results of this first simulation are then fed as an input to the robot Multi-Body Simulation that is presented in the next section.

3.2 Robot Multi-body Simulation

After the loading that is introduced by the material removal process is calculated the next step is to determine the impact of this load on the dynamic response of the robotic arm, in order to use it as an input for the dynamic simulation of the cell. For this purpose, a Multi-Body Simulation (MBS) of the robotic arm is utilized. The MBS considers both the flexibility of the joints and the links of the robot.

The joints are modeled as 1 Degree-of-Freedom (DoF) revolute joints with a spring-damper system along the rotation direction to capture the elastic deformation of the joints due to the machining loads.

For the modeling of the links the Component Mode Synthesis (CMS) method is utilized. This is an effective method to reduce the complexity of the model, while preserving the true static and dynamic behavior of the links. The Finite Element (FE) model of the links is developed and the interface points (i.e., the mating surfaces with the joints of the previous and next link) are defined. Next, the Craig-Bampton method is used to build the reduced order model of each link. From the full order FE model, only the interface DoFs are preserved, as well as some DoFs that are necessary to calculate the vibration modes of the link with a clamped-clamped boundary condition. As a result, the Craig-Bampton method retains both the static and vibration modes of the link and dramatically reduces the DoFs of the model, thereby its computational requirements. The result of this method is the creation of the so-called Superelements, which are comprised of the reduced mass and stiffness matrices of each link.

Finally, the superimposition of the joints and links models enables the development of the MBS that calculates the dynamic behavior of the whole robot during machining. By importing the pre-calculated cutting forces in the MBS, it is possible to estimate the final loads that will be transferred to the base of the robot and, ultimately, on the load bearing structure of the portable cell. These loads are utilized as an input for the next step of the analysis, which is the determination of the dynamic behavior of the cell. Figure 3 provides an overview of the robot MBS approach.

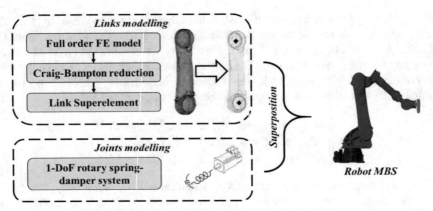

Fig. 3. Overview of the robot multi-body simulation

3.3 Simulation of the Robot Machining Cell

The final step of the whole approach is the determination of the structural and dynamic behavior of the load bearing structure of the portable cell. There are three key objectives that need to be fulfilled in order to create a successful design that will ensure a safe and reliable operation and transportation.

1. The load bearing structure of the cell should be able to withstand its own weight during transportation. For this purpose, a static analysis has been setup to ensure that the stresses during transportation are within the elastic limit
2. The deflections of the load bearing structure should be such, so that they do not impair the accuracy of the process. Ultimately, the main criterion to address this objective are the deflections between the mounting points of the robot and the mounting points of the workpiece. These can be determined through a harmonic response analysis. Since the loading during the milling process is harmonic, due to the intermittent cutting, the harmonic response analysis is a suitable tool to calculate the dynamic response of the cell in the whole frequency range of interest, which is determined by the capabilities of the machining spindle.

Based on the results of these three simulations a design of the load bearing structure of the cell can be performed, so that it is suitable for machining operations.

4 Case Study: A Portable Robotic Cell for Hybrid Manufacturing

The case study that has been selected to demonstrate the whole approach is for a portable robotic cell for hybrid manufacturing operations. The intended purpose of the cell is the in-situ repair and remanufacturing of industrial components. As a result, the main design requirements of the cell are: (a) a relatively low weight that can enable easy transportation between the different operating sites; (b) a dynamic behavior that will enable it to successfully complete milling operations; (c) manufacturability of the cell should also be considered. The robot that will be utilized is a Yaskawa GP225 industrial robot arm, along with a Yaskawa DK-500 2-DoF table, which will offer an additional flexibility to the machining and Additive Manufacturing processes, due to its redundant DoFs.

4.1 Architecture of the Portable Cell

The key elements of the main structure that affect its final design are related mainly with the layout of the various subsystems that is determined by the functional requirements of the cell. First of all, there is a need for two entrance points (one in the front and one in the side of the cell) to enable ingress and egress of the operators, as well as loading and unloading of the workpieces. Moreover, in the rear side of the cell, a shelf with an enclosure should be integrated, where the tools and process heads can be stored, when they are not in use. A retractable roof is installed on the top of the cell, so that it can be removed during transportation. Finally, a detachable platform where all the auxiliary equipment is stored (controllers, laser source, etc.) is installed on the rear side of the cell. All of these elements, dictate the areas where material can or cannot be placed in the load bearing structure, thereby affecting its design. Figure 4 provides an overview of the whole design.

Fig. 4. Overview of the portable cell architecture

4.2 Concept of the Load Bearing Structure

The part that is of interest for the simulation is the part that actually affects the machining process, due to its dynamic behavior. So, the load bearing structure can be isolated for further analysis and optimization of its design. As mentioned previously, the manufacturability of the cell is one of the key design requirements. Therefore, a welded construction based on square beams has been selected for the load bearing structure. Moreover, two C-section beams are welded on the bottom side of the load bearing structure in such a distance that will enable the transportation with a forklift. Shows the concept of the load bearing structure. Finally, vibration cancelling mounts have been selected two support the cell. Figure 5 provides an overview of the load bearing structure.

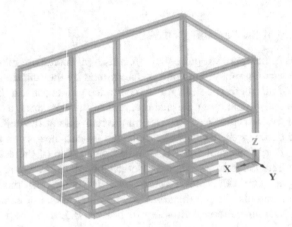

Fig. 5. Concept of the load bearing structure

4.3 Cutting Force Calculation

The milling processes that the robot will need to serve are mainly post-processing of Additive Manufactured components. In most applications where Additive Manufacturing is used for repair of industrial components, the target material is either a hardened steel or Heat-Resistant Alloys (HRAs). One of the HRAs that presents especially low manufacturability, leading to high cutting forces, is IN718. Moreover, IN718 is a very common material for Additive Manufactured parts. Therefore, in this case study milling of Additive Manufactured IN718 was considered. Based on the specific force coefficients for this specific material [2] the cutting forces that were calculated as a worst-case scenario with aggressive process parameters (Table 1) were $F_x = 1650N$, $F_y = 3130N$ and $F_z = 230N$.

Table 1. Process parameters for worst-case scenario

Process parameter	Value
Cutting tool diameter [mm]	50
Axial depth of cut [mm]	5
Radial depth of cut [mm]	45
Cutting speed [m/min]	450
Feed per tooth [mm]	0.16

4.4 Dynamic Behavior of the Robot

Next, the cutting forces were used as an input in the MBS of the robot to calculate its dynamic response. In order to also take into account the effects of the robot posture on its overall dynamic response, a large part has been designed and simulated, in order to evaluate the dynamic response over the whole toolpath (Fig. 6).

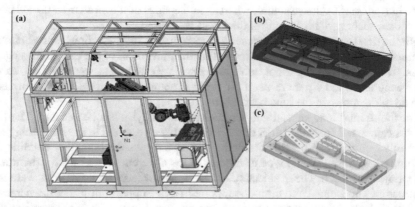

Fig. 6. (a) Toolpath used to evaluate the cutting forces; (b) close-up view of the toolpath; (c) close-up view of the workpiece and stock material

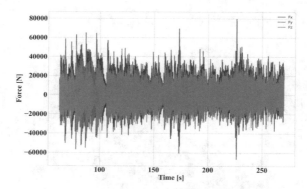

Fig. 7. Forces at the robot base during machining of IN718

Based on the results of the MBS the boundary conditions for the dynamic simulation of the cell can be determined. Figure 7 shows the MBS results regarding the forces at the base of the robot during machining of IN718. By identifying the amplitude of these forces, it is possible to create the input for the dynamic simulation of the cell.

4.5 Simulation of the Load Bearing Structure of the Portable Cell

Finally, the simulation of the load bearing structure of the portable cell can be performed. ANSYS Finite Element software has been utilized for this simulation. The grid that comprises the load bearing structure has been designed and the beams sections at each line of the grid have been inserted. Based on an iterative design process of simulation and re-design, 140 × 140mm steel square sections with 4mm wall thickness were used for the bottom part of the structure, whereas in the sides of the structure 100 × 100mm steel square sections with 4*mm* wall thickness were used.

For the first load case that was described in Sect. 3.3, the C-sections at the bottom of the structure were used as supports, restricting the movement of the cell on the negative Z-axis. The weights of the robot, table and auxiliary equipment were applied, as well as the weights of the structure itself. A static structural simulation was performed to estimate the stresses on the load bearing structure.

For the second load case, a harmonic response analysis has been setup. The force that was calculated from the MBS of the robot has been applied in the mounting points of the robot on the load bearing structure. Figure 8 shows the boundary conditions of the simulation. Based on the type of mounts at the different nodes of the structure, the appropriate restraint has been selected. For the nodes where simple mounts were used, the structure was considered simply supported (grey supports in Fig. 8), whereas for the nodes where the mounts are going to be anchored in the foundation of the floor, fixed supports were used (black supports in Fig. 8). The deflections between the robot mounting point and the table mounting point were monitored, in order to identify the effect of the dynamic behavior of the structure in the machining process.

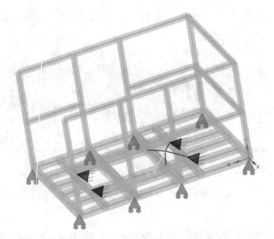

Fig. 8. Boundary conditions of the harmonic response analysis

5 Results and Discussion

The results of the static analysis are presented in Fig. 9. As it can be observed, the stresses on the structure are very low, indicating that the transportation of the cell can be performed without any issues.

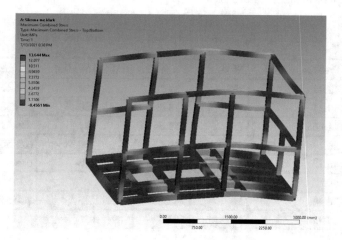

Fig. 9. Static analysis results

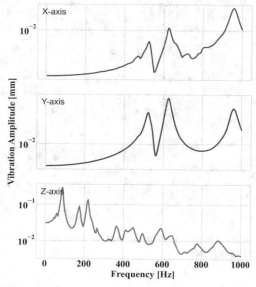

Fig. 10. Vibration amplitudes of the cell at the frequency range of interest

Figure 10 shows the semi-logarithmic plots of the vibration amplitudes over the frequency range of interest. As it can be observed, the main vibration component is that on the z-axis. This is something to be expected by the design of the load bearing component, since the beams are loaded in bending, due to the Z-axis component of the force, whereas in other two orthogonal axes they are loaded in tension-compression, where they present significantly higher stiffness (Fig. 11).

Table 2. Eigenfrequencies of the load bearing structure of the cell

Mode #	Frequency [Hz]	Mode #	Frequency [Hz]	Mode #	Frequency [Hz]
1	12.08	10	89.28	19	524.42
2	18.36	11	218.09	20	528.38
3	21.86	12	239.64	21	539.51
4	26.14	13	249.47	22	625.50
5	46.12	14	297.11	23	658.97
6	59.66	15	476.87	24	691.88
7	76.44	16	481.11	25	730.55
8	84.71	17	507.53	26	792.42
9	87.49	18	511.51		

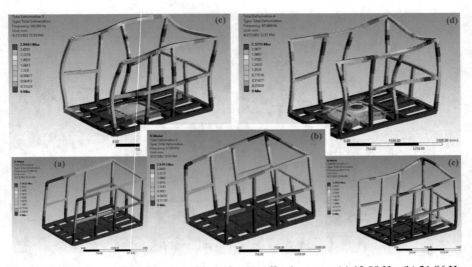

Fig. 11. Mode shapes of the 5 modes with the largest effective mass: (a) 12.08 Hz; (b) 21.86 Hz; (c) 46.12 Hz; (d) 87.49 Hz; (e) 89.28 Hz

The eigenfrequencies of the load bearing structure of the cell are presented in Table 2. Figure 12 depicts the mode shapes for the 5 eigenfrequencies that contain the largest effective mass, according to the modal analysis. It is also important to pay particular attention at the loading frequency of 87.49 Hz, where the highest peak in the Z-axis vibrations can be observed. Figure 12 presents the vibration amplitudes of the whole cell at this loading frequency. It can be observed that the highest deformations can be found at the top horizontal beam on the rear side and the two vertical beams on the front side. Again, this is a result to be expected, since these two beams are the least supported in the whole load bearing structure. Nevertheless, such a deformation is not worrying since it will not interfere with the accuracy of the machining process. By examining closely the bottom of the structure and especially the mounting points of the robot and table, it can be observed that the deformation in those points is in the 10^{-2} mm range. Such a result is perfectly acceptable, given the fact that in order to achieve the cutting forces that were calculated in Sect. 4.3, a roughing operation should be implemented. In such a case, an accuracy loss in that range is not detrimental for the process.

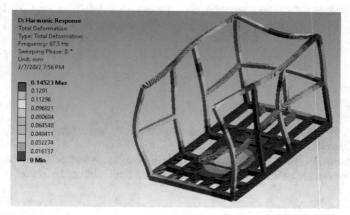

Fig. 12. Deformation of the cell at 87.5 Hz loading frequency

6   Conclusions

The scope of this work was to present an approach for the exploitation of the digital twin of the robot machining process, in order to design portable and structurally sound cells, towards in-situ repair and manufacturing. The whole approach consisted of an integrated machine-process modelling concept, where the machining process was firstly modelled, serving as an input for an MBS of the machining robot, which determined the boundary conditions for the simulation of the dynamic behavior of the robotic cell. Based on the results of the work the following conclusions can be drawn:

- The MBS of a machining robot can be a powerful tool during the design of its structural cell
- The use of the digital twin of the robotic machining process to build the boundary conditions of the simulation of the robotic cell can effectively lead to a virtual prototype

of the whole system that can reduce the need for physical prototyping and trial and error approaches

Future work should include the closure of the loop of this whole approach, where the effect of the dynamics of the portable cell are also taken into account in the MBS of the robot, yielding further accuracy in the whole simulation. Moreover, tools for automation and optimization of the design process should be developed, based on this digital twin approach, which will reduce the manual effort of setting and evaluating diverse simulations and can assist the engineers during the design phase. Finally, the validation of the actual dynamic behavior of the manufactured cell should be performed and used as a benchmark for the simulation.

Acknowledgement. This work has been co-financed by the European Regional Development Fund of the European Union and Greek national funds through the Operational Program Competitiveness, Entrepreneurship and Innovation, under the call RESEARCH – CREATE - INNOVATE (project code: T2EDK-03896).

References

1. Verl, A., Valente, A., Melkote, S., Brecher, C., Ozturk, E., Tunc, L.T.: Robots in machining. CIRP Ann. **68**(2), 799–822 (2019)
2. Stavropoulos, P., Bikas, H., Souflas, T., Ghassempouri, M.: A method for cutting force estimation through joint current signals in robotic machining. Procedia Manuf. **55**, 124–131 (2021)
3. Dávila, J.L., Neto, P.I., Noritomi, P.Y., Coelho, R.T., da Silva, J.V.L.: Hybrid manufacturing: a review of the synergy between directed energy deposition and subtractive processes. Int. J. Adv. Manuf. Technol. **110**(11–12), 3377–3390 (2020). https://doi.org/10.1007/s00170-020-06062-7
4. Stavropoulos, P., Bikas, H., Avram, O., Valente, A., Chryssolouris, G.: Hybrid subtractive–additive manufacturing processes for high value-added metal components. Int. J. Adv. Manuf. Technol. **111**(3–4), 645–655 (2020). https://doi.org/10.1007/s00170-020-06099-8
5. Jiménez, A., Bidare, P., Hassanin, H., Tarlochan, F., Dimov, S., Essa, K.: Powder-based laser hybrid additive manufacturing of metals: a review. Int. J. Adv. Manuf. Technol. **114**(1–2), 63–96 (2021). https://doi.org/10.1007/s00170-021-06855-4
6. Rahito, W.D., Azman, A.: Additive manufacturing for repair and restoration in remanufacturing: an overview from object design and systems perspectives. Processes **7**, 802 (2019)
7. Uriarte, L., Zatarain, M., Axinte, D., et al.: Machine tools for large parts. CIRP Ann. **62**(2), 731–750 (2013)
8. Durão, L.F.C.S., Christ, A., Zancul, E., Anderl, R., Schützer, K.: Additive manufacturing scenarios for distributed production of spare parts. Int. J. Adv. Manuf. Technol. **93**(1–4), 869–880 (2017). https://doi.org/10.1007/s00170-017-0555-z
9. Rauch, E., Dallinger, M., Dallasega, P., Matt, D.T.: Sustainability in manufacturing through distributed manufacturing systems (DMS). Procedia CIRP **29**, 544–549 (2015)

10. Allen, J., Axinte, D., Roberts, P., et al.: A review of recent developments in the design of special-purpose machine tools with a view to identification of solutions for portable in situ machining systems. Int. J. Adv. Manuf. Technol. **50**, 843–857 (2010)
11. Huo, D., Cheng, K., Wardle, F.: A holistic integrated dynamic design and modelling approach applied to the development of ultraprecision micro-milling machines. Int. J. Mach. Tools Manuf **50**(4), 335–343 (2010)
12. Altintas, Y., Brecher, C., Weck, M., Witt, S.: Virtual machine tool. CIRP Ann. **54**(2), 115–138 (2005)
13. Leonesio, M., Molinari Tosatti, L., Pellegrinelli, S., Valente, A.: Procedia CIRP **2**(1), 38–43 (2012)
14. Law, M., Rentzsch, I.S.: Developing of a dynamic substructuring framework to facilitate in situ machining solutions using mobile machine tools, Procedia Manuf. **1**, 756–767 (2015)
15. Law, M., Phani, A.S., Altintas, Y.: Position-dependent multibody dynamic modeling of machine tools based on improved reduced order models. ASME J. Manuf. Sci. Eng. **135**(2), 021008 (2013)
16. Checchi, A., Costa, G.D., Merrild, C.H., et al.: Offline tool trajectory compensation for cutting forces induced errors in a portable machine tool. Procedia CIRP **82**, 527–531 (2019)
17. Garnier, S., Subrin, K., Arevalo-Siles, P., et al.: Mobile robot stability for complex tasks in naval industries. Procedia CIRP **72**, 297–302 (2018)
18. Hazel, B., Côté, J., Laroche, Y., Mongenot, P.: A portable, multiprocess, track-based robot for in situ work on hydropower equipment. J. Field Rob. **29**(1), 69–101 (2012)
19. Wagner, H.J., Alvarez, M., Kyjanek, O., Bhiri, Z., Buck, M., Menges, A.: Flexible and transportable robotic timber construction platform – TIM. Autom. Constr. **120**, 103400 (2020)
20. Buchli, J., Giftthaler, M., Kumar, N., et al.: Digital in situ fabrication - challenges and opportunities for robotic in situ fabrication in architecture, construction, and beyond. Cem. Concr. Res. **112**, 66–75 (2018)
21. Zhu, Z., Tang, X., Chen, C., et al.: High precision and efficiency robotic milling of complex parts: Challenges, approaches and trends. Chin. J. Aeronaut. **35**(2), 22–46 (2022)
22. Stavropoulos, P., Gerontas, C., Bikas, H., Souflas, T.: Multi-Body dynamic simulation of a machining robot driven by CAM [Accepted for Publication]. In: Proceedings of 55th CIRP Conference on Manufacturing Systems, Lugano, CH, 29 June–1 July 2022 (2022)
23. Altintaş, Y., Lee, P.: A general mechanics and dynamics model for helical end mills. CIRP Ann. **45**(1), 59–64 (1996)

Development Process for Information Security Concepts in IIoT-Based Manufacturing

Julian Koch[✉], Kolja Eggers, Jan-Erik Rath, and Thorsten Schüppstuhl

Institute of Aircraft Production Technology, Hamburg University of Technology, Hamburg, Germany
julian.koch@tuhh.de

Abstract. Digital technologies are increasingly utilized by manufacturers to make processes more transparent, efficient and networked. Novel utilization elicits the challenge of preventing deployed information technology from compromising processual security. The digital enabling of formerly analog operation technology, the extensive use of information technology connectivity like MQTT, TCP/IP, Wi-Fi, and the deployment of IoT edge computing platforms create an application scenario for the Industrial Internet of Things (IIoT), which also introduces the associated vulnerabilities, which have been extensively exploited in the past. This paper introduces a development process for information security concepts designed for production scenarios based on the IIoT. This concept is then applied using an illustrative use case from aircraft production. The main contents of the development process include: Formulation of reasonable assumptions, system modelling, threat analysis including risk assessment, recommendation of countermeasures, reassessment after incorporating countermeasures. Specifically, a Data Flow Diagram as the model is developed, and a "risk first" variation of the STRIDE methodology is applied to identify threats and prioritize them. The aforementioned state-of-the-art methodologies are adjusted to our cyber-physical use case in the IIoT. The resulting concept aims to enable manufacturing processes to be digitized as sought. The adjustments to the methodologies are independent from our use case and may be suitable to a broad field of scenarios in the IIoT.

Keywords: Threat modelling · IIoT · STRIDE · Cyber-physical systems · Information security · Industry 4.0 · DFD

1   Introduction

In recent years, the amount and impact of cyberattacks on companies in the industrial sector has increased drastically and is expected to increase further [1, 2]. Due to rising danger cybersecurity has become a high priority for any party making use of Industrial Internet of Things (IIoT) environments [3]. Although there is no single definition of IIoT [4], there are certain recurring characteristics of the IIoT in the literature that are especially relevant for cybersecurity in manufacturing companies. In particular, the connection of a wide variety of cyber-physical systems to form a network should be mentioned here, which in turn places special requirements on connectivity, interoperability,

The original version of this chapter was revised: The incorrect affiliation of all the authors has been corrected. The correction to this chapter is available at
https://doi.org/10.1007/978-3-031-18326-3_40

© The Author(s) 2023, corrected publication 2023
K.-Y. Kim et al. (Eds.): FAIM 2022, LNME, pp. 316–331, 2023.
https://doi.org/10.1007/978-3-031-18326-3_31

scalability and data processing [5]. The increasing amount and impact of cyberattacks in the industrial context are attributable to the merging of the traditionally separated domains of Operation Technology (OT) and Information Technology (IT) into the IIoT [6]. Due to differences in scope, impact, and context of possible threats, securing IIoT systems is typically arduous, and differs substantially from securing both traditional IT systems and traditional OT systems [7]. The objective of this work is to develop a process to elaborate on information security concepts for digital manufacturing processes. To ensure that information security is integrated into the introduction of digital technologies, this approach particularly focuses on systems under development. For this objective, a system-driven [8], Security-By-Design [9] approach is chosen. In Sect. 2, the necessary background such as secure system development, data flow diagrams (DFDs) and STRIDE as well as related work will be considered. Section 3 describes the actual development methodologies as well as the proposed changes to the DFDs and the STRIDE method. Section 4 applies the methodology to a use case exploring the quality assurance of aircraft structure components and presents the subsequent findings. Last, Sect. 5 discusses the presented development process and provides an outlook on future work.

2   Related work and background

This section provides essentials and related work to facilitate a better understanding on the proposed methodologies and the respective adjustments.

2.1   Secure System Development

On the strategic level, secure system development considers the overall development process of secure systems, and is divisible into two approaches. The first approach aims to develop security measures for an existing system. The second approach integrates the security development into the actual system development process which aims to achieve a Security-By-Design approach. The first approach is followed by the BSI-security process, which in its description towards the development of a security concept, is applicable to existing processes [10]. This process starts with the specification of the scope, which is followed by a structural analysis of the underlying system and the definition of protection requirements, as well as the modelling of the system based on the prior steps. Based upon the model, the system's protection requirements are checked. If the protection requirements have not already been met, a risk analysis and subsequent risk consolidation is undertaken, which then triggers the next instance of protection requirement checks. This is an iterative process until the requirements are seen to be met and pertaining safeguards are implemented. Last, the process describes the maintenance and continuous improvement of the achieved results. However, this process provides inadequate guidance for Security-By-Design approaches, since the security development process should be integrated with the system development. Therefore, such approaches cannot be built on top of an existing system. For this reason, deviations from the BSI-process were developed which aim to make it suitable for Security-By-Design approaches as the said approach does not elaborate on the specific steps in the development process [11]. Publicly available use cases regarding end-to-end security development for industrial cyber-physical cases are rare, however the threat modelling use cases for industrial cyber-physical cases do exist [12]. Furthermore, the unadjusted application of methodologies from the cybersecurity domain does not sufficiently consider physical threats. Last, the proposed method collocates risk determination after the threat analysis, which tends to produce a high number of low-priority threats and is therefore a point of inefficiency.

2.2 System Modelling

Modelling approaches as a foundation for threat analyses, specifically also in IIoT contexts, vary, while the most common approach is to model the system as a data flow diagram (DFD) [13–19]. DFDs are based on the stages of digital data and model data-in-use as processes, data-at-rest as data stores and data-in-transit as data flows. Furthermore, DFDs may include trust boundaries which denote transitions of the respective trust assumptions between sections of the model [20]. The aptitude of DFDs regarding their use in threat analyses is a topic of scientific discussion and several enhancements to account for shortcomings exist [19, 21]. Regarding IIoT-systems, the incapability to model physical aspects will be more specifically considered and motivates the proposal of the adapted DFD notation. One aspect of enhancement included in the eSTRIDE methodology will be utilized as a reference [21].

2.3 Threat Analysis

STRIDE is the most common methodology for threat analyses, but due to its genesis in software security at Microsoft, suffers from shortcomings regarding use cases which increasingly differ from classical OS and software security [22]. However, STRIDE is used as a basis for threat analyses in the IIoT domain [14–19]. STRIDE provides six classes of common threats which facilitate the brainstorming process. The classes are "Spoofing", "Tampering", "Repudiation", "Information Disclosure", "Denial-Of-Service" and "Elevation-Of-Privilege" [20]. Deviations to account for challenges such as threat explosions and cyber-physical systems are discussed in varying literature [21, 22]. One of these approaches is called eSTRIDE and is relevant to our proposed methodology. eSTRIDE as a deviation to STRIDE applies a risk-first approach to the threat analysis, which otherwise is done after finding the threats via STRIDE [23].

3 Proposed Development Process and Methodologies

This section describes the derived development process for information security concepts (Fig. 1), including the adjusted DFD modelling and the adjusted application of the STRIDE methodology. The phases with their tactical steps of the development process will be laid out, placing greater emphasis on the proposed methodologies applied in these steps. The applied development process represents an adaptation of the BSI development process for a security concept on a strategic level, however, it is adapted for Security-By-Design approaches. The development process was devised for cyber-physical IIoT systems, while the adapted STRIDE methodology is applicable for any use case dealing with assets. The adapted DFD modelling was devised for IIoT systems with cyber-physical and physically distributed components.

3.1 Development Process

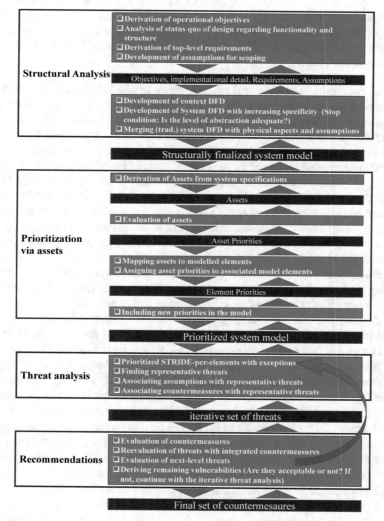

Fig. 1. Proposed development process with four phases including tactical steps.

The entire development process consists of four phases with tactical steps, which can be seen in Fig. 1. The tactical steps are shown in light blue and the results in dark blue. While in principle the development process follows the downstream flow seen in Fig. 1, the whole process is iterative in the sense that, for following the Security-By-Design approach, it must react to changes in the development of the system to be secured. Therefore, a change in system specifications triggers the upstream flow depicted in Fig. 1. In general the upstream flow might not only be triggered by a change of system specifications, but rather when the analyst assumes that previously executed steps have for whatever reason no longer valid results. The process is then gone through backwards in order to pinpoint those aspects which no longer hold, must be changed or must be added respectively. When the most upstream step of the process, which was affected by the change, has been adequately adapted, then the process is to be applied downwards to account for resulting changes. Within the following, the development process is detailed based on the main phases.

Structural Analysis: The structural analysis serves to provide a foundation for the subsequent modelling and threat analysis. Its execution depends on the state of the design phase and the degree to which implementational detail is known. In the conceptual phase, the objectives of the system are formulated with increasing detail in order to develop functional components, which can later serve as building blocks for the model. It should be noted that "components" describe conceptual parts of the system, which are not necessarily represented directly in the model, while "elements" are the defined building blocks of the DFD model. Therefore the modelling of the system as a DFD depends on the adequate representation of system components in the DFD using elements. Subsequently, already known implementational details are gathered. The produced set of implementational details is used in the last step of the modelling process in which the traditional DFD is merged with the implementational details. After this, the top-level requirements pertaining to the development of the information security concept for the underlying system under development are devised. These requirements should state what the information security concept should provide and what requirements the process itself must meet. Last, assumptions and scoping decisions are taken to guide the development. These assumptions will later be included in the model and the assessment of found threats. The gained knowledge of the underlying conditions developed in the prior steps are then utilized to develop an adjusted DFD to model the system. A conventional DFD model is devised with the DFD notation established in [20], differing only in that the processes are denoted with circles. When an adequate model of the system has been achieved, the changes to the DFD are made, which aim to better model systems such as the considered IIoT use cases (see Fig. 2). For this, implementational detail is added to the model. First, data flows are annotated with channel information such as communication protocols, much alike to those in eSTRIDE. Second, DFD elements which are to be implemented on the same device are aggregated in that regard. Third, system sections with multiple distributed instances are marked as such. Last, the resulting model is annotated with security relevant assumptions, divided into hard security assumptions (green), soft (yellow) and compromising (red). The assumptions are either specific to an element of the model or affect several elements indirectly.

Prioritization via Assets: The purpose of this phase is to provide the foundation for the subsequent prioritization of threats which facilitates a prioritized *risk-first* approach to the implementation of mitigations. As the origin of this prioritization the potentially endangered assets are utilized. The proposed process assumes that assets have already been described in the course of functional development of the IIoT system. The assets can be any kind of data produced and needed in IIoT-based manufacturing process. Examples for assets in this context are machine data, measurement data or intralogistics

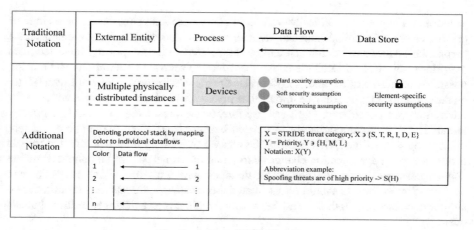

Fig. 2. Adjusted DFD Notation.

data. The derivation of assets can be performed by various techniques. For example, the system specifications can be used to create an Entity-Relationship model from which the resulting data can be derived. The assets are mapped onto the associated DFD elements, developed in phase 1. The assets are prioritized into low, medium and high priority in the categories of the STRIDE methodology regarding possible impact and exposure. Notably, this step takes place before finding threats, while in traditional STRIDE application a risk assessment or prioritization of threats of any kind is not an inherent part of the methodology and is therefore often applied afterwards, therefore the approach described in this paper is denoted as a *risk-first* approach. For this assessment, expertise regarding the underlying system and information security are required, which can be integrated by executing the assessment in collaboration with an expert of the underlying system. It is important to note that this assessment does not include *attack scenarios*, but only considers the generic STRIDE threats as basis for the assessment. The analyst is free to choose a specific existing methodology for the assessment. Examples for standardized qualitative methodologies for this task can be found in [24, 25]. When the prioritization is done, the DFD elements inherit the highest possible priority of their associated assets. It should be noted that the prioritization of assets and the inheritance of the resulting priorities by the DFD elements is a novel approach.

Threat Analysis: The threat analysis provides the threats to the system. Furthermore, the threat analysis aims to provide representative threats which group threats together if they are sufficiently similar. This serves to reduce the analysis effort. The STRIDE methodology is applied with the limiting assumptions and scope settings in order of the prioritized elements of the DFD. While the STRIDE methodology is conventionally applied to a model without prioritization, the approach in this work provides a prioritization to the model, which aims to result in focusing the threat analysis on higher priority threats. This aspect makes this a *risk-first* approach. The general approach is based on the STRIDE-per-Element variant. However, threat scenarios are generally not limited to single elements, therefore the "per-Element" notion is not regarded as absolute, rather as more of a guiding principle. Therefore, in addition to assessing the elements in their prioritized order, the threat analysis examines scenarios in which multiple elements must partake in the threat execution. Threats are associated with their priority inherited from the priority of affected DFD elements, the affecting relevant assumptions taken prior, possible countermeasures and lastly similar threats which form a set from which later on representative threats may be drawn. The priority is used to determine in which order mitigations are considered. High priority threats are considered first and low priority threats last. The assumptions are associated because they affect the possible and recommended mitigations. This enables the analyst to directly consider affected threats if due to system specification changes certain assumptions cannot be upheld. Possible countermeasures should be associated to threats to have a set from which a selection can be done. Similar threats should be associated, because those threats may be sufficiently similar to merge certain threats into representative threats as a means to reduce analysis effort.

Countermeasure Recommendation: The countermeasure recommendation represents the mitigation of threats found in the prior phase and produces a set of countermeasures which constitute the information security concept. The potential countermeasures from the threat analysis must be evaluated in a holistic manner. This includes their mitigation potential on the respective threat category, their possible impact on other threat categories, the necessary effort of implementation and their role and interdependence in the system-wide mitigation effort. The evaluation of a potential countermeasure in these categories is carried out with expertise regarding the system. A specific evaluation methodology is not considered in this paper, but exists in standardized form e.g. in [24] under the term *consolidation*. Based on this first evaluation, a set of countermeasures is selected. Upon this selection, a reevaluation of the found threats is executed, now including the selected mitigations, and in addition, considering new threats introduced by the selected countermeasures. This triggers the second of possibly more iterations starting at the 1st step of the 3rd phase of the described process resulting in a set of first and higher-level countermeasures. These countermeasures form the recommendations representing the aspired information security concept.

4   Application of the Methodology and Results

The use case and the application of the described development process onto the use case will be illustrated in this section. For better comprehension and topical focus representative aspects of the steps described prior are presented.

4.1   Description of the Use Case

The use case for the application of the described development process evolves from the digitization of a Quality Assurance (QA) process for aircraft structure components. The original QA process requires the inspectors to examine features like steps and gaps of fuselage elements or heights of rivet heads based on an inspection plan in paper format distributed to the inspectors. The inspection itself is executed manually with analog measurement tools (e.g. calipers) and the inspection results are to be written in paper form. From there, the resulting documentation reports are sent to be manually digitized by office staff.

The described QA process suffers from several drawbacks. First, manual unassisted inspections based on individual worker-skill are not sufficient with narrow tolerances. Second, tolerances for every feature must be extracted from the physical inspection plan. Third, a lack of information transparency regarding the state of the inspection may lead to duplicate work. Last, measurements taken manually cannot be directly integrated into higher-level data management systems.

To improve the QA process regarding the described shortcomings, several objectives were developed. Among those is the deployment of digitally enhanced inspection tools for the seamless integration of measurement data attained from the inspection of the device under test. The measurement data is forwarded to an automated documentation process, which produces a final report from incoming measurement data. This report is then stored in the data base, and appropriately forwarded for print, to be signed for legal reasons. Furthermore, an automated orchestration (flow generator) process is intended to distribute the inspection plans (workflows) and prior documentation data to the inspectors. To provide the digitally distributed data to the inspector and to provide assistance in the inspections, assistance systems such as in [26] are deployed which provide information to the inspector. Last, an operational administrator has to organize the data base for which a digital entry point is needed (Admin relay). All data in the system (flow generator, documentation, assistance, inspection) is centrally managed by the above-mentioned database system which is accessed by the respective processes. System components already known to have to store data (at least an intermediate) are assigned temporary memory. The overall system described is currently still under development, so not all subsystems have been completely defined and rolled out. However, the implementation of the project is based on typical technologies and protocols of the IIoT. In concrete terms, this means that the measurement tool represents a cyber-physical system and that different protocols such as MQTT and HTTP are employed to transmit data between the different systems used in the process. Thus, this project constitutes a suitable use case for applying the development process for information security concepts proposed here.

4.2 Structural Analysis

First, objectives were developed with decreasing degree of abstraction and reaching an implementational approach. This aspect has already been performed in the use case description.

Second, implementational detail already known in the design process is gathered. For example digitized inspection tools are to be included to facilitate direct integration of measurement data with higher-level data management systems.

Subsequently, requirements are developed. Exemplary the QA process assures the quality of structure components therefore preventing any compromising impact regarding this assurance is representative of a top-level requirement. To guide the ensuing process assumptions are taken. One assumption maintains that Wi-Fi channels are assumed to be secured via WPA technology. The assumption is not absolute since e.g., configuration management influences how well the employment of such technology translates into tighter security.

Based on the prior steps, the first traditional DFD is developed (Fig. 3) following the methodology illustrated in Fig. 1. First, a context diagram is constructed, which contextualizes the digital quality assurance process in the manufacturing process. Based on the developed objectives the DFD is then iteratively constructed by decomposing model elements into more specific elements until an adequate level of abstraction is achieved. This DFD is altered in respect to the described adjustments (see Sect. 3). If several DFD elements are on the same device this is denoted (e.g., the documentation process and the Flow Generator), also implementationally known communication protocols like the Bluetooth link between inspection process and documentation process are added. Furthermore, the section of the DFD which represents several distributed instances is also denoted. Last, the assumptions affecting individual elements of the DFD are integrated into the model as locks (see Fig. 4), for example the soft assumption (yellow) that Wi-Fi connections employ WPA3. Hard security assumptions are denoted in green and compromising assumptions in red. However, some assumptions may not be clearly associated with only one element and must therefore be considered implicitly for the whole system. This can be seen in the assumption that an industrial shopfloor is generally not accessible to the public.

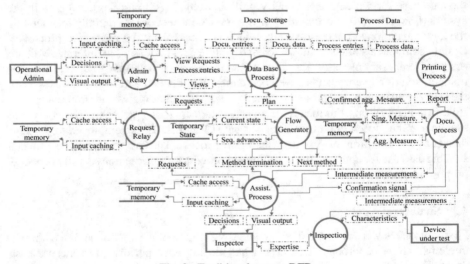

Fig. 3. Traditional system DFD.

4.3 Prioritization via Assets

Assets of the system are developed based on a preexisting Entity-Relationship model developed for the system, which followed the methodology described in [27] to cope with extensive, heterogeneous, unstructured data. Regarding the selected exemplary objective of digitized inspection tools, the exemplary asset "intermediate measurement data" is considered (see Fig. 3 data from inspection to documentation process). This asset describes measurement data which was generated in the sensors of the digitally enhanced inspection tools, but was not yet confirmed by the inspector as the correct value.

All assets are evaluated regarding severity of the STRIDE threat categories if a threat of said category was to be executed successfully. Exemplary, the intermediate measurement data is assessed to have a high priority regarding spoofing and tampering threats, which results directly from the formulated top-level requirement that any compromise of the QA process results is of high priority.

The developed assets are mapped to their associated DFD elements. In the case of the intermediate measurement data this includes the inspection process as well as the data flow from there to the documentation process amongst others. The associated DFD elements subsequently inherit the resulting priorities, which renders the deviating model where resulting priorities are assigned to a selected set of elements (Fig. 4).

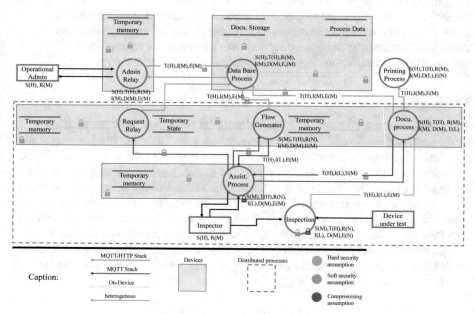

Fig. 4. System DFD with proposed adjustments and inherited priorities.

4.4   Threat Analysis

The STRIDE-per-element threat analysis was executed in the order of the assigned priorities. As an example, the tampering threats regarding the data flows associated with intermediate measurement data were among the first to be analyzed. Deviating from the traditional STRIDE-per-element execution, the found threat scenarios were merged into representative threat scenarios in the described manner. All representative threat scenarios were assessed regarding made assumptions in the structural analysis and potential countermeasures were assigned. Amongst others, this results in a higher importance of configuration management for employed Wi-Fi technology than in the technical aspects of possible spoofing or tampering threats affecting Wi-Fi connections since WPA3 was assumed to be deployed.

4.5 Countermeasure Recommendation

The potential countermeasures from the threat analysis were evaluated in the manner described in Sect. 3.1 "countermeasure recommendation". For all countermeasures which were assessed to be promising, the threat analysis is executed again to assess if new threats are introduced. If so, those second-level threats are treated the same as first-level threats. This iterative process was executed until residual vulnerabilities were considered acceptable. Exemplary, spoofing threats from human actors in the system are aimed to be mitigated with an authentication scheme. As a second-level threat, the authentication scheme might require only weak passwords or degrading back-up authentication. These common threats are assessed to be mitigated by employing state-of-the-art authentication schemes implementing concepts described for example in [28]. The set of countermeasures in its totality forms the information security concept.

4.6 Results

The analysis of the adjusted DFD model with its prioritized sections resulted in 20 representative threats with 13 high-priority threats (see Table 1). The countermeasure recommendation provided respective mitigations for almost all found threats. Further, it produced residual vulnerabilities such as the possibly compromising capabilities of digitized inspection tools regarding secure device authentication schemes and potential architectural improvement regarding security by transferring interactions with "feature tolerance" assets to the inspectors instead of the operational administrators. It should be noted that the vulnerability arising from the inspection tool capabilities is a Security-By-Design related aspect since this can either be specified or considered solved when more implementation detail becomes known. As key results the mentioned representative high priority threats are presented.

Table 1. Representative high priority threats.

#	Threat description
1	Spoofing of the operational admin and manipulating documented measurement values and feature tolerances
2	Admin repudiates against illegitimate manipulation of data, e.g. altering feature tolerances
3	The inspector is spoofed to the inspection process and produces illegitimate measurement data
4	The inspector is spoofed to the assisting process and to the inspection process which enables the adversary to produce measurement data and confirm it
5	Documented values or a report contain problematic entries and no inspector claims responsibility
6	Documentation process is tampered with as a means to manipulate measurement data after it has been confirmed
7	Spoofing the documentation process to the printing process and printing illegitimate reports
8	Spoofing inspection process to documentation process and sending illegitimate measurement data
9	Spoofing the documentation process to its temp data store to manipulate cached confirmed measurement values
10	Inadequate authorizations and consequentially usability issues undermining security policies
11	Tampering with the aggregated measurement values transmitted from documentation process to DB process
12	Information tampering threat on dataflow from inspection process to documentation process
13	Tampering with the documentation data store of the DB to alter reports or with the association data store to alter tolerances

5 Discussion and Future Work

In summary, this paper presented a strategic development process utilizing adjusted DFDs and an adjusted STRIDE methodology. IIoT use cases have distinct aspects which justify the adjustments. The aspects are: their cyber-physical nature, the Security-By-Design approach and lastly their size regarding modelled elements and assets.

5.1 Summary

The presented strategic development process provides a modular framework for Security-by-Design threat modelling approaches in the IIoT domain and is integrated with the development of the underlying system. Objectives, functional requirements, assumptions, implementational detail and assets result from the functional system development and are integrated with security-oriented requirements, scoping decisions and asset prioritizations. Based on the strategic phases 1 and 2 and the outlined prioritized model, the threat analysis can be executed in 3 and followed up by the recommendation of countermeasures in 4.

The proposed deviation from the traditional DFD modelling integrates the description encompassing data flow communication channels from physical to application layer. Additionally, it introduces the notion of devices into the DFD, which provides context for the threat analysis. Last, it includes notation for centralized and singular system sections and vice-versa for decentralized sections with various instances, which provides information affecting exposure and impact of attacks.

The adapted STRIDE method applies a risk-first approach where the deviation from the eSTRIDE method consists of transferring the perspective from the evaluated assets back to the DFD elements. This deviation makes it possible to apply the risk-first approach with more traditional STRIDE variants like STRIDE-per-element and STRIDE-per-interaction. Similar to eSTRIDE, assets are identified beforehand and evaluated. In the proposed method this evaluation is based on the STRIDE threat categories. The resulting priorities are assigned to the assigned DFD elements and in the prioritized order threats are analyzed with the STRIDE-per-element variant of the STRIDE method.

5.2   Discussion About the Methodology and Its Application

The application of the strategic development process with its modularization into phases and tactical steps onto the use case described in Sect. 4 systematically produced an information security concept covering all found threats or describing system aspects, which could not be concluded upon due to the design stage of the considered system under development. The process allowed for swift adaption in the event of a change in prior phases or tactical steps and its results present a solid starting point for the continuous development in the sense of Security-By-Design. It thereby successfully adapts the BSI-security process to Security-By-Design development. Furthermore, the process is clearly modularized regarding the purpose of the described phases and tactical steps. Last, it provides the strategic perspective for risk-first threat analysis approaches and embeds them into the overall development process.

Regarding the adapted DFD modelling approach, the time consumption of the additional aspects of the model proved negligible in comparison to the traditional system modelling, while the device notation enabled the analysis to consider threats aimed at whole devices rather than pure cyberattacks aimed at traditional DFD elements. By using the protocol annotation for data flows, the produced threats become more specific in comparison to the threats associated with more generic DFD elements. Annotating distributed system components with multiple instances facilitates assessments of exposure and impact of possible attacks. While the analysis benefitted from annotating the assumptions, it became apparent that due to on the quantity of assumptions and their difference in specificity (to one element) and generality (affecting all elements), they may struggle to be manifested through a visualized format. Given that assumptions may be very peculiar, it also did not seem reasonable to proceed in unison to the asset priorities and let the DFD elements inherit a general level of security based on taken assumptions. Therefore, a selection of assumptions to present visually might be made.

The prioritized and adapted STRIDE method is based on the evaluation of assets. This was considered positive, given that the assets were conceptually known while the implementational detail was still only partly defined. The categories to evaluate the assets may be improvable, seeing as the evaluation based on STRIDE threat categories pushes the analyst to consider threats before the actual threat analysis. This may cause confusion regarding the otherwise clearly separated phases. A solution would be to use security properties as the basis of evaluation (e.g., Confidentiality, Integrity, Availability, Accountability and Authenticity). Another notable aspect is that the risk-first approach prioritizes the threat analysis such that high priority threats are found faster, while in traditional STRIDE threats are found without regard to their possible priority. Many of those unprioritized threats may then be discarded afterwards when a risk assessment renders a low priority. This however, means that the effort that was spent finding them was spent inefficiently [eSTRIDE case studies "Finding security threats that matter"]. Furthermore, the possibility to combine the risk-first approach and thereby integrating its benefits with the well-established and documented STRIDE-per-element variant is considered the most important beneficial take-away. Notably, it serves as an alternative to eSTRIDE if the prior development steps render many assets in comparison to the number of DFD elements. In such a situation, the described approach might reduce analysis effort.

With regard to the manufacturing domain, the presented process with the associated methodologies can be used to examine IIoT-based applications for critical aspects of security even during their development phase. Using the adjusted DFD, a graphical representation of the use case with critical aspects is modeled, fostering a common understanding between manufacturing and security experts. Impacts of design decisions, such as the choice of a particular communication protocol, can be quickly captured and evaluated, and countermeasures to potential threats can be developed in parallel with the overall application. This work thus represents a contribution to the enablement of secure IoT applications in manufacturing.

5.3 Future Work

Future work should investigate modelling approaches to cyber-physical systems which do not depend as heavily on the perspective of digital data as DFDs. Furthermore, efforts should be made to find standardized manners of integrating physical system aspects into the modelling process. Additionally, approaches for the piecewise integration of implementational detail into models for iterative Security-by-Design approaches may be explored. Regarding the application of the STRIDE methodology in cyber-physical systems further research should consider options to better include physicality; either in modelling or in the analysis itself. The same holds true for the threat analysis of architectural systems, where the promising research might exist in efficient approaches to threat modelling of systems of higher abstraction.

This work provides a development process for information security concepts containing three novel approaches which enable the efficient and effective integration of security into the development process of IIoT-systems, thereby minimizing security risks.

References

1. Morgan, S.: Global Cybercrime Damages Predicted To Reach $6 Trillion Annually By 2021(2020). https://cybersecurityventures.com/annual-cybercrime-report-2020/. Accessed 08 Feb 2022
2. CSIS. https://www.csis.org/programs/strategic-technologies-program/significant-cyber-inc idents. Accessed 08 Feb 2022
3. Yu, X., Guo, H.: A survey on IIoT security. In: 2019 IEEE VTS Asia Pacific Wireless Communications Symposium (APWCS). IEEE (2019)
4. Boyes, H., et al.: The industrial internet of things (IIoT): an analysis framework. Comput. Ind. **101**, 1–12 (2018). https://doi.org/10.1016/j.compind.2018.04.015
5. Bostjancic Rakas, S., et al.: Industrial Internet: architecture, characteristics and implementa- tion challenges. In: 2021 20th International Symposium INFOTEH-JAHORINA (INFOTEH): 17–19 March 2021, Jahorina, East Sarajevo, Republic of Srpska, Bosnia and Herzegovina: Proceedings, pp. 1–4. IEEE, Piscataway, NJ (2021)
6. Ani, U.P.D., He, H., Tiwari, A.: Review of cybersecurity issues in industrial critical infras- tructure: manufacturing in perspective. J. Cyber Secur. Technol. **1**, 32–74 (2017). https://doi. org/10.1080/23742917.2016.1252211
7. Tsochev, G.: Some security problems and aspects of the industrial internet of things. In: 2020 International Conference on Information Technologies (InfoTech), pp. 1–5 (2020)
8. Burnap, P.: Risk Management & Governance: Knowledge Area (2021). https://www.cybok. org/media/downloads/Risk_Management_Governance_v1.1.1.pdf. Accessed 08 Feb 2022
9. Santos, J.C.S., Tarrit, K., Mirakhorli, M.: A catalog of security architecture weaknesses. In: 2017 IEEE International Conference on Software Architecture Workshops (ICSAW), pp. 220–223. IEEE (2017)
10. Bundesamt für Sicherheit in der Informationstechnik BSI-Standard 200–2 - IT-Grundschutz Methodology (2017)
11. Eckert, C.: IT-Sicherheit: Konzepte, Verfahren, Protokolle, 10. Auflage. De Gruyter studium. De Gruyter Oldenburg, München (2018)
12. Mohamed Shibly, M.U.R., Garcia De Soto, B.: Threat modeling in construction: an example of a 3D concrete printing system. In: Proceedings of the 37th International Symposium on Automation and Robotics in Construction, ISARC 2020: From Demonstration to Practical Use - To New Stage of Construction Robot, pp. 625–632 (2020)
13. Shevchenko, N., et al.: Threat Modeling: A Summary of Available Methods (2018)
14. AbuEmera, E.A., ElZouka, H.A., Saad, A.A.: Security framework for identifying threats in smart manufacturing systems using STRIDE approach. In: 2022 2nd International Conference on Consumer Electronics and Computer Engineering (ICCECE), pp. 605–612. IEEE (2022)
15. Borgaonkar, R., et al.: Improving smart grid security through 5G enabled IoT and edge computing. Concurr. Comput. Pract. Exper. **33**, 1–16 (2021). https://doi.org/10.1002/cpe. 6466
16. Danielis, P., Beckmann, M., Skodzik, J.: An ISO-compliant test procedure for technical risk analyses of IoT systems based on STRIDE. In: 2020 IEEE 44th Annual Computers, Software, and Applications Conference (COMPSAC). IEEE, pp. 499–504 (2020)
17. Empl, P., Pernul, G.: A flexible security analytics service for the industrial IoT. In: Gupta, M., Abdelsalam, M., Mittal, S. (eds.) Proceedings of the 2021 ACM Workshop on Secure and Trustworthy Cyber-Physical Systems, pp. 23–32. ACM, New York, NY, USA (202)
18. Khan, R., et al.: STRIDE-based threat modeling for cyber-physical systems. In: 2017 IEEE PES Innovative Smart Grid Technologies Conference Europe (ISGT-Europe), pp. 1–6. IEEE (2017)

19. Yampolskiy, M., et al.: Systematic analysis of cyber-attacks on CPS-evaluating applicability of DFD-based approach. In: 2012 5th International Symposium on Resilient Control Systems, pp. 55–62. IEEE (2012)
20. Shostack, A.: Threat Modeling: Designing for Security. Wiley, Indianapolis (2014)
21. Tuma, K., et al.: Finding security threats that matter: two industrial case studies. J. Syst. Softw. **179**, 111003 (2021). https://doi.org/10.1016/j.jss.2021.111003
22. Shevchenko, N., Frye, B.R., Woody, C.: Threat Modeling for Cyber-Physical System-of-Systems: Methods Evaluation. Carnegie Mellon University Software Engineering Institute, Pittsburgh, United States (2018)
23. Tuma, K., Scandariato, R., Widman, M., Sandberg, C.: Towards security threats that matter. In: Katsikas, S.K., et al. (eds.) Computer Security. LNCS, vol. 10683, pp. 47–62. Springer, Cham (2018). https://doi.org/10.1007/978-3-319-72817-9_4
24. Bundesamt für Sicherheit in der Informationstechnik BSI-Standard 200-3 - Risk Analysis based on IT-Grundschutz (2017)
25. Joint Task Force: Risk management framework for information systems and organizations. National Institute of Standards and Technology, Gaithersburg, MD (2018)
26. Müller, R., et al.: The Assist-By-X system: calibration and application of a modular production equipment for visual assistance. Procedia CIRP **86**, 179–184 (2019). https://doi.org/10.1016/j.procir.2020.01.021
27. Koch, J., Lotzing, G., Gomse, M., Schüppstuhl, T.: Application of multi-model databases in digital twins using the example of a quality assurance process. In: Andersen, A.-L., et al. (eds.) Towards Sustainable Customization: Bridging Smart Products and Manufacturing Systems. LNME, pp. 364–371. Springer, Cham (2022). https://doi.org/10.1007/978-3-030-90700-6_41
28. Grassi, P.A., et al.: Digital identity guidelines: authentication and lifecycle management. National Institute of Standards and Technology, Gaithersburg, MD (2017)

Augmented Virtuality Input Demonstration Refinement Improving Hybrid Manipulation Learning for Bin Picking

Andreas Blank[1]([⊠]) [ID], Lukas Zikeli[1], Sebastian Reitelshöfer[1], Engin Karlidag[2], and Jörg Franke[1]

[1] Institute for Factory Automation and Production Systems (FAPS), Friedrich-Alexander-Universität Erlangen-Nürnberg (FAU), Erlangen, Germany
{andreas.m.blank,georglukas.zikeli,joerg.franke}@fau.de
[2] Digital Industries (DI) Factory Automation (FA), Siemens, Nuremberg, Germany

Abstract. Beyond conventional automated tasks, autonomous robot capabilities aside human cognitive skills are gaining importance in industrial applications. Although machine learning is a major enabler of autonomous robots, system adaptation remains challenging and time-consuming. The objective of this research work is to propose and evaluate an augmented virtuality-based input demonstration refinement method improving hybrid manipulation learning for industrial bin picking. To this end, deep reinforcement and imitation learning are combined to shorten required adaptation timespans to new components and changing scenarios. The method covers initial learning and dataset tuning during ramp-up as well as fault intervention and dataset refinement. For evaluation standard industrial components and systems serve within a real-world experimental bin picking setup utilizing an articulated robot. As part of the quantitative evaluation, the method is benchmarked against conventional learning methods. As a result, required annotation efforts for successful object grasping are reduced. Thereby, final grasping success rates are increased. Implementation samples are available on: https://github.com/FAU-FAPS/hybrid_manipulationlearning_unity3dros

Keywords: Bin picking · Machine learning · Mixed reality · Human-in-the-loop

1 Motivation

Short product life cycles, an increasing amount of product variants and more complex goods pose challenges to the manufacturing industry. Flexible automation involving robot systems contributes to improve the situation. However, conventional automation reaches limitations in scenarios with uncertainties. These include manipulation and grasping operations in bin picking for material supply and machine feeding [1].

Deep Reinforcement Learning (RL) is an enabler for autonomous robot skills able to cope with complex grasping operations. However, implementation of RL within industrial applications is limited to specific and low demanding use cases. This is caused by

© The Author(s) 2023
K.-Y. Kim et al. (Eds.): FAIM 2022, LNME, pp. 332–341, 2023.
https://doi.org/10.1007/978-3-031-18326-3_32

the complexity of learning environment setups [2]. Deep Imitation Learning (IL) represents an alternative, whereby human cognitive skills are involved in the learning process. Thus, less parametrization is required. As demonstrations formulate an explicit, often intuitive learning-objective, more manipulation scenarios are covered [3].

Limiting factors of IL are the restriction to human demonstrations and the required effort to generate a sufficient annotations amount [4]. The utilized Human Machine Interface (HMI) represents another factor for IL performance. Existing approaches lack a real-time complementary exploitation of multiple sensor and semantic data sources.

In this context, the contribution of this paper is to propose an Augmented Virtuality (AV)-based input demonstration refinement method. The method enables efficient hybrid learning for manipulation operations. Hybrid learning combines known RL and IL algorithms by formulating weighted objective functions within shared constraints. In computer science, AV refers to the augmentation of the Virtual Reality (VR) with real-world elements, enriching the user experience [5]. The overall objective of the method is to reduce required adaptation efforts to new and changing scenarios. In addition, the improved annotation quality through demonstrations increases grasping success rates. The hybrid learning method is further characterized by flexible iterative learning. In addition, successive AV-based dataset refinement and fault interventions during system ramp-up enable application tuning up to operational productive deployment.

2   Related Work: Learning Strategies in Industrial Bin Picking

While fully autonomous robots have not yet been proven to be deployable to the shopfloor, industrial application of partly autonomous robot capabilities is an active field of research. Skill- and behavior-based abstracting architectures as a flexible design paradigm for robot software find application across industrial research domains [6].

Industrial bin picking is a subdomain of manipulation of the aforementioned setting. Stereo vision-based object recognition and pose estimation based on Convolutional Neural Networks (CNN) improve success rates of bin picking applications [7, 8].

However, the underlying manipulation skill is often a source of grasping failures, still inhibiting success rates [9]. Current research focuses on flexible grasping strategies with increased activity in the RL domain [1, 2]. Markovian Q-learning within Actor-Critic neural networks like Soft Actor Critic (SAC) can enable the application of RL to complex robot scenarios [10]. As a result, high success rates are achieved for simple handling [11]. Through Policy Gradient methods like Proximal Policy Optimization (PPO), collision avoidance can be integrated into learning strategies [12].

In [11] and [12] manipulation tasks are either presented in a simplified manner or unsolved grasping attempts still remain. RL is affected by specific hyperparametrization and environment stochasticity [13]. In general, heterogeneous manipulation tasks benefit from more abstract approaches instead of case-specific RL implementations.

IL, as part of the learning by demonstrations domain, is an alternative Machine Learning (ML) paradigm suitable for complex manipulation. In particular, it provides symbiotic integration with AV teleoperation and offers a more abstract characteristic enabling a wider robotic application range [14]. IL is mostly applied in scenarios requiring sequences of specific state transitions to reduce search space complexity [15]. Since algorithms like Behavioral Cloning (BC) and Generative Adversarial Imitation Learning (GAIL) require multiple high-quality demonstrations, IL is combined with Human-in-the-Loop (HuITL) [3, 16]. Here, failure intervention through teleoperation realizes dataset refinement [4]. This strategy is costly in terms of input data generation and more or less overfits when utilized in easy-to-solve bin picking tasks.

Since RL and IL share the theoretical paradigm of Markovian Decision Processes, scientific approaches combining their target functions exist [17]. Thereby, initial learning from demonstrations is proposed to enable faster positive reward collection. The combined RL and IL approach serves for policy improvement and diversification. This hybrid learning strategy is referred as reward-consistent Imitation Learning [17].

Consequently, although IL and HuITL approaches for teleoperated intervention as well as hybrid RL and IL concepts are already considered in research, more in-depth R&D is required for industrial manipulation scenarios (e. g. bin picking). Related methods are either tailored to specific setups and manipulation scenarios or do not explicitly address IL for manipulation bottlenecks. Furthermore, IL potentials often remain unexploited due to inappropriate VR-HMIs. These are inferior compared to an AV-based real-world environment reconstruction deployed for dataset refinement.

3　Augmented Virtuality-Based Hybrid Manipulation Learning

In the following section, the hybrid RL-IL method involving AV-based input refinement for bin picking scenarios is introduced (see Fig. 1). Subsequently, Fig. 2 describes an architectural concept for method integration along a suggested ramp-up process.

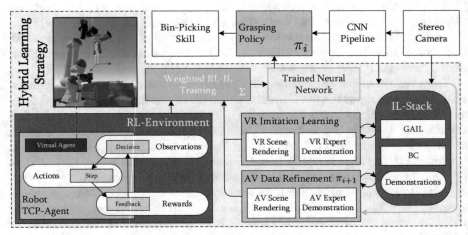

Fig. 1. Hybrid Learning Strategy for weighted Reinforcement- and reward consistent Imitation Learning improving the bin picking grasping policy by utilizing AV-based input refinement.

The proposed hybrid learning strategy (see Fig. 1) is designed to improve grasping policies underlying to autonomous bin picking skills iteratively. In the upper right section of Fig. 1, sensor data obtained from a 3D-RGB camera is processed by a CNN for object localization and pose estimation. This data serves as input for the grasping policy (top).

As hybrid learning strategy (dashed outlined box), DL implemented as weighted RL-IL training (Fig. 1, center, green box) is utilized. The initial training is either performed via conventional RL (Fig. 1, lower left) or via hybrid learning. RL, PPO and SAC algorithms are utilized for training. For hybrid learning, a VR-based IL environment (Fig. 1, bottom, center) serves during virtual commissioning. The latter enriches the initial dataset with further human demonstrations for subsequent training iterations.

A simple RL environment for industrial bin picking serves for digital grasping failure simulation (Fig. 1, lower left). It consists of: a virtual agent (blue) with a collision model, a virtual Small Load Carrier (SLC), the collision environment and virtual objects within the SLC (top, left). The physics engine of the VR is utilized for realistic random multi-object arrangement and filling of the SLC with virtual grasping objects.

The Tool-Center-Point (TCP) of a simulated gripper represents the virtual agent. Continuous actions in form of single translations or rotations within global cartesian space are taken with each step of an episode. Thereby, an episode end is reached as soon as the agent either surpasses a maximum number of steps taken or by receiving a sparse reward. Positive sparse rewards are triggered by the collision of the TCP with grasping areas of an object to grasp. Negative rewards, on the other hand, are triggered by the collision with any environmental element. Optionally, dense rewards are awarded each time the agent frame approximates the grasping area of a target object.

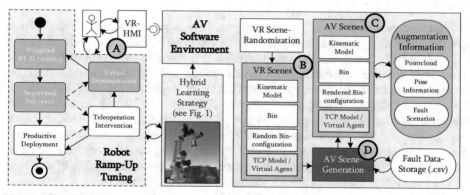

Fig. 2. Method utilization along ramp-up process (left). Sample architecture for method integration involving scene-generation for fault reproduction and a demonstration-HMI (right).

Once initial data generation and virtual refinement are completed (Fig. 1), the grasping policy is deployed to the robot system. In case of failed autonomous grasping attempts during ramp-up or subsequent productive operation, an AV teleoperation interface serves as fault intervention mechanism. Hence, human fault-solving capabilities serve as demonstration input for subsequent weighted RL-IL-retraining.

Human demonstrations (Fig. 2, (A)) are captured during virtual commissioning (VR scenes, (B)) and also during online intervention of occurring grasping failures (AV scenes, (C)). The AV serves as the scene for HuITL input data refinement involving online or recorded offline sensor data. Search space complexity is aimed to be reduced by utilization of an initial demonstration dataset generated in randomized scenarios.

The AV scene generator (D) provides the required storing, respectively snapshotting of virtual as well as real-world robot scenes. This enables in addition asynchronous offline input data refinement (C). Therefore, a digital twin is generated involving stored raw and processed sensor data from a defined area of interest. This includes the point cloud of the SLC area, the environment configuration as well as related component specific information (e. g. derived object classifications, localizations and six Degrees of Freedom (6DoF) pose estimates).

Both VR and AV share in principle the same 3D rendering engine. In AV mode, however, the real-world robot environment is rendered by a soft real-time capable environment reconstruction pipeline. The latter operates with multiple sensor data inputs and their subsequent processing and characterization (e. g. object localization, pose estimation and knowledge augmentation) [5]. The IL stack (Fig. 1, right) used for VR and AV scenes employs reward-consistent IL utilizing BC and GAIL algorithms.

4   Setup and Procedure of Experiments

A demonstrator and a digital twin are set up for method validation. The repository is provided on: https://github.com/FAU-FAPS/hybrid_manipulationlearning_unity3dros.

4.1   Demonstrator Setup

The method as well as the architecture proposed are implemented using Unity3D as a physics simulation engine running on an Industrial PC (IPC) equipped with a NVIDIA RTX 2080 GPU. Unity ML-Agents is utilized as the Deep Learning API for episode design and runtime environment for demonstrations, training and inference. An HTC Vive Pro serves as HMI, thereby the SteamVR Unity plugin and OpenVR are used [5]. On a second IPC, the Robot Operating System (ROS) is installed for communication with the robot. Motion commands generated throughout the HMI are sent to a teleoperation middleware [18]. Point clouds are gathered by a robot wrist mounted stereo camera. Pose estimates of grasping objects are provided by a combined image processing pipeline. Here, the DL-based Frustum PointNets algorithm is complemented by the fifth release of the You Only Look Once (YOLO, v5) algorithm for region proposal.

Regarding robot and grasping components, industrial standard systems and semi-finished goods facilitate comparability. Thereby, a YASKAWA HC10 six-joints articulated robot equipped with a conventional electro-mechanical two-finger gripper is utilized. As major benchmark component a shifting rod from a lorry's limited-slip differential is chosen (see Fig. 3). The method is proven adaptable to components with differing characteristics, as shown and described within the github-repo documentation.

4.2   Procedure of Experiments

For evaluation, three scenes within Unity3D are implemented: Scene A for AV fault-virtualization and -demonstration, Scene B for training based on a defined set of hyperparameters and Scene C for virtual or real robot inference (see Fig. 3).

Fig. 3. Evaluation workflow involving AV fault virtualization, demonstration and training as well as virtual inference and ROS-trajectory export; in addition: real-world bin picking setup with exemplary grasping component "shifting rod" and YASKAWA HC10 articulated robot.

Reward accumulation during training serves as evaluation metric for comparison of:

RL not requiring human demonstrations, weighted RL-IL with 350 initial demonstrations in randomized scenarios as well as weighted RL-IL based on 100 initial demonstrations and 250 fault scenario demonstrations as input refinement. The latter will be referred to as weighted REF.

For every learning method, ten training runs are initiated for statistical validation of results. Each run consists of 3×10^7 steps. Weightings of objective functions are adapted during runs involving IL: BC algorithm is active with a weighted objective strength of 20% until step 1×10^7, whereas GAIL is active with a strength of 10% throughout an entire run. Dense reward functions are active until step 2×10^7. Fault scenario demonstrations are performed in 25 real-world bin picking intervention scenarios caused by failed RL- or RL-IL-based component grasping. For each individual fault scenario, ten subsequent clearance demonstrations are performed. Three volunteers experienced with the system performed the teleoperated demonstration process.

In a second experiment, grasping success rates achieved and resulting grasping durations during inference with the virtual and the real robot system are compared. This is performed for RL, RL-IL and REF. To this end, a sample size of $N = 51$ runs is chosen. The sample size is validated through calculation of the according p-values for RL, RL-IL and REF. For each method, the networks with the most representative accumulated reward achieved during the first experiment is chosen. Every failed grasping in the virtual scene is also counted as a failure for the real-world robot system. Resulting trajectories are not exported in order to prevent collision damage.

RL, RL-IL and REF share the same configuration of hyperparameters. The experiments do not focus on optimization of hyperparameters, hence one configuration leading to satisfying learning has been chosen and remains unchanged for valid comparison. The utilized hyperparameter configuration files will be provided within the repository.

5 Results and Discussion of the Hybrid Learning Evaluation

Good results for all methods are achieved with PPO over SAC. For graphs in Fig. 4 (A), the final mean accumulated reward for RL has a value of 0.765 (Standard Deviation (SD): 0.016), for weighted RL-IL it equals 0.833 (SD: 0.005) and for weighted REF 0.865 (SD: 0.006) is calculated. With these networks, mean grasping success rates during virtual inference over 1×10^7 steps of 76.61% for RL, 82.45% for RL-IL and 84.38% for REF is obtained. While RL learns in a shorter time span, it converges faster.

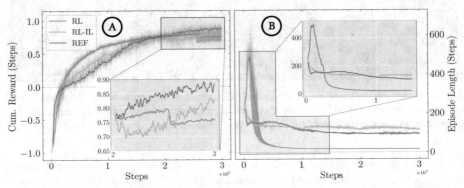

Fig. 4. Mean reward value for cumulative reward (A) and episode length across each step (B) with 2.5% and 97.5% quantiles. Each for five trainings in three learning sets.

For the graphs shown in Fig. 4 (B), the mean episode length given in steps for RL is 34.0 (SD: 0.44), for RL-IL 117.8 (SD: 3.56), and for REF 97.6 (SD: 2.07).

Virtually achieved grasping success rates (see Fig. 5 (A)) show significance as p_{RL} = 0.0078, p_{RL-IL} = 0.0024 and p_{REF} = 0.0064 are calculated for ten inference runs over the sample number N chosen. Experiments using the real robot system show a drop in grasping success rate for RL, which is not as drastic for RL-IL and REF. On the other hand, the drop is smaller for REF in comparison to RL-IL.

Mean grasping durations of 1.1 s (SD: 0.01) for RL, 5.9 s (SD: 0.61) for weighted RL-IL and 4.9 s (SD: 0.63) for weighted REF are obtained. The measured superiority of RL within Fig. 5 (B) matches observations in Fig. 4 (B). While the obtained distribution of grasping durations is more homogeneous for RL-IL and REF, the values are closer for RL. This improves slightly for REF over RL-IL, while some outliers larger than the median are measured. Grasping duration of all methods is adjustable by scaling the trajectory execution. The grasping success itself is not influenced by this.

Graphs plotted in Fig. 4 (A) reveal improved cumulative reward due to hybrid learning in virtual training environments. REF increases rewards even further. Advantages of RL over RL-IL and REF with regard to productivity are in a faster learning between steps 0.3×10^7 and 1×10^7 as well as in shorter episode length (Fig. 4 (B)). The increased episode length in RL-IL and REF is explainable by the more elaborate and tentative nature of the human grasping, being imitated. The superior performance of REF aligns with a shorter episode length during fault scenario demonstration. An underlying reason

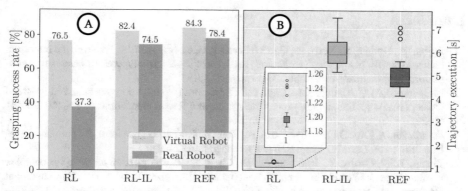

Fig. 5. Results for grasping success rate (A) and grasping duration (B) within the real robot environment in comparison to virtual inference before trajectory export to ROS

therefore could be human routine over repeated demonstrations. Considering computed quantiles across both graphs within Fig. 4, a more stable and reliable learning of RL-IL and REF compared to RL is concluded.

Figure 5 (A) verifies observations made with regard to accumulated reward. During inference, RL-IL and REF reveal a higher grasping success rate compared to RL. The grasping success rate drops while inference for RL utilizing the real robot. This is explained due to simplifications within the training environment. As RL, in contrast to IL, is optimizing manipulation movements, some trajectories learned under a simplified setting do not lead to success in real-world. Nevertheless, RL-IL and REF almost replicated success rates to the real environment. It is concluded that even with more simple but efficient learning models, sufficient results are achieved by IL. In particular, this applies for input refinement by demonstration within intervention scenarios (REF).

6 Conclusion and Outlook

In this work, a method for AV input demonstration refinement improving hybrid manipulation learning is described. Within a bin picking experimental setup involving standard components and systems, the method proves to considerably reduce required component adaptation effort for successful grasping. Thereby, grasping success rates for the application are noticeably increased. Major enablers are the weighted enrichment of RL with IL as well as the successive reward-consistent demonstration within the immersive AV (exploiting human cognitive skills). In contrast to solely RL-based learning, the hybrid strategy is less affected by the domain gap between virtual commissioning and reality. Compared to pure IL, the required effort to generate a sufficient number of annotations for autonomous operation is considerably reduced.

Even though the hybrid learning strategy shows promising outcomes, further R&D is required. Future work will therefore investigate optimization of hyperparameters. In addition, iterative continuous AV input demonstration refinement along the ramp-up process will be emphasized. Further research is required regarding transferability of human demonstrations to similar use cases through higher levels of abstraction.

References

1. Fujita, M., Domae, Y., Noda, A., Garcia Ricardez, G.A., Nagatani, T.: What are the important technologies for bin picking? technology analysis of robots in competitions based on a set of performance metrics. Adv. Robot. **33**(17), 1–15 (2019)
2. Mohammed, M.Q., Chung, K.L., Chyi, C.S.: Review of Deep Reinforcement Learning-Based Object Grasping: Techniques, Open Challenges, and Recommendations **8**, 178450-178481 (2020)
3. Mandlekar, A., Xu, D., Martín-Martín, R., Zhu, Y., Fei-Fei, L., Savarese, S.: Human-in-the-Loop Imitation Learning using Remote Teleoperation (2020)
4. Zhang, T., McCarthy, Z., Jow, O., Lee, D., Chen, X.: Deep imitation learning for complex manipulation tasks from virtual reality teleoperation. In: Lynch, K. (Hrsg) IEEE International Conference on Robotics and Automation (ICRA). 21–25 May 2018, S 5628–5635 (2018)
5. Blank, A., Kosar, E., Karlidag, E., Guo, Q., Kohn, S., Sommer, O.: Hybrid environment reconstruction improving user experience and workload in augmented virtuality teleoperation. Procedia Manufacturing **55**, 40–47 (2021)
6. Pane, Y., Mokhtari, V., Aertbelien, E.: Autonomous Runtime Composition of Sensor-Based Skills Using Concurrent Task Planning. Rob. Aut. Lett. **6**(4), 6481–6488 (2021)
7. Wang, H., Situ, H., Zhuang, C.: 6D Pose estimation for bin-picking based on improved mask R-CNN and DENSEFUSION. In: 2021 26th IEEE International Conference on Emerging Technologies and Factory Automation (ETFA). IEEE, S 1–7 (2021)
8. Lee, S., Lee, Y.: Real-time industrial bin-picking with a hybrid deep learning-engineering approach. In: Lee, W.: (Hrsg) 2020 IEEE International Conference on Big Data and Smart Computing. 19–22 February 2020, Busan, Korea : proceedings, S 584–588 (2020)
9. Yan, W., Xu, Z., Zhou, X., Su, Q., Li, S., Wu, H.: Fast object pose estimation using adaptive threshold for bin-picking. IEEE Access **8**, 63055–63064 (2020)
10. Brito, T., Queiroz, J., Piardi, L., Fernandes, L.A., Lima, J., Leitão, P.: A machine learning approach for collaborative robot smart manufacturing inspection for quality control systems. Procedia Manufacturing **51**, 11–18 (2020). https://doi.org/10.1016/j.promfg.2020.10.003
11. Marchesini, E., Corsi, D., Benfatti, A., Farinelli, A., Fiorini, P.: Double Deep Q-Network for Trajectory Generation of a Commercial 7DOF Redundant Manipulator. In: International Conference on Robotic Computing. 25–27 February 2019, Naples, Italy. IEEE, pp. 421–422 (2019)
12. Lim, J., Lee, J., Lee, C., Kim, G., Cha, Y.: Designing path of collision avoidance for mobile manipulator in worker safety monitoring system using reinforcement learning. In: Ohara, K., Akiyama, Y.: (Hrsg) IEEE-ISR2021 (online). 2021 IEEE International Conference on Intelligence and Safety for Robotics : 4–6 March 2021, Nagoya, Japan, pp. 94–97 (2021)
13. Fan, L., Zhu, Y., Zhu, J.: SURREAL: Open-source reinforcement learning framework and robot manipulation benchmark. In: Conference on Robot Learning, pp. 767–782 (2018)
14. Luebbers, M.B., Brooks, C., Mueller, C.L., Szafir, D., Hayes, B.: Using augmented reality for interactive long-term robot skill maintenance via constrained learning from demonstration. In: 2021 IEEE International Conference on Robotics and Automation (ICRA), (2021)
15. Kim, H., Ohmura, Y., Kuniyoshi, Y.: Gaze-based dual resolution deep imitation learning for high-precision dexterous robot manipulation. Rob. Aut. Lett. **6**(2), 1630–1637 (2021)
16. Ablett, T., Marić, F., Kelly, J.: Fighting Failures with FIRE: Failure Identification to Reduce Expert Burden in Intervention-Based Learning (2020)

17. Kawakami, D., Ishikawa, R., Roxas, M., Sato, Y., Oishi, T.: Learning 6DoF Grasping Using Reward-Consistent Demonstration (2021)
18. Blank, A., Karlidag, E., Zikeli, G.L., Metzner, M., Franke, J.: Adaptive Motion Control Middleware for Teleoperation Based on Pose Tracking and Trajectory Planning. Annals of Scientific Society for Assembly, Handling and Industrial Robotics, Hannover (2021)

Manufacturing Process Optimization via Digital Twins: Definitions and Limitations

Alexios Papacharalampopoulos and Panagiotis Stavropoulos[(⊠)]

Laboratory for Manufacturing Systems and Automation (LMS), Department of Mechanical Engineering and Aeronautics, University of Patras, 26504 Rio Patras, Greece
`pstavr@lms.mech.upatras.gr`

Abstract. Manufacturing process real-time optimization has been one of the main digital twins' operations. It is of utmost importance to the processes, since it enables the feedback of a digital twin towards the real world. However, it is quite difficult to be implemented, since it requires modelling of the process, adaptivity of both the model and the process, real-time communication and link to other functionalities. Under the framework of formalizing such activities, the current work attempts to categorize the types of manufacturing process real-time optimization and show their limitations. For the sake of simplicity, generic process models are adopted and then the requirements for the process control are given, driving the aforementioned definitions. Specific numerical examples are used to illustrate the definitions, while the latter presented herein span all categories of real-time optimization as well as all manufacturing performance indicators. Finally, both mathematically and physics-wise, the limitations are discussed.

Keywords: Manufacturing process real-time optimization · Digital twins

1 Introduction

Manufacturing process real-time optimization (MPRTO) [1] is a concept, that even if it can be considered as well-defined, the emerging Industry 4.0 Key Enabling Technologies have helped in its reappearance into spotlight. Also, it is a fact that the documentation of the respective challenges and aspects that need to be addressed has started some years ago (in 1995) [2], linking the manufacturing process control requirements to the implementation (i.e. communication protocols) and the application itself (manufacturing process addressed).

Furthermore, it can be easily found that it is closely related to classical control theory; if one considers a dynamic (linear) system S of state x, its exact controllability or complete controllability [3] can be defined as follows: "The system S is controllable on an interval $[t_0, t_1]$ if $\forall x_0, x_1 \in R_n$, $\exists$ controllable function $u \in L_2([t_0, t_1]: R_m)$ such that the corresponding state space satisfying $x(t_0) = 0$ also satisfies $x(t_1) = x_1$". This statement means that we can define the behavior of the system, at least for the second time point.

© The Author(s) 2023
K.-Y. Kim et al. (Eds.): FAIM 2022, LNME, pp. 342–350, 2023.
https://doi.org/10.1007/978-3-031-18326-3_33

An additional concept that is highly useful, is that of structural controllability [4, 5], implying that it is the physical structure of the dynamic system, and not its occasional parameter that define the controllability. It is worth noting that these two concepts cannot directly be applied in the manufacturing process area, as this will be shown below.

After having defined the concept of control, a wide area of branches can be defined, such as static controllers, PID controllers, adaptive controllers [6] in the case where the system is unknown, empirical controllers [7] that bypass theoretical calibration, even robust controllers [8] that address the case of systems with uncertainties, either structural or functional. Furthermore, there are secondary types, like the Fuzzy Control which can extend other concepts (i.e. the PID concept [9]).

Modelling plays an important role in the application of this. As shown in literature [8, 10], the uncertainties or the non-linearities can affect the model as well as the controller that will have to be chosen. Also, the manufacturing processes modelling is not straight-forward, as the criteria (i.e. quality) can affect the choice of the I/O model, requiring the definition of the so-called Intermediate Variables [10]. These primarily constitute the first challenges come across in the MPRTO procedure.

Another useful concept can be the stochastic control (i.e. through polytopes) [11], that must not be confused with statistical control. Optimal control [12], next, can be used to address the various extra (manufacturing) criteria, such as energy efficiency, while the practicality and the generality can be addressed through some intelligent control approach, such as neural modelling, or even neural control [13, 14].

Multiple-Input-Multiple-Output (MIMO) control from the classical control theory [15] can be also highly useful in this attempt to address manufacturing process control. This consecutive integration of definitions could be summarized by the depiction of Fig. 1. This is a (non-unique) approach, aiming at describing the digital-twin (DT) based manufacturing process control.

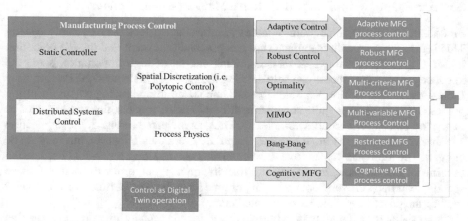

Fig. 1. Framework integrating control theory into DT-driven MPRTO.

Moreover, additional issues, such as security/safety and certification [16] and cloud services could be added in the requirements, leading to the depiction of Fig. 2. These are side-challenges that may appear in the procedure. Additionally, digital twins, being the result of extensive integration [17, 18] are in any case umbrellas of various technologies [19], exploiting to a large extent Artificial Intelligence techniques, among others, so that they are able to handle the data management. The complexity management, thus, may also hinder the modelling and the controlling procedures.

Thus, after reviewing the potential challenges set by the above mentioned technologies, the practicality of control as real-time optimization should be investigated. This paper introduces some concepts expected to be highly useful as a future reference for digital twin design frameworks (due to small execution time). In addition, the MPRTO will be benefited from the integration of control theory techniques (Figs. 1 and 2).

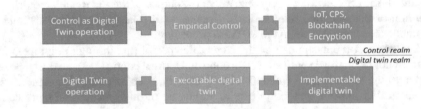

Fig. 2. Additional requirements for digital twin driven MPRTO.

2   Method for Integrating Manufacturing Process Control in DTs

2.1   Input-Output Approach: Adopting a Meta-modelling Framework

Modelling for application of control theory requires the existence of a dynamic system. This is also the case with manufacturing process control, except that the I/O definition is not always straight-forward; metamodelling can be of high importance, since the knowledge about the process can help defining the Process Parameters, the Performance Indicators, as well as Key Variables that are implicated in optimizing a manufacturing process [10]. As there is quite some variability in process models, this is a expected to be a good policy, as it reduces complexity. Thus, as indicated in Fig. 3, the application and the criteria define the Input Key and Output Key variables. However, this may not be enough. The reference values have to be defined. This may have a two-fold implication in terms of objectives: either to perform tracking on a signal, or for the output to have specific features (i.e. overshoot for quality or time scale for process time, etc.). The latter renders the procedure a rather complicated one.

It is crucial for a digital twin to be able to manipulate such information, thus making tacit knowledge on aspects quality or any other manufacturing attribute relevant. In addition, the interpretation of such knowledge into information on the selection of Input Key and Output Key Variables is essential to the whole procedure. Setting the criteria also requires the existence of a specific database, while the communication in real-time needs to handle the monitoring signals as well as the control signals. Finally, the extraction of knowledge to assist explicability is an extra criterion, however, this exceeds the purposes of the current work. It is noted, in any case, that linearization is the primary route to control. Generic non-linear approaches exceed the purposes of this work.

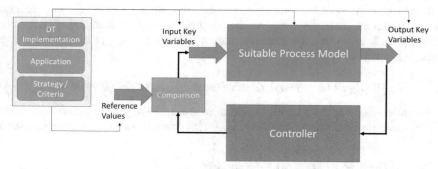

Fig. 3. An Input-Output model of a closed loop MPRTO.

2.2   Theoretical Definitions Facilitating DT-Driven Process Control

The first type of control that would need to be discussed under the concept of manufacturing processes control, would be the Direct Process Control. This is particularly for the case of operation in manufacturing. In this case, under classical control theory, given an integro-differential operator T, an output key variable (signal) y and an input key variable u, without loss of generality, the following systemic description stands.

$$Ty = x \tag{1}$$

It is noted that the operator T can include either a first-order temporal derivative (e.g. in the linear area of laser processes) or a second-order temporal derivative (e.g. in the elastic region in the case of mechanical processes).

In any case, direct control, given a desired output y_0, would be to select an input $x_0 = Ty_0$. However, this is not always applicable, since it implies the use of fields x and y as functions of space. An alternative would be what could be defined as Boundary Direct Control, taking into account the boundary conditions of the problem, e.g. in the case of laser incidence on metals, which is modeled as flux [7]. A mixed type could also be defined, given the fact that in some cases, the hybrid triggering of the field through both types may be needed. An example would be Beer-Lambert penetration of photons due to material-laser combination (i.e. CO_2 of 10.6um wavelength vs. SiO_2/Al_2O_3 material [20]) or due to scattering in powder [21].

Additional requirements could be added on these types of control, leading to Restricted Control, extending the Empirical Control that has been aforementioned. In this case, the key input variable may be subject to extra requirements, such as the positivity, and since formulation may not be applicable or the feasibility of the optimization problem may not be always the case, Heuristic Restricted control may be needed. An example for this case is also given in Sect. 3.1 below.

An additional definition that would be highly relevant in the case of process design is given herein. Given the structural controllability presented in the introduction, it is quite normal to assume that the system structure cannot always originate from a systemic approach, due to the distributed (in space) character of the processes.

Subsequently, in lumped approaches, such as in the case of mechatronics, the structural controllability is of high relevance. To this end, we are looking for the minimum type of system invasions (Δ_1, Δ_2,) so that the final system (A, B, C, D) is (state) controllable (Eq. 2, given that the ρ operator denotes the rank of a matrix and that n is the dimension of square matrix A).

$$\rho[B+\Delta_2, (A+\Delta_1)(B+\Delta_2), (A+\Delta_1)^2(B+\Delta_2), .., (A+\Delta_1)^{n-1}(B+\Delta_2)] = n \tag{2}$$

In addition, it would be highly useful to connect this to graphs annotation, so that the network interventions are also visualized. This could be defined as invasive control.

3 Numerical Results and Discussion

3.1 Application of DT-Driven Control Integrating New Concepts

The current application is related to enforcing direct process control in a simple process. The term simple implies the consideration of a linear area (i.e. Equation 4), while a specific spatio-temporal temperature profile has to be tracked (Eq. 3 describing heating up). There is a first-order temporal derivative, implying an adoption of thermal processes. Figure 4 depicts the outcome of the direct control application, as well as a potential experimental configuration; a metal sheet (blue surface) heated by a distribution of heating/cooling elements (yellow surfaces).

$$\nabla^2 T + Q = k\frac{\partial T}{\partial t} \tag{3}$$

$$T = 500e^{-x^2}(0.2y + 1)\frac{1}{t^2 + 1} \tag{4}$$

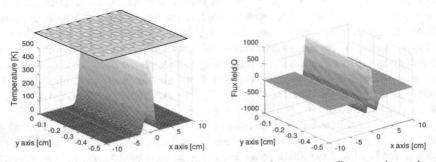

Fig. 4. Application of direct manufacturing process control: a corresponding experimental setup (planar surfaces), the desired spatial profile (left) and the flux required (right) at a given time.

Next, it would be quite interesting to also apply empirical control. In the case below, the temporal profile of Eq. 5 needs to be tracked for the model of Eq. 6 (an ordinary differential equation). The experimental configuration corresponding to this case could be 1D motion control of a machine tool. The requirement is that the input needs to be pulsed (restriction of the motor). After some trial and error (manually), this is achieved, as shown in Fig. 5. The oscillation is imposed by the dynamics of the equation, which is a crucial limitation of control.

$$T = 3te^{-t} \tag{5}$$

$$y(x) + y''(x) + 0.1y'(x) = u(x) \tag{6}$$

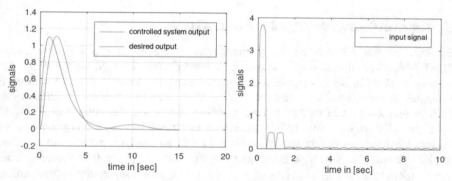

Fig. 5. Empirical direct control: Control performance (left) and input signal (right).

In addition, one could apply also restricted control. This would address machine-related restrictions, on top of aforementioned process-related ones, such as the very dynamics discussed above. For the model of Eq. 6, the control has to fulfil the approximate criterion of Eq. 7, where input has been assumed to be positive, as an extra restriction ($q_n^2 > 0$), using a specific set of pulses, denoted in Eq. 8 and 9 (duration of pulse is t_d and H being the Heaviside function). Results are shown in Fig. 6.

$$f = min_q \sum_{n=1}^{N_G} w_n [T(x_n) + T''(x_n) + 0.1T'(x_n) - q_n^2]^2, T = 60t^3 e^{-2t} \tag{7}$$

$$p(t, t_0) = (H(t - t_0) - H(t - t_0 - t_d))Sin^{0.1}(2\pi \frac{t - t0}{2t_d}) \tag{8}$$

$$u_c(t) = \sum_{n=0}^{N_d} |q_{n+1}|p(t, (n+1)t_d) \tag{9}$$

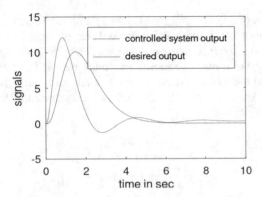

Fig. 6. Empirical restricted control applied.

3.2 Repercussions on the Digital Twins' Workflows

Empirical models and thus empirical manufacturing process control (or equivalently MPRTO based on closed loop schemas) have been proved to be inherently useful for many cases of manufacturing process control, however, the digital twin architecture that is adopted needs to be able to support all this. Thus, it can be claimed that human-in-the-loop optimization of manufacturing processes modelling and control or AI-based digital twins' workflows are of high importance towards achieving such operations. To this end, and to address robust manufacturing, it would be crucial to integrate monitoring capabilities towards identifying the actual thermal model of the part manufactured. This implies taking into account uncertainties in modelling, like temperature dependencies, or spatial variations due to material characteristics. This could be addressed by profiling the surficial temperature distribution at different part planes, and extracting the internal status of the part; implying further integration of AI in the digital twin.

4   Conclusions and Future Outlook

Digital twins are quite hard to be implemented, with control theory interpreted into MPRTO being an integral part of their set of operations. As such, the empirical use of process control needs take into account several aspects. At the same time, the application of empirical process control is promising for a variety of applications, especially in cases of low digital maturity. This can be integrated either in the form of trial and error (automatically or using a human-in-the-loop scenario), or through a specific optimization procedure.

In addition, in the case of designing, invasive control can be a useful definition. This would need, in any case, further elaboration, before suggesting a framework that is implementable. With respect to future outlook, specific data driven models should be elaborated for testing theoretically and practically the use of manufacturing process control, seemingly requiring a long series of tests, for a variety of processes and configurations, as well as AI-based monitoring techniques elaboration, as aforementioned.

Acknowledgements. This work is under the framework of EU Project AVANGARD, receiving funding from the European Union's Horizon 2020 research and innovation program (869986). The dissemination of results herein reflects only the authors' view and the Commission is not responsible for any use that may be made of the information it contains.

References

1. Chryssolouris, G.: Manufacturing Systems: Theory and Practice, 2nd edn. Springer, New York, NY, USA (2006). https://doi.org/10.1007/0-387-28431-1
2. Homepage for critique on National Research Council's Manufacturing Process Controls for the Industries of the Future. https://frankfurtrights.com/Books/Details/manufacturing-pro cess-controls-for-the-industries-of-the-future-17058745. Accessed 1 Feb 2022
3. George, R.K., Iist, T.: Controllability, observability, stability and stabilizability of linear systems, IIT Bombay, Lecture notes (2015)
4. Blackhall, L., Hill, D.J.: On the structural controllability of networks of linear systems. IFAC Proceedings **43**(19), 245–250 (2010)
5. Shields, R., Pearson, J.: Structural controllability of multiinput linear systems. IEEE Trans. Autom. Control **21**(2), 203–212 (1976)
6. Zhou, M., Jing, H., Yang, J., Yao, S., He, L.: An extended adaptive control system for EDM. Procedia CIRP **68**, 672–677 (2018)
7. Papacharalampopoulos, A., Stavropoulos, P., Stavridis, J.: Adaptive control of thermal processes: laser welding and additive manufacturing paradigms. Procedia CIRP **67**, 233–237 (2018)
8. Stavropoulos, P., Papacharalampopoulos, A., Michail, C.K., Chryssolouris, G.: Robust additive manufacturing performance through a control oriented digital twin. Metals **11**(5), 708 (2021)
9. Vasu, S.: Fuzzy PID based adaptive control on industrial robot system. Materials Today Proceedings **5**(5), 13055–13060 (2018)
10. Stavropoulos, P., Foteinopoulos, P., Papapacharalampopoulos, A.: On the impact of additive manufacturing processes complexity on modelling. Appl. Sci. **11**(16), 7743 (2021)
11. Ning, Z., Zhang, L., Mesbah, A., Colaneri, P.: Stability analysis and stabilization of discrete-time non-homogeneous semi-Markov jump linear systems: a polytopic approach. Automatica **120**, 109080 (2020)
12. Papacharalampopoulos, A., Stavridis, J., Stavropoulos, P., Chryssolouris, G.: Cloud-based control of thermal based manufacturing processes. Procedia CIRP **55**, 254–259 (2016)
13. Wong, W.C., Chee, E., Li, J., Wang, X.: Recurrent neural network-based model predictive control for continuous pharmaceutical manufacturing. Mathematics **6**(11), 242 (2018)
14. Kumar, S.S.P., Tulsyan, A., Gopaluni, B., Loewen, P.: A deep learning architecture for predictive control. IFAC-PapersOnLine **51**(18), 512–517 (2018)
15. Jin, X.: Adaptive fixed-time control for MIMO nonlinear systems with asymmetric output constraints using universal barrier functions. IEEE Trans. Autom. Control **64**(7), 3046–3053 (2018)
16. Putz, B., Dietz, M., Empl, P., Pernul, G.: Ethertwin: Blockchain-based secure digital twin information management. Inf. Process. Manage. **58**(1), 102425 (2021)
17. IBM homepage. https://developer.ibm.com/articles/what-are-digital-twins/. Accessed 1 Feb 2022

18. Gunasegaram, D.R., Murphy, A.B., Matthews, M.J., DebRoy, T.: The case for digital twins in metal additive manufacturing. J. Physics: Materials **4**(4), 040401 (2021)
19. Alexopoulos, K., Nikolakis, N., Chryssolouris, G.: Digital twin-driven supervised machine learning for the development of artificial intelligence applications in manufacturing. Int. J. Comput. Integr. Manuf. **33**(5), 429–439 (2020)
20. Lawrence, J., Minami, K., Li, L., Edwards, R.E., Gale, A.W.: Determination of the absorption length of CO_2, Nd: YAG and high power diode laser radiation for a selected grouting material. Appl. Surf. Sci. **186**(1–4), 162–165 (2002)
21. Schneider, M.: Laser cladding with powder, effect of some machining parameters on clad properties. University of Twente. Doctoral dissertation (1998)

The Circular Economy Competence of the Manufacturing Sector — A Case Study

Nillo Adlin[1], Minna Lanz[1], and Mika Lohtander[2]([envelope])

[1] Tampere University, Tampere, Finland
{nillo.adlin,minna.lanz}@tuni.fi
[2] Turku University of Applied Science, Turku, Finland
mika.lohtander@turkuamk.fi

Abstract. Circular economy refers to the intention to overcome the problems in the current production and consumption model. The current model is based on continuous growth and an increasing efficient resource utilization among the industry. Within the circular economy, an organizations are expected to minimize material and energy use, and by design and action reduce the environmental deterioration without restricting economic growth or social and technical progress. This development has been undertaken in many industries, especially in consumer-related businesses, many examples are well documented. However, there are only few studies available concerning the current circular economy activities in the manufacturing industry or their potential. The main goal of this paper is to identify and validate the circular economy readiness between communication and actual action in the circular economy business in manufacturing companies in a regional context. The results of this study will show the capabilities of the manufacturing industry to do business in regional circular economy activities. Two conclusions can be made. Firstly, the results of regional circular economy activities bring valuable information for academics, policymakers, and manufacturing companies. Secondly, the regional baseline of the circular economy can help highlight the current situation and identify the business areas which the next actions should be targeted at.

Keywords: Circular economy · Manufacturing capability

1 Introduction

The balance between industrial evolution and production, and environmental impacts and disposal is crucial for the business performance [8]. Environmental impacts have constantly increased pressure on industrial evolution to avoid things called waste. Eventually, when trying to bend the linear economy to a circular economy it is essential to remember that manufacturing itself is consuming fossil resources. So, even we can extract all used material and transform them into new raw material, we have to also transform our manufacturing in a way that we use only circulated energy sources. Manufacturing is an indispensable element of the innovation chain; it enables technological innovations to be applied in goods and services, making new products affordable and

© The Author(s) 2023
K.-Y. Kim et al. (Eds.): FAIM 2022, LNME, pp. 351–360, 2023.
https://doi.org/10.1007/978-3-031-18326-3_34

accessible to a multitude of consumers and, thus, increasing societal and economic benefits [9]. However, increased competition for access to scarce or critical resources has become another major concern for the manufacturing industry in addition to fulfilling the obligations of environmental legislation at a minimum cost [8]. By promoting the adoption of closing-the-loop production patterns within an economic system, the circular economy aims to increase the efficiency of resource use, with special focus on urban and industrial waste, and to achieve a better balance and harmony between the economy, environment and society [4].

The European Commission adopted the new circular economy action plan (CEAP) in March 2020. It is one of the main building blocks of the European Green Deal, Europe's new agenda for sustainable growth. The EU's transition to a circular economy (CE) will reduce pressure on natural resources and will create sustainable growth and jobs. It is also a prerequisite to achieve the EU's 2050 climate neutrality target and to halt biodiversity loss. The new action plan announces initiatives along the entire life cycle of products. The CEAP promotes CE processes, encourages sustainable consumption, aims at preventing waste and keeping resources in the EU economy for as long as possible [2, 3]. Similarly on national level, several European countries have launched R&D programs to advance different aspects of the circular economy. For example, in Finland, the Finnish innovation fund Sitra released a national roadmap towards the circular economy in 2016 [12]. At the regional level, smart specialization strategies have further defined and implemented EU priorities that strive towards a smart, sustainable and inclusive economy with high employment, productivity and social cohesion by the year 2020 [15]. Aligning with the European Commission's intentions, various regions have also increasingly raised the circular economy as part of their smart specializations.

2   Motivation and Research Question

The circular economy is understood well from an environmental sustainability perspective and is very visible in the waste management, recycling and bioeconomy industries. However, there is a lack of visibility in existing circular economy business activities. The circular economy conceptual frameworks are either too general to address the circular economy in manufacturing or they are too focused on a certain aspect related to the circular economy, such as remanufacturing or maintenance, and do not consider the relevance of the circular economy concept as a whole. The the main goal of this paper is to identify and validate a circular economy readiness between communication and actual action in the circular economy business in manufacturing companies in a regional context. This kind of readiness level study is a useful tool for public authorities and manufacturing companies to better understand existing situation to communicate fact based information and facilitate circular economy business development. We set the following research questions: 1.)What are the circular economy baselines in different regions in manufacturing? 2.) How to separate the existing capability from wishes?

3   Literature Review

Traditionally, the circular economy has been seen as a waste management approach. However, as recent studies such as Ghisellini et. al [4] show, the view is becoming wider by Leider [8]. Repurposing, remanufacturing or even repair could be the best strategy based on King et. al. [9] studies. From the perspective of production and operations research, the theoretical framework of sustainable manufacturing is slowly emerging to be part of the circular economy.

According to the Ellen MacArthur Foundation [1], the circular economy is restorative and regenerative by design. It can be viewed an economic strategy that suggests new ways to transform the current and predominantly linear system of consumption into a circular one while achieving economic sustainability with much needed material savings. By relying on a system-wide innovation, the circular economy aims not only to redefine products and services by minimizing negative impacts but, like Leider [8] explain, also to reduce solid waste, landfills and emissions through such activities as reuse, remanufacturing and/or recycling.

Circular economy business models fall in two groups. The first group encourage reuse and extended service life through repair, remanufacturing, upgrades and retrofits. The second group turn old goods into new resources by recycling the materials. Such a model would change the economic logic because it replaces production with sufficiency: reuse what you can, recycle what cannot be reused, repair what is broken and remanufacture what cannot be repaired [13, 14]. Lieder et al., [8] shows how concurrent engineering will take bigger role in modern engineering. Prieto-Sandoval et al., [10] proposed new kind of thinking for circular economy. As Korhonen et. al, [7] explain the circular economy concept has been created mainly practiners and policymakers. In general, it is accepted that the circular economy is seen as a transformation from a traditional linear economy to circular [6, 7, 10]. In this world view the linear and circular economies are not seen as opposites. Instead, they complement each other. Already in manufacturing, we can see different levels of circularity taking in place when old equipment like engines and powerelectronic components are remanufactured and repurposed. The circular economy can potentially close the loop in materials consumption at the end of the product lifecycle. Ideally, it enables a continuous cycle of resources without the need for disposing used material and extracting new resources to product lifecycles Signh [11]. Recently, the idea of a closed loop concept, especially in manufacturing, has been extended to other kind of resources, such as information and energy Ghisellini [4].

4   Research Methodology

This research is a qualitative study, literature review of circular economy baselines and data analysis based on the expert interviews and data analysis. The literature review was done mainly with Scopus database in 2019–2021. Searching phrases included terms like; circular economy, circular economy in manufacturing, modern manufacturing, material consumption, distributed manufacturing, reuse, repurpose and recycle. The case research included 4 main steps 1) data collection (incl. Collection of secondary data of manu.companies from available sources); 2) data mapping and interpretation (incl. Mapping regional CE acitivites among manufacturing companies based on qualitative analysis).3) Case study among two regions, and 4) analysis and further recommendation. The data search was related to the companies who had participated to university collaboration in the recent 5 years and were located in Tampere region and/or the South-Karelia region [15, 16].

The research data used was i) general information of company specific data; ii) the positions in value chain classifying companies within the 10 different roles of circular economy; and ii) the levels of activity in 15 different application domains both in the linear and the circular economy. In relation to the utilized circular economy conceptual framework, repurpose was dropped out in the early phase of data collection, as there was no sign of its industrial relevance. The application domains of design, manufacture and disposal were added to the data collection to indicate the more traditional establishments of the studied companies. In addition, recovery water treatment and energy recovery were distinguished separately (Table 1).

Table 1. Principles for data collection

Principle	Description
Scope of the mapping	All of the identified value chain and circular economy activities have to be done within a specified scope on a macro level, such as a regional, national or city level. The chosen companies have to have some activity, such as a business line, within the chosen macro area
The data sources for collecting company information	Official company websites, company registries, annual reports, newspapers and online news. In newspaper and online news, national and regional sources are emphasized
Company specific data	Name of the company, location of company's head office/regional office, sector using national TOL2008, compatible with NACE coding, turnover, number of personnel, input materials in production, and output materials/products in production (EC, 2017a)
Position in the circular value chain	Company can locate in several roles in the value chain framework, but the regional primary role has to be indicated for the purpose of clarifying the visualization
Circular economy level of activity within the application domains	**1 point:** Company acknowledges the existence of the specific application domain. Relevant word or concept is mentioned, but there is no proper indication on utilizing the application domain or the activity is outsourced **3 points:** Company utilizes the application domain **9 points:** Company utilizes the application domain in its core business or key product, or it shows to invest in or develop the specific application domain

Position in value chain \ Application domain	Design	Manufacturing	Share	Maintenance	Repair	Modernization	Reuse	Refurbish	Remanufacturing	Recycle close loop	Recycle open loop	Recovery	Water treatment	Energy recovery	Disposal	Σ
Gatherer of Core Resources	0	0	0	0	0	0	0	0	0	0	0	0	0	0	0	0
Primary Material Processor	0	0	0	0	0	0	6	0	0	24	24	28	18	21	21	142
Parts manufacturer(Suppliers)	81	99	0	27	10	16	0	9	9	3	2	0	0	0	0	256
Product manufacturer (OEM)	138	104	27	92	20	43	11	3	15	10	6	0	0	0	0	469
Packaging and Distributor	9	9	27	39	15	10	21	0	0	3	0	0	0	3	0	136
Service provider (engineering)	102	48	60	39	3	21	0	0	6	9	0	0	0	0	0	288
Service provider (life cycle)	57	47	51	93	43	36	14	12	15	9	0	0	0	3	0	380
User	0	0	54	21	12	9	18	0	0	0	0	0	0	0	0	114
Collector	3	9	18	21	9	0	15	0	0	43	51	28	18	15	31	261
Disposer	0	0	0	3	3	0	6	0	0	34	42	19	0	12	31	150
Σ	390	316	237	335	115	135	91	24	45	135	125	75	36	54	83	

Fig. 1. An example of circular economy activity in value chain positions

Figure 1 illustrates the matrix structure to combine the collected data. In this study, the Eurostat nomenclature of territorial units for statistics NUTS level 3 was used. As a result, a map, called the regional circular economy profile, illustrates macro-level activity. The rows indicate the different roles in the value chain, which are linked to columns that represent the different application domains in the linear and circular economy. Primary data presentation indicates the accumulation of regional circular economy activity. In this approach, all of the collected data from different companies are summed up to make visible the overall levels of circular economy activity in each application domain.

5   Results

For the Tampere region, we collected data from 62 companies in the manufacturing sector or directly linked to manufacturing value chains; 49 of the companies are registered in the Tampere region and 13 had side offices or active business lines in the region. The companies can be further classified as follows: 8 primary material processors, 12 part manufacturers, 16 product manufacturers, 5 distributors, 13 service providers in engineering design and software, 15 lifecycle service providers, 6 business to business users, 12 collectors and 7 disposers. Machinery and equipment, and waste collection, treatment and disposal activities—the material recovery (C28, C38 in Nace coding) sector was represented by 17 companies (NACE, 2010), Fig. 2. This sector is specifically strong in this region. The second largest sector represented was wholesale trade, excluding motor vehicles and motorcycles. This was closely followed by computer programming, consultancy and related activities by 6 companies. The latter sector is also one of the region's traditionally strong sectors in relation to other regions in Finland.

In the mapping of Tampere region's manufacturing sector—the manufacture of Based on the collected data, we found the regional distinctive profiles of the manufacturing sector's circular economy activity, which is illustrated in Fig. 4. Results indicates that the Tampere region has strong levels of design and manufacturing, which aligns with our original expectations. Overall, the most active circular economy domains were identified in maintenance, share and recycling. Concerning the individual circular economy application domains, the key role involves the product manufacturers. Within this domain,

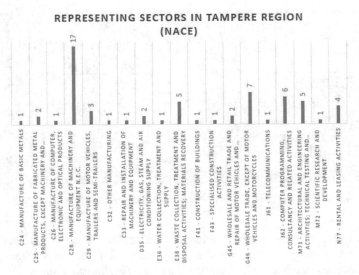

Fig. 2. Sectoral allocation of companies in the Tampere region study

maintenance business is the most active application domain and the area of investment; however, in many occasions, companies seem to include refurbishment, repair and remanufacturing as an integral part of their maintenance service.We assume that for this reason, refurbishing, repairing or remanufacturing may not be visible, although each can be actively used. In addition, product manufacturers are moderately active in modernizing their existing products. This is especially visible in the case of machinery products that have relatively long lifecycles. Accordingly, part manufacturers in general seem not to utilize circular economy actively, although few exceptions do apply (Fig. 3).

Service providers (engineering) were highly active in the share application domain and moderately active in maintenance. Most of the identified activity in practice concerned new digital solutions or services, which add value to the existing machinery through better capacity, improved utilization rates and anticipatory maintenance. Therefore, in many cases, share and maintain application domains emerged. Correspondingly, we identified the service providers (lifecycle) are highly active in the maintenance domain and moderately active in the share, repair and modernization domains. Lifecycle service providers tend to have these two services separated. Packagers and distributors are highly active in the maintenance, share and reuse domains and are moderately active in the repair and modernization domains, as they tend to provide one-stop shop services for customers to run the distributed products and redistribute old products. All the users in the case study are business-to-business renting companies, and, therefore, the share domain is highly active. In addition, the maintenance, reuse and repair domains are moderately active, as the companies benefit for longer lasting products, and later they can resell old machinery.

For the South Karelia region analysis listed 68 companies mainly comprising of machinery companies around the region's capital city of Lappeenranta; 12 of the companies are large enterprises, and 56 represent small and medium sized companies. Looking

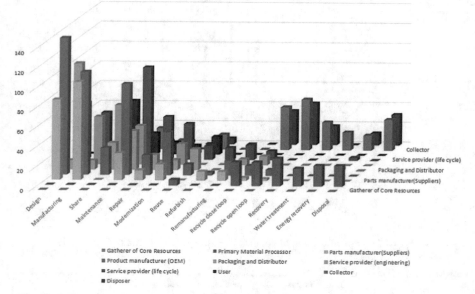

Fig. 3. Summary of Tampere region's circular economy profile in the manufacturing sector

at the region's sectoral allocation, most of the manufacturing companies, 19 in total, represent the manufacturing sector of fabricated metal products, except machinery and equipment (C25 in NACE). The second largest list of representatives were from the sector manufacturing machinery and equipment (C28 in NACE). Sectoral allocation in relation to NACE is illustrated in Fig. 4. What is notable in this sectoral allocation is the importance of sectors C16—manufacturing of wood products, except furniture; manufacturing of straw articles and plaiting materials—and C17, the manufacturing of paper and paper products. The large enterprises in these sectors spread across a wide range of different positions in the circular value chain. According to the identified regional profile in Fig. 5, this region has a strong level of manufacturing activity, and almost half of the companies have in-house design. In general, the most active circular economy domains are in maintenance, recycle close loop and reuse. The bio-economy refers to wood-based and biodegradable products, and it, seemingly, has a great importance in the region's circular economy context. The most active positions in the circular economy value chains are part manufacturers and engineering consulting service providers. Both are moderately active in the maintenance business.

Accordingly, the primary material processors are moderately active in the recycle close loop, maintenance and reuse domains. Surprisingly, product manufacturers in general seem to be only lightly active in the given circular economy domains, although the companies are high in number in the region. Service providers in for the product lifecycle are low in numbers and only moderately active in maintenance. Gatherers of core resources are highly active in the recycle close loop, reuse and maintenance domains

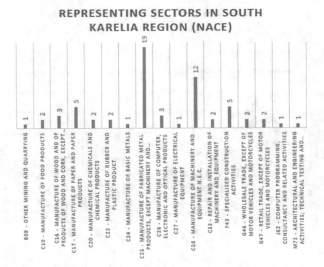

Fig. 4. Sectoral allocation of companies in the South Karelia region's study

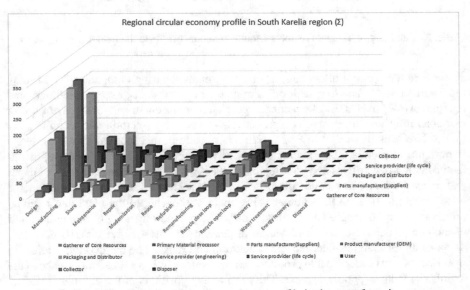

Fig. 5. South Karelia region's circular economy profile in the manufacturing sector

and moderately active in the share and repair domains. Collectors are highly active in the closed loop recycling, reuse and maintenance domains and moderately active in share and repair domains. Packagers and distributors and users (business-2-business) represent the same forest industry companies and are highly active in the maintenance, reuse, recycle in close loop, share and repair domains. Finally, disposers that are represented mainly by the two large enterprises from forest industry are highly active in the maintenance, repair, reuse and recycle close loop domains, as discussed in the 9R approach from Jawahir et. al. [5].

6 Findings

The research provided answers to RQ1 "what are the circular economy baselines in the region" by looking at the regional circular economy profiles and comparing them, we can make generalizable remarks. In the Tampere region, machinery companies are naturally orienting themselves towards maintenance business. There is a growing interest towards service business in general, which extends the current traditional maintenance and repair shop activities towards more digitally advanced and proactive prevention of product failure. In the Tampere region, the cross-sectoral possibilities of combining regional strengths in machinery production, information and communication technology (ICT) and engineering design make the sharing business models a fruitful area to invest in. In Tampere region waste management, water management and recycling sectors, a connection between waste and the circular economy has raised attention. Additionally, remanufacturing is gaining business potential, as there are few pioneering companies in the region.

The South Karelia region had fewer large or medium sized enterprises and fewer original equipment manufacturing companies. Thus, the region is more dependent on part manufacturers, which are dependent on subcontracts. Therefore, there is relatively little orientation towards industrial services. Instead, the region's manufacturing sector still focuses on investment in production technologies. In the South-Karelia region there is a strong dependence on the few large forest companies, which characterize the regional manufacturing sector. This makes the region naturally orientate towards material flows in recycling. The importance of biomaterials is explicit in South Karelia's baseline. South Karelia, the increasing consumption of packaging materials, both plastics and bio-based alternatives, makes the area of packaging reuse and recycle an especially important area to focus on.

For the RQ2 "How to separate existing capability from wishes" the partial answer was reched. The visualisation of circular economy activities in manufacturing companies. Based on the experience gained from these case studies, many of the mapped circular economy application domains are relevant to manufacturing. It has to be noted that in general, manufacturing companies do not yet identify themselves as practicing circular economy business models. For manufacturing companies, the most relevant and clear application domains were maintenance, repair, modernization, reuse (resales) and remanufacture. Based on the finding from the research, material recycling is evidently not a main stream business model within the collected data context.

Therefore, it is deduced that circular economy is in very early phase among companies. Research institutes, universities and news paper are ahead of reality. Based on realized and practical information the gap between the circular economy theory and real manufacturing are far from each other.

References

1. Ellen Macarthur Foundation: Delivering the circular economy - A toolkit for policymakers, p. 177 (2016)
2. European Commission: A new Circular Economy Action Plan, COM(2020) 98 final (2020)
3. European Commission: Communication from the commission to the european parliament, the council, the european economic and social committee and the committee of the regions - Closing the loop - An EU action plan for the Circular Economy. COM/2015/0614 final (2015)
4. Ghisellini, P., Cialani, C., Ulgiati, S.: A review on circular economy: the expected transition to a balanced interplay of environmental and economic systems. J. Clean. Prod. **114**, 11–30 (2016)
5. Jawahir, I.S., Bradley, R.: Technological elements of circular economy and the principles of 6R-based closed-loop material flow in sustainable manufacturing. Procedia CIRP **40**, 103–108 (2016)
6. Kirchherr, J., Reike, D., Hekkert, M.: Conceptualizing the circular economy: an analysis of 114 definitions. Resour. Conserv. Recycl. **127**, 221–232 (2017)
7. Korhonen, J., Honkasalo, A., Seppälä, J.: Circular economy: the concept and its limitations. Ecol. Econ. **143**, 37–46 (2018)
8. Leider, M., Rashid, A.: Towards circular economy implementation: a comprehensive review in context of manufacturing industry. J. Clean. Prod. **115**, 36–51 (2016)
9. Manufuture High level group: Manufuture Vision 2030: Competitive, Sustainable And Resilient European Manufacturin (2018)
10. Prieto-Sandoval, V., Jaca, C., Ormazabal, M.: Towards a consensus on the circular economy. J. Cleaner Production (2017)
11. Singh, J., Ordonez, I.: Resource recovery from post-consumer waste: important lessons for the upcoming circular economy. J. Clean. Prod. **134**, 342–353 (2016)
12. Sitra: Leading the cycle – Finnish road map to a circular economy 2016–2025. Sitra Studies 121 (2016). https://media.sitra.fi/2017/02/28142644/Selvityksia121.pdf. Accessed 12 May 2022
13. Stahel, W.R.: The business angle of a circular economy for higher competitiveness, higher resource security and material efficiency. In: Proceedings of the A New Dynamic and Effective Business in a Circular Economy. Ellen Macarthur Foundation (2013)
14. Szira, Z., Alghamdi, H., Othmar, G., Varga, E.: Analyzing waste management with respect to circular economy. Hungarian Agric. Eng. **30**, 75–86 (2016)
15. Tampere Region: Smart Specialization Strategy (2017). Accessed 2 Sept 2017
16. Region, S.-K.: Smart specialization strategy (2017). Accessed 24 Sept 2018

A Framework for Manufacturing System Reconfiguration Based on Artificial Intelligence and Digital Twin

Fan Mo[1(✉)], Jack C. Chaplin[1], David Sanderson[1], Hamood Ur Rehman[1],
Fabio Marco Monetti[2], Antonio Maffei[2], and Svetan Ratchev[1]

[1] Institute for Advanced Manufacturing, University of Nottingham,
Nottingham NG8 1BB, UK
fan.mo@nottingham.ac.uk
[2] Department of Production Engineering, KTH Royal Institute of Technology,
Brinellvägenn 68, 114 28 Stockholm, Sweden

Abstract. The application of digital twins and artificial intelligence to manufacturing has shown potential in improving system resilience, responsiveness, and productivity. Traditional digital twin approaches are generally applied to single, static systems to enhance a specific process. This paper proposes a framework that applies digital twins and artificial intelligence to manufacturing system reconfiguration, i.e., the layout, process parameters, and operation time of multiple assets, to enable system decision making based on varying demands from the customer or market. A digital twin environment has been developed to simulate the manufacturing process with multiple industrial robots performing various tasks. A data pipeline is built in the digital twin with an API (application programming interface) to enable the integration of artificial intelligence. Artificial intelligence methods are used to optimise the digital twin environment and improve system decision-making. Finally, a multi-agent program approach shows the communication and negotiation status between different agents to determine the optimal configuration for a manufacturing system to solve varying problems. Compared with previous research, this framework combines distributed intelligence, artificial intelligence for decision making, and production line optimisation that can be widely applied in modern reactive manufacturing applications.

Keywords: Artificial intelligence · Multi-agent programming · Digital twin · Process simulation

1 Introduction

Digital transformation is the integration of digital technologies in a company to improve performance and productivity. Digital transformations can be applied to

Supported by DiManD Innovative Training Network (ITN) project funded by the European Union through the Marie Sklodowska-CurieInnovative Training Networks (H2020-MSCA-ITN-2018) under grant agreement no. 814078.

K.-Y. Kim et al. (Eds.): FAIM 2022, LNME, pp. 361–373, 2023.
https://doi.org/10.1007/978-3-031-18326-3_35

any area of the business, and impacts the manufacturing shop floor, the business processes and models used, and the overall customer experience [1]. Digital twins and simulations are increasingly used to plan and optimise processes [2], and given the large volumes of data generated by these, there is a push toward using smart data analytics tools to analyze the stream of data and let managers take responsible, rapid decisions to regulate and improve productivity [3]. Machine Learning (ML) algorithms provide solutions for processing large volumes of data [4], for example, processing and making predictions with sensor signals [5].

By applying simulation and digital twins, it is more realistic to implement reconfigurable systems and customized, demand-based production. The application of artificial intelligence to the digital twin-based streams of data allows companies to analyze and react to sudden changes in demand or disruptions [2]. The reconfiguration of production processes in response to external changes is a difficult challenge. The scope of system reconfiguration includes system layout, process parameters, operation time of multiple assets, sequence of the operations, and material handling systems.

This paper proposes a new framework to enable the system to do different kinds of reconfiguration through the application of existing artificial intelligence algorithms, such as the genetic algorithm and particle swarm optimization (PSO), or newly developed algorithms. With this framework, the manufacturing system reconfiguration can be simulated at first in the virtual environment, and then transferred to the actual production plant. We developed a digital twin environment that simulates the manufacturing process with multiple industrial robots performing multiple tasks. A data pipeline with an API that integrates artificial intelligence is built in the digital twin. With the help of artificial intelligence and the digital twin, the performance of the decision-making process and the reconfiguration process is improved. Finally, a multi-agent program to communicate between different agents and share the negotiation status that determines an optimal configuration for the manufacturing system is made. Section 2 presents a brief overview of works related to the same topic. Section 3 introduces a new framework for manufacturing system reconfiguration and a framework of the reconfiguration engine. Section 4 describes an application with multiple robots, which applies the previously mentioned reconfiguration framework. Section 5 presents the conclusion and suggestions for future research.

2   Related Work

Increasing market demand for mass customization is forcing manufacturers to adopt rapid reconfiguration capabilities, allowing for the sudden change of system configuration – whilst maintaining full system effectiveness – in the event of unpredictable customer demands, line failures or need for maintenance [6]. There is a large body of research on automation agent architectures – distributed artificially intelligent programs that collaborate to solve problems – that allow for 'on the fly' reconfiguration of manufacturing systems, exploiting asset flexibility and control to change the behaviour of the manufacturing system [7]. Multi-agent systems and the concept of cyber-physical systems (CPS) were key contributors to research in the reconfiguration of manufacturing environments at the beginning of Industry 4.0 [8,9].

Emerging artificial intelligence techniques are now being applied to the problem of manufacturing reconfiguration [10]. Human-machine collaboration is an increasingly important concept in flexible and reconfigurable manufacturing systems, and artificial intelligence is also useful here, as well as considerations of safety in such environments [11].

Another important topic that emerged from smart manufacturing's adaptability to demands and the consequent issues of decision-making based on data is the self-repair ability of smart systems; a recent study handled it again using an artificial intelligence approach to select the best strategy, based on product and module swapping, operation rescheduling and reconfiguration, to significantly reduce the capacity loss [12]. The volume of data coming from this knowledge base is a challenge to handle, but also represents an opportunity exploitable for the development of the Digital Twin (DT) concept, a virtual representation of a physical asset where both counterparts are connected to each other and are dynamically evolving through the whole life cycle [13]. DT emerged as a concept for monitoring processes and is evolving towards being an instrument used for reconfiguration. They can be used in parallel with an artificial intelligence approach to reconfiguring human-robot collaborative assembly lines [14,15]; they model the real world and exchange data back and forth with the various assets of a manufacturing environment to be able to affect the system's decisions and behaviour [16]; and they can represent the base on which to build smart, fast and responsive manufacturing systems based on robotics assets [17,18].

An implication of these proposed approaches is that integrating complex adaptive systems with artificial intelligence techniques, combined with flexible multi-functional manufacturing assets such as robotics, will result in a continuously evolving and changing knowledge base. This volume of data in this knowledge base is both a challenge, but also an opportunity for enabling real-time learning of new reconfigurations and behaviours in manufacturing areas such as production, logistics, and assembly. All the different methods of artificial intelligence aim to achieve the best possible performance in terms of scheduling, efficiency, and productivity, helping to apply autonomous technologies and behaviour to robots, manipulators, and eventually the whole system. Given the increasing availability of data, and the ability to be iteratively applied, this approach is also self-improving [19].

3   Framework

3.1   Overview of Manufacturing System Reconfiguration System

This section proposes a framework for the intelligent optimisation of manufacturing system reconfiguration management. It also provides a concept of the reconfiguration engine with multi-agent system integration. This reconfiguration framework covers two levels; the manufacturing execution level, and the controller level. The manufacturing execution level interacts in real-time with the controller level to deal with the customer's orders and make decisions. The controller level will receive the information from the manufacturing execution level and send the related control command to the physical device. The reconfiguration engine is a digital environment with multi-agent system integration and enables the system to determine the optimum reconfiguration based on the selection criteria it receives from the decision engine and the reconfiguration data bank (Fig. 1). Layout, cost, time, sequence of the operation, and others can be used as the selection criteria. This framework consists of the following steps:

(1) **Product requirement**

The first step in the framework is to generate the product requirements, as these will specify the operations the system must perform. The bill of resources and the bill of processes will be generated, breaking the requirements into the equipment needed, and the order of processes required to manufacture the product. This step can be done with several different approaches, including the customer specifying exactly the bill of resources and bill of the process themselves, manually creating the bill of resources and process, or utilising a semantic model and matching algorithm to generate these automatically.

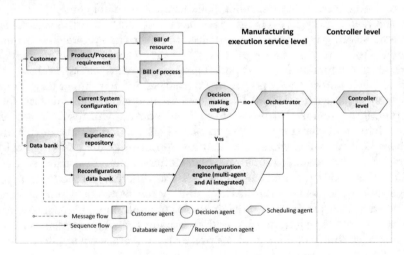

Fig. 1. Overview for system reconfiguration

(2) **Experience data bank**

The data bank is a database for storing data from (i) the current system configuration, (ii) prior experience and configurations, and (iii) stored algorithms and simulations required to support reconfiguration decision making. The current system configuration data (i) consists of information about the current system, such as process information, equipment information, layout information, machine runtime condition, and value-chain information.

Standard data models are used to represent the current system configuration information. For instance, asset administration shells can be used to represent system configuration information [20]. The experience repository (ii) is a data bank of historical data for helping the system's decision making. For instance, configuration information from previous system reconfigurations and processes and their performance. The reconfiguration data bank (iii) contains the offline data components of digital twins for the current use case, the simulation part of the digital twin optimisation algorithms, simulation results from the previous system configurations, and libraries of PLC code for updating the control programs. To store the offline data of the digital twin, multiple approaches are used. For example, the simulation model, the interfaces to access data in the live physical system, and other information such as device information, operation information, and signal information are all stored in the database for future use.

(3) **Decision-making engine**

The information from the bill of resource/process, system configuration, and the experience repository are analyzed by the decision-making engine. The decision-making engine determines if a reconfiguration is needed based on different aspects, such as cost, time, utilization of the equipment, and productivity. If a reconfiguration is needed, the decision-making engine will give recommendations via an agent to the reconfiguration engine about the reconfiguration approach to be used based on previous successful experiences. The reconfiguration approach can be an algorithm or a combination of algorithms that can be best utilized.

(4) **Reconfiguration engine**

The reconfiguration engine is used by the decision-making engine to begin the reconfiguration process. Depending on the type of system being reconfigured, different approaches are selected for use, such as inbound reconfigurable transportation systems [21], system layout problems [22], reconfigurable manufacturing system configuration selection [23], or planning & scheduling in reconfigurable manufacturing systems [24, 25]. The optimized approach will be sent to the orchestrator after the reconfiguration has been made. Additionally, the optimized approach will also be stored in the data bank for future reference.

(5) **Orchestrator**

The orchestrator has a similar function to an ERP (Enterprise resource planning system) and can work with existing ERP systems. The orchestrator is responsible for arranging, controlling, and optimizing workloads in production. The orchestrator will allocate manufacturing resources, plan human resources, and plan and schedule production processes, based on the information received from the decision-making and reconfiguration engines.

(6) **Multi-agent integration**

To make the system resilient and robust, the multi-agent approach is used. There are five types of agents in this framework: customer agent, database agent, decision agent, reconfiguration agent and scheduling agent (see Table 1).

Table 1. Agent description

Agent type	Functionalities
Customer agent	The customer agent handles new product requests and may read the information from the current system and get the necessary information from the data bank to generate a bill of process
Database agent	The database agent is responsible for generating data models and storing information. It responds to requests from the decision making engine to supply the necessary data for the decisions
Decision agent	The decision agent will submit requests for relevant information from the database agent and customer agent and will decide if a reconfiguration is needed based on the decision criteria
Reconfiguration agent	The reconfiguration agent responds to requests from the decision agent for changes to the system configuration. It invokes the artificial intelligence algorithms as required based on different optimization scenarios
Scheduling agent	The scheduling agent receives the output from the decision agent and the reconfiguration agent and determines how the system should respond to enact the new configuration at the control level

3.2   Reconfiguration Engine Algorithm Based on Digital Twin

The reconfiguration engine can select multiple algorithms (either provided by the user or previously stored in the data bank) to determine the optimum system configuration for a given product. In complex robotic manufacturing use cases, many types of optimisation decisions are needed to specify the reconfiguration. These include robot locations, robot paths, and the operation sequence. A digital twin is needed to simulate the process to test multiple scenarios and optimise it. Here we present a framework for utilising digital twins and multi-objective optimisation algorithms. The framework consists of three parts: the *digital twin*, the *training gym*, and the *communication block*.

The *digital twin* environment will be taken from the experience repository. Digital twins themselves are a combination of three primary components: a virtual twin, a physical twin (i.e. the physical manufacturing system), and the data flow component. The data flow component is used to feed data from a physical twin to its virtual twin and returns information from the virtual twin.

The digital twin and the training gym share data via the *communication block*. The communication can be sockets, OPC-UA, MQTT, TCP/IP or other communication protocols depending on the application domain, and can be dynamically chosen by the reconfiguration engine depending on the system context. For instance: if a cloud-based environment is needed, then the MQTT approach could be used to enable communication.

In the *training gym*, different artificial intelligence algorithms and approaches are used to optimize the simulation in the digital twin to meet the different optimization targets. For example, if the optimization target is multi-objective, then multi-objective optimization algorithms are selected. If the manufacturing system looks to optimise a single parameter with the new product – such as faster operation, or lower cost – then reinforcement learning can be used in the training gym. For highly complex or difficult problems, deep reinforcement learning will be suggested by the decision engine. After the optimized result is found and evaluated successfully in the simulation environment, the optimized result (such as the optimized robot program, robot path, and the structure of the production line) will be applied to the physical device via the orchestrator. The orchestrator may make direct changes to PLC code or robot controllers to instantiate the change or make recommendations to human operators to make the changes manually. The orchestrator will receive change requests via the communication protocol suggested by the reconfiguration engine.

4   Application

This section describes a demonstrator used to verify the effectiveness of the proposed approach in achieving system reconfiguration. The following example illustrates the reconfiguration of a pick and weld production station, using Siemens Tecnomatix Process Simulate as the digital twin enabler. As shown in Fig. 2, this use case consists of two KUKA KR2700ultra robots, one conveyor for transporting the product, one welding station, and one work table for storing the parts after welding. For the two KUKA robots, one robot (robot 1) is armed with the pick and place end effector. The other robot (robot 2) is equipped with an arc welding gun.

Normally, robots are drilled into concrete or mounted to metal plates, so cannot quickly be moved. This application applies a modern reconfigurable approach that allows the robot to find the optimized position at first in the simulation environment and then applied to the real physical device. This is either done before the cell is commissioned to determine the optimal location, for situations where the robot can be moved to different locations with a reconfigurable flooring solution (which can be seen in Fig. 2), or the robot will (depending on payload restrictions) be placed on an automated ground vehicle (AGV) and moved to the new position after it receives the optimized location. The process consists of five steps:

(1) The product (yellow part) is placed on the conveyor by the previous process in the production line.
(2) The part will be transported via the conveyor to robot 1 for picking.
(3) Robot 1 will pick the part up and move it to the welding station.
(4) After the part is successfully put on the welding station, robot 2 will perform arc welding on the product.
(5) After welding, robot 1 will pick the product up and put it on the worktable, and return to the home position.

In this use case, it is assumed that the customer wants a welded cubic product and the decision-making engine has already decided that a system layout reconfiguration is needed. Both robots need to be relocated to their optimal locations (based on the suggestions from the reconfiguration engine) in order to optimize the production time of the process. Optimizing process time is a common industrial requirement, for example, to minimize cycle time or meet a takt time. An additional constraint on the optimisation is ensuring that neither robot collides with other parts during operations.

The application for the framework of the system reconfiguration engine for this use case is shown in Fig. 3. The digital twin environment will send operation time and collision situation to the training gym via a socket (C#) based on Tecnomatix .NET API. Tecnomatix .NET API is connected to the Tecnomatix Process Simulate simulation environment with Tecnomatix .NET viewers.

In the training gym, after optimization via the selected artificial intelligence approach, the updated robot location will be sent to the digital twin, to get a new operation time and check for collisions. After all the iterations, an optimised location of the two robots will be found. The results (locations of the robots) will at first be validated in the digital twin. Then the final optimized robot locations will be sent to the orchestrator. It is the role of the orchestrator to communicate the robot locations to the relevant resource that will make the change - e.g. the human worker to arrange the robot move, or the AGV the robot is mounted on. Finally, the physical KUKA robots will receive the new optimized robot location via PLC. In this example, the CEE (Cyclic Event Evaluator) simulation mode in Tecnomatix Process Simulate is used. The CEE, which functions as a PLC, is used inside the Tecnomatix Process Simulate to control how a typical robotics simulation progresses using logic. Once the start signal is true, the simulation will start. Originally, there are no robot move relocation functions defined. With Tecnomatix .NET API, the move operation can be generated in each simulation. In the first iteration, the training gym will send random coordinates of both robots to the simulation environment, and then two object flow operations in Tecnomatix Process Simulate will be generated and be linked with other operations.

Fig. 2. Layout of the simulation environment

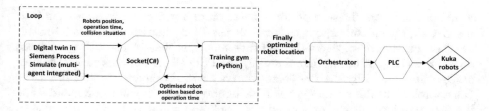

Fig. 3. Framework for reconfiguration engine

To let the result converge faster, penalties are given to situations where a robot collides with other parts or cannot complete the operation due to reachability. As a multi-objective optimization problem, multiple different artificial intelligence approaches can be applied to this use case. The reconfiguration engine can choose which approach is the best solution based on past experience. In this use case, two different types of optimization algorithms will be used to do robot location optimization. One is the global Particle Swarm Optimization (PSO) approach. The other is a Genetic Algorithm (GA). The sequence flow of these two algorithms is listed below (Fig. 4). PSO is a bio-inspired algorithm that searches for an optimal solution through iterative improvement. In this use case, the robot's initial random locations will be set in the initialization swarm, and progressively optimized based on the simulation time received from Tecnomatix Process Simulate. GA is a search heuristic inspired by the theory of natural evolution. This algorithm reflects the process of natural selection where the fittest individuals are selected for reproduction in order to produce offspring for the next generation. The hyperparameter for the PSO and GA are listed below (Table 2).

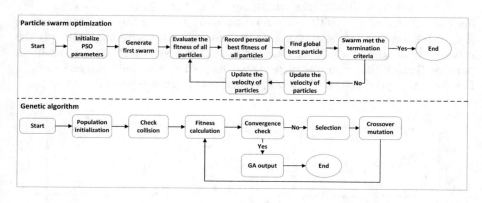

Fig. 4. Flow diagram for PSO and GA algorithms

After the training gym has finished the optimization, the optimized location of the robots will be found. The key performance indicators in this example are the cell's total operation time and the number of iterations the algorithm took to find an optimal operation time. The comparison of these two approaches is listed below (Fig. 5 and Fig. 6). The reconfiguration engine will choose the scenarios based on different performances. From Fig. 5 and Fig. 6, we find that the GA algorithm converges quicker than the PSO algorithm approach for this use case. Furthermore, the optimized time found by the GA algorithm is almost the same as the optimized time found by the PSO algorithm approach. The best process time found by the PSO algorithm is 23.72 s compared to 23.74 s for the GA approach. In this situation, the reconfiguration engine will recommend using the GA algorithm as the optimized approach and can save this information in the data bank for future reference.

Table 2. Hyperparameters for Genetic algorithm and PSO algorithm

Hyperparameters	PSO algorithm	Genetic algorithm
Population size		10
Number of genes		6
Number of parents mating		4
Swarm size	10	
Acceleration coefficient	c1 = 1.5, c2 = 1.5	
Inertia weight	w = 0.5	
Pick robot's coordinates	([−2000, 2000], [−500, 500], [0, 0])	
Welding robot's coordinates	([−2000, 2000], [−500, 500], [0, 0])	
Number of iterations	50	
Penalties	50	

There are some limitations to the approach used in the applications. For instance, if the initial position of the robot is too far from the work-piece where they can't execute the operations, the result will not easily converge – if the robots can't execute the operations, the training gym will always receive a penalty (bigger than the operation time) instead of the valid operation time. Convergence speed is also highly dependent on the chosen penalty value.

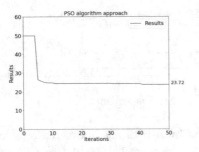

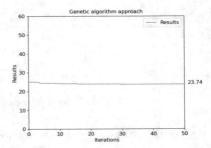

Fig. 5. PSO approach   **Fig. 6.** Genetic algorithm approach

5 Conclusions and Future Work

The design and operation of manufacturing systems are increasingly changeable as the markets require quicker responses to new products, supply disruptions, and volume demands. Reconfiguration and optimisation of the production process in response to external changes is a difficult challenge for complex and flexible systems. This paper proposes a new framework to enable the system to find optimized operation parameters and configurations autonomously. With this framework, the manufacturing system reconfiguration can be enabled at first in the simulation environment and then deployed to the physical system. Though currently applied to a simulation environment, our next step is to apply the framework to our physical robotic manufacturing cells. More key performance indicators will be introduced as optimisation criteria for this framework. Lastly, more complicated use cases will be considered to use this framework. Specifically, applications where manufacturing is not limited to one workstation.

References

1. Westerman, G., Calméjane, C., Bonnet, D., Ferraris, P., McAfee, A.: Digital transformation: a roadmap for billion-dollar organizations. MIT Center Digit. Bus. Capgemini Consult. **1**, 1–68 (2011)
2. da Cunha, C., Cardin, O., Gallot, G., Viaud, J.: Designing the digital twins of reconfigurable manufacturing systems: application on a smart factory. IFAC-PapersOnLine **54**(1), 874–879 (2021)
3. Davis, J., Edgar, T., Porter, J., Bernaden, J., Sarli, M.: Smart manufacturing, manufacturing intelligence and demand-dynamic performance. Comput. Chem. Eng. **47**, 145–156 (2012)
4. Torayev, A., Schultz, T.: Interactive classification of multi-shell diffusion MRI with features from a dual-branch CNN autoencoder (2020)
5. Li, T., Sun, S., Bolić, M., Corchado, J.M.: Algorithm design for parallel implementation of the SMC-PHD filter. Signal Process. **119**, 115–127 (2016)
6. Koren, Y., et al.: Reconfigurable manufacturing systems. CIRP Ann. **48**(2), 527–540 (1999)
7. Lepuschitz, W., Zoitl, A., Vallée, M., Merdan, M.: Toward self-reconfiguration of manufacturing systems using automation agents. IEEE Trans. Syst. Man Cybern. Part C Appl. Rev. **41**(1), 52–69 (2011)
8. Rehman, H.U., et al.: Cloud based decision making for multi-agent production systems. In: Marreiros, G., Melo, F.S., Lau, N., Lopes Cardoso, H., Reis, L.P. (eds.) EPIA 2021. LNCS (LNAI), vol. 12981, pp. 673–686. Springer, Cham (2021). https://doi.org/10.1007/978-3-030-86230-5_53
9. Estrada-Jimenez, L.A., et al.: Integration of cutting-edge interoperability approaches in cyber-physical production systems and industry 4.0. In: Rea, P., Ottaviano, E., Machado, J., Antosz, K. (eds.) Design, Applications, and Maintenance of Cyber-Physical Systems, chap. 7, pp. 144–172. IGI Global (2021). http://services.igi-global.com/resolvedoi/resolve.aspx?doi=10.4018/978-1-7998-6721-0.ch007

10. Schwung, D., Reimann, J.N., Schwung, A., Ding, S.X.: Self learning in flexible man-
    ufacturing units: a reinforcement learning approach. In: 9th International Confer-
    ence on Intelligent Systems 2018: Theory, Research and Innovation in Applications,
    IS 2018 - Proceedings, pp. 31–38 (2018)
11. El-Shamouty, M., Wu, X., Yang, S., Albus, M., Huber, M.F.: Towards safe human-
    robot collaboration using deep reinforcement learning. In: Proceedings - IEEE
    International Conference on Robotics and Automation, pp. 4899–4905 (2020)
12. Epureanu, B.I., Li, X., Nassehi, A., Koren, Y.: Self-repair of smart manufacturing
    systems by deep reinforcement learning. CIRP Ann. 69(1), 421–424 (2020)
13. Barricelli, B.R., Casiraghi, E., Fogli, D.: A survey on digital twin: definitions, char-
    acteristics, applications, and design implications. IEEE Access 7, 167653–167671
    (2019)
14. Tsarouchi, P., Michalos, G., Makris, S., Athanasatos, T., Dimoulas, K., Chrys-
    solouris, G.: On a human-robot workplace design and task allocation system. Int.
    J. Comput. Integr. Manuf. 30(12), 1272–1279 (2017). https://www.tandfonline.
    com/doi/full/10.1080/0951192X.2017.1307524
15. Kousi, N., et al.: Digital twin for designing and reconfiguring human-robot col-
    laborative assembly lines. Appl. Sci. 11(10), 4620 (2021). https://www.mdpi.com/
    2076-3417/11/10/4620
16. Kousi, N., Gkournelos, C., Aivaliotis, S., Giannoulis, C., Michalos, G., Makris, S.:
    Digital twin for adaptation of robots' behavior in flexible robotic assembly lines.
    Procedia Manuf. 28, 121–126 (2019). https://linkinghub.elsevier.com/retrieve/pii/
    S2351978918313623
17. Magnanini, M.C., Tolio, T.A.: A model-based digital twin to support responsive
    manufacturing systems. CIRP Ann. 70(1), 353–356 (2021). https://linkinghub.
    elsevier.com/retrieve/pii/S0007850621000676
18. Zhang, C., Xu, W., Liu, J., Liu, Z., Zhou, Z., Pham, D.T.: Digital twin-enabled
    reconfigurable modeling for smart manufacturing systems. Int. J. Comput. Integr.
    Manuf. 34(7-8), 709–733 (2021). https://www.tandfonline.com/doi/full/10.1080/
    0951192X.2019.1699256
19. Chen, Q., Heydari, B., Moghaddam, M.: Leveraging task modularity in reinforce-
    ment learning for adaptable industry 4.0 automation. J. Mech. Des. Trans. ASME
    143(7) (2021)
20. Cavalieri, S., Salafia, M.G.: Asset administration shell for plc representation based
    on IEC 61131-3. IEEE Access 8, 142606–142621 (2020)
21. Carpanzano, E., et al.: Design and implementation of a distributed part-routing
    algorithm for reconfigurable transportation systems. Int. J. Comput. Integr. Manuf.
    29(12), 1317–1334 (2016)
22. Yamada, Y., Ookoudo, K., Komura, Y.: Layout optimization of manufacturing
    cells and allocation optimization of transport robots in reconfigurable manufactur-
    ing systems using particle swarm optimization. In: Proceedings 2003 IEEE/RSJ
    International Conference on Intelligent Robots and Systems (IROS 2003) (Cat.
    No. 03CH37453), vol. 2, pp. 2049–2054. IEEE (2003)
23. Koren, Y., Shpitalni, M.: Design of reconfigurable manufacturing systems. J.
    Manuf. Syst. 29(4), 130–141 (2010)

24. Li, A., Xie, N.: A robust scheduling for reconfigurable manufacturing system using petri nets and genetic algorithm. In: 2006 6th World Congress on Intelligent Control and Automation, vol. 2, pp. 7302–7306. IEEE (2006)
25. Yu, J.-M., Doh, H.-H., Kim, J.-S., Kwon, Y.-J., Lee, D.-H., Nam, S.-H.: Input sequencing and scheduling for a reconfigurable manufacturing system with a limited number of fixtures. Int. J. Adv. Manuf. Technol. **67**(1–4), 157–169 (2013)

AI-Based Engineering and Production Drawing Information Extraction

Christoph Haar[✉], Hangbeom Kim, and Lukas Koberg

Fraunhofer Institute for Manufacturing Engineering and Automation IPA,
Stuttgart, Germany
christoph.haar@ipa.fraunhofer.de

Abstract. The production of small batches to single parts has been increasing for many years and it burdens manufacturers with higher cost pressure. A significant proportion of the costs and processing time arise from indirect efforts such as understanding the manufacturing features of engineering drawings and the process planning based on the features. For this reason, the goal is to automate these indirect efforts. The basis for the process planning is information defined in the design department. The state of the art for information transfer between design and work preparation is the use of digital models enriched with additional information (e.g. STEP AP242). Until today, however, the use of 2D manufacturing drawings is widespread. In addition, a lot of knowledge is stored in old, already manufactured components that are only documented in 2D drawings. This paper provides an AI(Artificial Intelligence)-based methodology for extracting information from the 2D engineering and manufacturing drawings. Hereby, it combines and compiles object detection and text recognition methods to interpret the document systematically. Recognition rates for 2D drawings up to 70% are realized.

Keywords: Drawing · Manufacturing drawing · Image recognition · Text recognition · Process planning

1 Introduction

Most components have been manufactured on the basis of 2D drawings up to now. In addition to the pure geometry, these 2D drawings contain a lot of additional information like e.g. surface tolerances, dimensional tolerances, and a heat treatment that have to be read out and described semantically. This information is called Product Manufacturing Information (PMI) [1].

State-of-the-art transmission of information from design to process planning and on to production are enriched data formats from software such as CATIA, INVENTOR, Pro/E, SolidWorks, NX, etc., or software-independent exchange formats like STEP, JT, 3D PDF, and STL [2]. Some of these formats like STEP AP 242 include the documentation of the described additional information based on the ISO 10303-242:2014. On this basis, the exchange between disciplines is

© The Author(s) 2023
K.-Y. Kim et al. (Eds.): FAIM 2022, LNME, pp. 374–382, 2023.
https://doi.org/10.1007/978-3-031-18326-3_36

theoretically possible today. Despite these prerequisites, many manufacturing companies use or receive 2D manufacturing drawings. This is because the reuse of existing drawings, the use of 2D CAD tools, and the not standardized annotation of manufacturing-specific information on 3D models are demanded. The digitalization and semantic description of 2D drawings are the basis for the automation of work planning and digitalized test procedures in production. The combination of this information and technology data form a knowledge database that can be used to derive rules for the automation of the work planning process. This paper focuses on AI (Artificial Intelligence)-based methodology for the extraction of non-geometric information which contains a high information content that is essential for rough pricing and work planning. In further work, it is planned to combine non-geometric and geometric information extracted from 2D manufacturing drawings. This combined information can be merged with 3D models to build a sufficient base for fully automated detailed work planning.

2   State of the Art

A literature review shows that there are several approaches to extracting non-geometric information from technical drawings. Prabhu et al. [3] propose a system, called the AUTOFEAD algorithm, that is able to extract non-geometric information from manufacturing drawings using Natural Language Processing (NLP) techniques. To search for dimensions and their attributes, a heuristic search procedure is developed. Scheibel et al. [4] describe a method to extract dimensional information from pdf manufacturing drawings. The text and position are extracted into HTML format. By clustering multiple text elements by position, the dimensional information is extracted. The authors suggest that the extracted information can be used to optimize a quality control system. Elyan et al. [5] develop an end-to-end framework to process and analyze engineering drawings. To interpret the drawings deep learning methods are used to detect and classify symbols.

To recognize text from images Optical Character Recognition (OCR) technology can be used. In the past years, a lot of research has been done on OCR methods. In [6,7], a rule-based algorithm is developed for text and graphics separation from engineering drawings. OCR method is used to recognize text from the separated areas. Jamieson et al. [8] propose a deep learning-based approach for text detection and recognition from engineering diagrams. The model is capable to recognise horizontal and vertical text.

Object detection is one of the most fundamental problems in computer vision techniques for locating instances of objects in images. A deep convolutional neural network is able to learn robust and high-level feature representations of an image. In the deep learning field, object detection can be categorized into two main groups "one-stage detection" (e.g. YOLO [9], SSD [10]) and "two-stage detection" (e.g. Faster R-CNN [11], Mask R-CNN [12]), where the former is regarded as "complete in one step" while the latter is called as a "coarse-to-fine" process [13]. The object detection technology is applied to understand the class and the location of symbols in [5].

The challenge to transfer OCR and AI-based object detection to manufacturing drawings remains and is addressed in this paper.

3   Methodology

Information to be extracted out of manufacturing drawings can be categorized into 5 categories. The dimensions, geometry, tolerances, general information, and additional manufacturing information (Fig. 1).

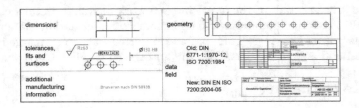

Fig. 1. Information in production drawings

The focus of the work is to recognize the non-geometric information. Figure 2 describes the process of AI-based drawing information extraction. The input drawing is divided into text and symbol information. The OCR method is used to interpret the text from the drawing image. Simultaneously, the object detector allows delivering of the classified objects with bounding boxes. Then, the extracted information is handled by matching algorithms. Based on this, PMIs are read and visualized.

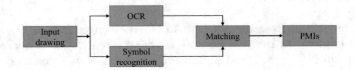

Fig. 2. Phases of the AI-based information extraction system

In the following three sections, the text recognition, the symbol recognition, and the compilation of the information are described in more detail.

3.1   Text Recognition

There are numerous open-source libraries and cloud solutions available for text recognition, also called OCR. The most commonly used systems are MMOCR, EasyOCR, Google Vision, Keras-OCR, and so on. Multiple evaluations of different tools show different results for this system. These results also arise from the use of a wide variety of image data [14]. A basic analysis of some available solutions showed the superiority of the Google Vision system. Due to the goal of open source and python integration, the Python EasyOCR library is chosen. It is composed of 3 main components: feature extraction (Resnet) and VGG, sequence labeling (LSTM), and decoding (CTC). Due to the poor recognition of vertically oriented text, the images are rotated at intervals of 90° for text recognition.

3.2 Symbol Recognition

You Only Look Once(YOLO), a well-known and single-stage target detection algorithm, is used as our basic architecture [9]. The network divides the image into regions and predicts bounding boxes and probabilities for each region. These bounding boxes are weighted with predicted probabilities. As a result, YOLO achieves high detection performance and outstanding inference speed.

Dataset Generation. Synthetic data is an approach to producing datasets that meet specific needs [15]. It enables humans to lessen labor for labeling or generating data manually. The synthetic data is generated to make up for the lack of data. 15 basic 2D drawing documents are used to enlarge the dataset with the information relating to the class and the location of symbols for the symbol recognition. For this purpose, 17 symbols like surface, edge, arrow, and tolerance are cropped from the basic drawings and randomly added to the empty background of basic drawings with different rotations and sizes (Fig. 3). The associated information about the class and the location of the symbols is stored in YOLO labeling format. The YOLO model is trained and tested with 1000 synthetic images. 80% of the dataset is used for training and validation and the rest of them is utilized for the test set.

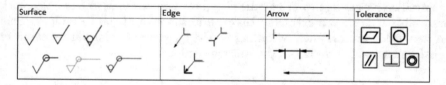

Fig. 3. 17 different extracted symbols, arranged by super-classes, for synthetic data

Object Detection Method. The latest version: YOLOv5 consists of multiple varieties of pre-trained models such as YOLOv5s, YOLOv5m, YOLOv5x, and so on. The difference between them is the size of the model. The lightweight model version YOLOv5s is not for accurate predictions but for fast inference time. Therefore, the YOLOv5x model is considered the main architecture since accuracy is the most significant factor to analyze 2D drawings. The SGD optimizer is used for training with the 1e-2 initial learning rate. Then the model is trained with 16 batch sizes and 640 image sizes. Figure 4 illustrates the symbol recognition sample with an actual 2D drawing document.

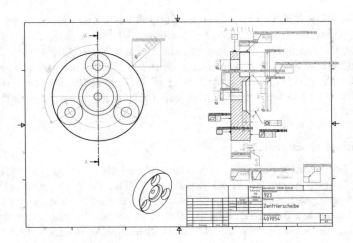

Fig. 4. Inference image using the trained model on the test set of the actual data

3.3   Matching of Symbol and Text

After the text recognition and the symbol recognition are completed, the result data is merged to extract the relevant information in each symbol. Tests have shown that any text with less than 50% recognition accuracy was mostly not recognized correctly, these are all filtered out. When it comes to symbol recognition, confidence and intersection over union(IoU) thresholds are set at 0.25 and 0.45, respectively. Then intersections of bounding boxes of text and symbols are searched for. If there is an intersection between two bounding boxes, the text is regarded as a related text for the symbol.

Specific characteristics of the drawing are used to analyze the title block. First, all text fields are extracted that are in the area of the title block. Then the text fields are assigned according to their geometric position or based on their format. Table 1 describes the rules we define:

Table 1. Rules for title block-matching

Feature	Rule
Title	Next to "Benennung"
Drawingnumber	Number with 6 digits
Material	Next to "Werkstoff"
Materialnumber	Next to "number"
Date	Search for data format

4   Experimental Results and Discussion

In this chapter, the results of the implemented method are presented and discussed. The strengths and weaknesses of the methods used are shown.

4.1   Text Recognition

Due to the focus of OCR systems on horizontal texts, the text images are analyzed both horizontally and rotated by 90°. Only texts with a confidence level over 0.5 are included in the analysis. The score for the evaluation of the text recognition is created per correctly recognized character and correctly recognized text field. 5 drawings with together 278 text fields are analyzed.

- 68% of the characters are recognized correctly
- 62% of the conveyors are recognized correctly

The difficulties in recognition are with mathematical special characters and with texts that, are positioned at an angle and position close to other forms (Fig. 5).

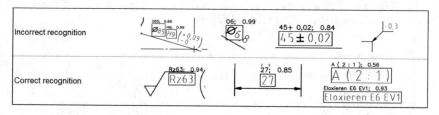

Fig. 5. Examples for incorrect and correct recognized characters, above the bounding boxes the recognized text, and the confidence level are shown

4.2   Symbol Recognition

We use the test set of the synthetic data and the test set of actual 2D drawings to estimate the trained model. The test set of synthetic data is created from the same 15 drawings as the training set but the symbols are positioned in different positions. When it comes to the actual 2D drawings test set, they are original documents and unseen for training the model. The model is evaluated by the detection mean average precision (mAP) since this is a common evaluation metric for object detection. The average precision is calculated with two different IoU thresholds: mAP and AP50. The intersection over union is a similarity measure between the bounding box of ground truth and the predicted detection. The mAP indicates the average of all 10 average precisions with the increment of the 0.05 IoU threshold steps from 0.5 to 0.95. And AP50 is the average precision at the IoU 0.5. The model achieves 0.927 accuracies with AP50 and 0.876

Table 2. The mAP values of the YOLOv5 on the test set of synthetic and actual data

Dataset	Test set (synthetic)	Test set (actual)
all	0.876	0.364
surface_all (6 classes)	0.870	0.537
edge_all (4 classes)	0.911	0.480
arrow_all (2 classes)	0.792	0.215
tolerance_all (5 classes)	0.888	0.207

accuracies with mAP in the test set of the synthetic data. We list the mAP of the final results of the model in each category in Table 2.

Most of the symbols are detected as their original label on the test set of synthetic data. Especially, the model shows outstanding predictions for edge classes with 0.911 accuracies. The average of the arrow classes results in relatively low accuracy at 0.792. This is because some lines are drawn across the arrow symbols and numbers are placed inside the bounding box of the arrows. On the other hand, the model shows difficulties in predicting the symbols on the actual dataset since the model relies on only small amounts of sample symbols and drawings. For this reason, we recommend training the model with abundant data for precise detection of symbols (Fig. 6).

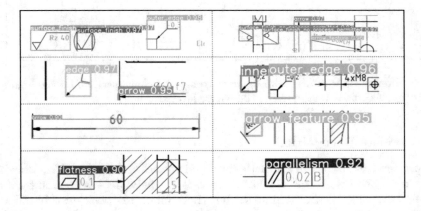

Fig. 6. Samples of detected symbols in 2D drawing documents

4.3   Matching of Symbol and Text

To find a score for the matching algorithm, the number of correct matches from the recognized texts and symbols is calculated. Only recognized features are used, so that recognition issues are not included in the score. 21 drawings are analyzed, and a total of 72% correct assignments are found. Problems arise especially when the bounding box is too small and the text is further away from the symbol. The actual quality of the assignments depends to a large extent on the text and symbol recognition. From the title box, 88% of the information could be extracted correctly. The defined rules are reliable but must be adapted to other drawing types.

5 Conclusion

In this work, the flexibility of machine learning-based systems is adapted to the use case of production drawings recognition. Thus a system could be developed, which is able to read out information from production drawings. Based on 15 test drawings an accuracy of more than 70% is achieved. There is still potential for optimization in each of the described fields text recognition, symbol recognition, and merging. In the field of text recognition, the orientation of the texts and the recognition of special characters is a weakness, which can be eliminated by training new models. In the area of symbol recognition, the greatest potential lies in the extension of the training data set, especially to non-standard drawings. In the area of merging, the extension of the set of rules for semantic processing of the recognized texts and symbols has great potential. All in all, the realized approach represents an expandable basis for the recognition of information from 2D drawings. The biggest advantage of the presented method is the easy extension of the logic due to the use of machine learning-based approaches.

Acknowledgement. The results are obtained as part of the project "SE.MA.KI - Self-learning control of cross-technology matrix production by simulation-based AI", funded by the German Federal Ministry of Education and Research (BMBF) under grant number L1FHG42421.

References

1. Lipman, R., Lubell, J., Hedberg, T., Freeney, B., Frechette, S.: MBE PMI validation and conformance testing project-NIST. Technical report, NIST (2013)
2. Nzetchou, S., Durupt, A., Remy, S., Eynard, B.: Review of CAD visualization standards in PLM. In: Fortin, C., Rivest, L., Bernard, A., Bouras, A. (eds.) PLM 2019. IAICT, vol. 565, pp. 34–43. Springer, Cham (2019). https://doi.org/10.1007/978-3-030-42250-9_4
3. Prabhu, B., Biswas, S., Pande, S.: Intelligent system for extraction of product data from CADD models. Comput. Ind. **44**, 79–95 (2001)
4. Scheibel, B., Mangler, J., Rinderle-Ma, S.: Extraction of dimension requirements from engineering drawings for supporting quality control in production processes. Comput. Ind. **129**, 103442 (2021)
5. Elyan, E., Jamieson, L., Ali-Gombe, A.: Deep learning for symbols detection and classification in engineering drawings. Neural Netw. **129**, 91–102 (2020)
6. Kulkarni, C.R., Barbadekar, A.B.: Text detection and recognition: a review. Int. Res. J. Eng. Technol. (IRJET) **4**(6), 179–185 (2017)
7. Lu, Z.: Detection of text regions from digital engineering drawings. IEEE Trans. Pattern Anal. Mach. Intell. **20**(4), 431–439 (1998)

8. Jamieson, L., Moreno-Garcia, C.F., Elyan, E.: Deep learning for text detection and recognition in complex engineering diagrams. In: 2020 International Joint Conference on Neural Networks (IJCNN), pp. 1–7 (2020)
9. Bochkovskiy, A., Wang, C.-Y., Liao, H.-Y.M.: Yolov4: optimal speed and accuracy of object detection, arXiv preprint arXiv:2004.10934 (2020)
10. Liu, W., et al.: SSD: single shot multibox detector, CoRR, vol. abs/1512.02325 (2015)
11. Ren, S., He, K., Girshick, R.B., Sun, J.: Faster R-CNN: towards real-time object detection with region proposal networks, CoRR, vol. abs/1506.01497 (2015)
12. He, K., Gkioxari, G., Dollár, P., Girshick, R.B.: Mask R-CNN, CoRR, vol. abs/1703.06870 (2017)
13. Zou, Z., Shi, Z., Guo, Y., Ye, J.: Object detection in 20 years: a survey, CoRR, vol. abs/1905.05055 (2019)
14. Smelyakov, K., Chupryna, A., Darahan, D., Midina, S.: Effectiveness of modern text recognition solutions and tools for common data sources. In: CEUR Workshop Proceedings, pp. 154–165 (2021)
15. Nikolenko, S.I., et al.: Synthetic Data for Deep Learning. Springer, Cham (2021). https://doi.org/10.1007/978-3-030-75178-4

Generation of an Intermediate Workpiece for Planning of Machining Operations

Dušan Šormaz$^{(\boxtimes)}$ ⬥, Anibal Careaga Campos ⬥, and Jaikumar Arumugam

Ohio University, Athens, OH 45701, USA
{sormaz,ac985015}@ohio.edu

Abstract. Digital twins in manufacturing plays a key factor for the digital trans-formation. A necessary component of any digital twin in manufacturing is a geo-metric model of a workpiece as it is processed through steps. DT requires solid 3d models, machining features, and information regarding machines, tools, and its constraints such as initial setup, machining direction, etc. The objective of this paper is to generate alternate feature interpretations to identify geometric con-straints, machine and tool requirements, and stock materials to generate flexible manufacturing plans that fit a defined criterion. In this study we propose using the IMPlanner system to retrieve a 3d model from a CAD software, read its geomet-ric features and convert them into possible machining features. This information along with information from the database of stock materials, tools, machines, and tolerances, the system generates several feature interpretations, thus offering a more flexible manufacturing plan.

Keywords: Manufacturing planning · Manufacturing modeling · Machining features

1 Introduction

Manufacturing companies strive to streamline their processes to reduce operational costs and minimize lead times. The constant pursues to integrate computer aided design CAD and computer aided process planning CAPP resulted in several feature-based solutions. However, this integration proved challenging due to the complexity of the features for each part, and different interpretations of features in each step. Several solutions have been presented to overcome such challenges, which will be explained in the following section. The goal of this paper is to propose a new methodology to generate alternate feature interpretations using a digital twin model which creates intermediate solid mod-els. This way the design and planning processes use the same data types, and the system has the ability to exchange information bidirectionally between the two. We use the IMPlanner systems that connects to a Siemens NX™ CAD software's API and retrieves the geometric model information which is used to generate a digital twin of the desired part and the features included. The different solutions proposed in the literature are out-lined in Sect. 2. Section 3 will explain the procedure to generate the alternate features, a machining process sequence and intermediate workpiece model. An example using a complex part is presented in Sect. 4, and finally the conclusion along with prospect future work to complement this research is presented in Sect. 5.

© The Author(s) 2023
K.-Y. Kim et al. (Eds.): FAIM 2022, LNME, pp. 383–393, 2023.
https://doi.org/10.1007/978-3-031-18326-3_37

2 Literature Review

Computer aided process planning CAPP and its integration with computer aided design CAD has received significant attention since the 1980s [1]. CAD and CAPP integration became the main focus area due to the versatility of options provided by feature-based integration [1]. Yang et al. [2] reviewed the different types of product modeling explaining the advantages of feature-based product modeling over solid product modeling. Their reasoning states that feature-based modeling maintains data integrity across several domains and it's easier to transfer. Additionally, they explain the role of those modeling types in newer modelling techniques such as object-oriented product modeling. Ma et al. [3] explain how features can be used as information units to avoid incompatible data structures amongst different design stages. Their unified modeling scheme combines both geometric and non-geometric attributes to determine the inter or intra-relation between features. Kritzinger et al. [4] define digital twin as mirroring an asset with advanced technology to reproduce them in a virtual setting. This practice allows for further product development and more efficiency. Goopert et al. [5] emphasize the benefits of digital twins in modern manufacturing practices. As they explain, the connection between physical and digital assets enables smooth data transfer and easier adaptations. Liu and Wang [6] presented a feature sequencing method using knowledge-based rules and geometric reasoning to generate feature sequencing rules considering setups on tool-approaching directions. To expand on this study, Mokhtar et al. [1] discuss the machining precedence of STEP-NC features. Their approach is to detect interacting features considering their geometric constraints and machining constraints. Using both they determined the precedence of the features and generated several machining plans. Adding to this work, Dipper et al. [7] propose a set of algorithms to generate a feature interaction detection system for STEP-NC data models. Their algorithms, written in the programming language C++, calculate the volumes of all the features and faces that are interconnected in a part. However, this method is limited to simple faces. Further efforts considered more complex features. Zheng and Mohd Taib [8] developed an algorithm for feature recognition of inner loops. This methodology focuses on the Solidworks models and converts shapes such as islands, through pockets and pockets into machining features. Ma et al. [3] explained the importance of collaborative feature-based engineering systems to improve the product development process. The authors outline the issues related to this approach, including different definitions of features and their data structures. Therefore, they proposed a feature-based unification to integrate all the features into suitable applications. Different approaches with more unconventional algorithms were taken. Houshmand et al. [9] proposed a flower pollinating with artificial bees (FPAB) technique to generate machinable features. Their approach is to develop elementary volumes and then merge them to generate machinable volumes which are considered as candidates and their tool evaluates them to generate possible sequences of machining operations. Liu et al. [10] proposed a semantic feature language that is compatible across multiple domains in concurrent engineering. Their approach to ease integration between CAD and CAM systems separates features hierarchically. Different hierarchies are required for different domains. Therefore, this approach simplifies the communication and exchange of machining features across different platforms. Mokhtar and Xu [1] complemented this work with a proposed a rule-based system for machining precedence relations and make it compatible with STEP-NC models. Their system considers the interaction between all the features and includes more complex faces such as pockets, and through holes. To include sequence considerations, Khoshnevis et al. [11] proposed a planning system using feature reasoning and space search-based optimization. Their approach is to upstream integration with design systems and downstream integration with scheduling considering dynamic constraints such as machine and tool availability to produce more efficient process plans. Applying this methodology to manufacturing, Šormaz et al. [12] designed and proposed the IMPlanner system. A set of algorithms and rules that use XML data representation to generate process plans considering all the mentioned constraints. Expanding upon the previous work, Sormaz et al. [13] proposed a rule based process selection of milling processes considering GD&T requirements. This approach considers different requirements which outputs new manufacturing routes for more optimality in the entire process. Guo et al. [14] proposed a hybrid feature recognition method using graph and rule based methods. Their approach uses traditional graph techniques such as boundary representation, and a set of algorithms to recognize individual features. It then loads the information into a matrix to find the relationships and extract the required information for the following process. Throughout the years there has been extensive work done in the area of feature recognition and interpretation for manufacturing planning. Earlier work focused on simpler faces and pockets, and later work expanded to more complex shapes. The industry 4.0 movement increased the digitalization of the manufacturing processes using data driven methods. Our work expands on a systematic approach to generate alternate feature interpretations. The previous research considers feature recognition is several ways, however, this methodology uses a novel approach of creating a digital twin model for a more extensive feature interpretation to generate intermediate solid models for further analysis and application in manufacturing planning and processing.

3 Methodology

This section expands the process to generate intermediate workpiece from a CAD model via an interface with a CAD system. After a brief overview, previously built modules are briefly described and modules for intermediate workpiece generation explained.

3.1 Overview

Figure 1 shows a complete map of the process to retrieve CAD model and use its information to develop a machining processing plan. This task is accomplished using the IMPlanner system and the CAD software Siemens NX. The IMPlanner system is a software tool developed at Ohio University comprised of several algorithms for the purpose of CAPP. It connects to the CAD system's API and retrieves the desired model. Next, the system completes feature mapping and feature solid creation operations. Then, a feature interaction analysis is performed to check the interaction between features, this analysis considers face neighborhood, feature neighborhood, feature interaction and feature precedence. It considers geometric factors concerned with tolerance, technological factors concerned with machine feasibility, and economic factors concerned with cost.

This information helps separate required features from optional features, then generate a precedence network which then is used for the process sequencing stage. All the information is uploaded to the feature-based CAPP. This system then generates optimal machining operation sequences for all the volumes and processes them into the manufacturing assembly model.

The major steps of the procedure are feature mapping and feature interaction analysis, process selection and process sequencing, feature volume generation, intermediate workpiece model (digital twin) generation. Some of the steps were reported in the earlier work and they will be just briefly explained. The steps for feature volume generation and intermediate workpiece are subject of this paper and they are explained in more details.

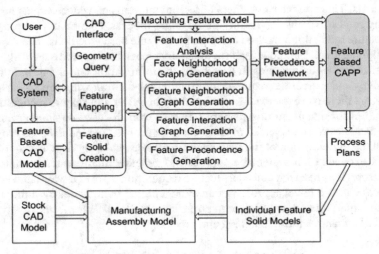

Fig. 1. Map of manufacturing assembly model

3.2   Feature Mapping and Interaction Analysis

A manufacturing feature is a collection of related geometric elements, which can be associated to a manufacturing process. Features of mechanical components have been selected as the communication medium for the integration of CAD/CAM systems [15]. The features are usually viewed differently in different stages of the product design and manufacturing process. Design features differ from manufacturing features. Figure 2 shows the different interpretation for a hole as a design feature and a manufacturing feature. A CAD software shows a simple hole with a specified radius and depth inside a block, whereas a cylinder removed from a block showing an approach axis and depth is considered a machining feature. Feature mapping considers all the faces and generates a coordinates index array which contains the axis points, reference points and the attributes of the feature such as radius, length, distance among others. Figure 3 outlines the approach for feature mapping of a general slot. The design features and their parameters are retrieved from the CAD model. The program converts the model features into appropriate machining features by finding its outline location and setting it as a reference position, then projects a normal vector from it and uses the placement outline to create profiles for the new volume. Finally, the floor offset sets the bottom distance. This approach changed the design features into machining features that satisfy approachability criteria. After the machining features are mapped, their interactions are analyzed using geometric and neighborhood information from the CAD model. This results in the Feature Precedence Network (FPN) which is a set of constraints on the order of manufacturing individual features. The details of the procedure are outside the scope of this paper, but its details can be found in [16].

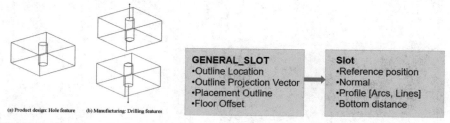

(a) Product design: Hole feature   (b) Manufacturing: Drilling features

Fig. 2. Design feature hole versu hole drilling feature

Fig. 3. Process of feature mapping

3.3 Process Selection and Process Sequencing

The interaction analysis determines required and optional features. Features which volumes can only be removed by machining processes associated with it are considered required. On the other hand, optional features can be removed while machining another feature. This aspect is crucial for process and sequence selection. In this step the system verifies GD&T requirements, process capability, and tools and machine availability. Then implements a process selection procedure based on specific and generic rules. Specific rules include cutting operation and quality while generic rules include compatible machines, process capability, and feature relations. The optimization algorithm generates process instances, then checks for process capabilities. If a match is complete, then the process is accepted for the feature, then tools and machine is specified and finally the machining time and cost are considered guarantee optimality (see Fig. 4). Process sequencing uses a space search-based optimization to generate alternate sequences. All those decisions are made based on knowledge in rules and facts inferred from the CAD model. Details of these steps have been reported in earlier work and, therefore, are out of the scope for this paper.

3.4 Feature Volume Generation

Feature volume generation is based on a stock volume and parameters of each feature. Using process information which may have several access directions thus can be approached from several different positions, we can generate volume for each possible feature. Considering each non-floor face of a feature as the floor face for a possible alternative feature allows to project the face on the stock for the part, then the generated volume is a new possible feature. Features are presented in a parametric way, and it is necessary to convert those parameters into solid volumes. These volumes depend on the size and shape of the stock. Figure 5 shows the parameters used to map out the slot and how they are derived to generate a solid volume which is minimal for material removal.

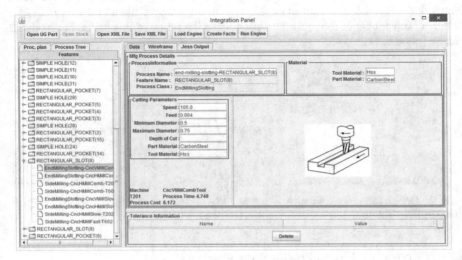

Fig. 4. IMPlanner UI after process selection

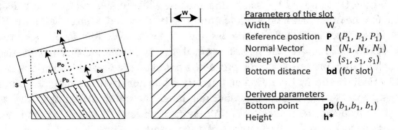

Parameters of the slot
Width   W
Reference position   P (P_1, P_1, P_1)
Normal Vector   N (N_1, N_1, N_1)
Sweep Vector   S (s_1, s_1, s_1)
Bottom distance   bd (for slot)

Derived parameters
Bottom point   pb (b_1, b_1, b_1)
Height   h*

Fig. 5. Volume generation of slot

3.5 Intermediate Workpiece Generation

Upon the completion of process planning procedure, an intermediate workpiece of the manufacturing process is created. The IMPlanner system takes the manufacturing part model and the stock material as input, both of them in the form of CAD models. Based on recommendation from Siemens NX [17] an assembly file that contains all components of the digital twin is created. The components of that assembly file are (see Fig. 6): the part model, the stock model, delta volume, and incremental manufacturing model. The part model and stock model are input for a given design, the delta volume is a CAD model of all machining features (or set difference between stock and part). An incremental manufacturing model corresponds to removal of individual features from stock. Based on the process sequence, as individual feature is machined, its volume is subtracted from the stock volume and intermediate model is created. Those intermediate models can be saved in order to explore forces and stress during machining. An example of this procedure will be shown later in Sect. 4. All these steps are performed from IMPlanner's digital twin module, but they generate and modify CAD models using Siemens NX Java API.

Fig. 6. Manufacturing assembly file

4   Case Study

In order to illustrate the intermediate workpiece generation procedure, we will use an example of a mechanical part called slider. Figure 7 shows the drawing from a CAD model of the part with all its specifications.

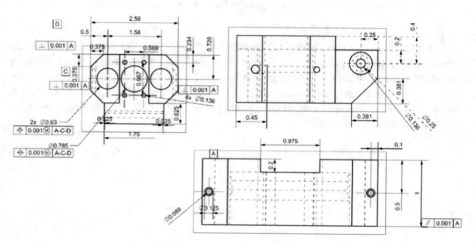

Fig. 7. Drawing of slider example

The IMPlanner system loads the part from Siemens NX and reads all its geometric features, Fig. 8 shows the results of feature mapping process with a 3D view for verification. Process selection and process sequencing procedures are executed within IMPlanner with generation of alternate processes and their sequencing according to the FPN. The results of those steps are shown earlier in Fig. 4. The complete process plan is loaded into the Digital twin module to finalize the CAD Models creation steps. The individual feature volumes and the whole delta volume are created first as shown in Fig. 9. Incremental volume can be created in two modes: individually by selecting features in the feature tree, or automatically using the process sequence. The result after completing some features is shown in Fig. 10 in NX (three features) and in Fig. 11 in IMPlanner (nine features).

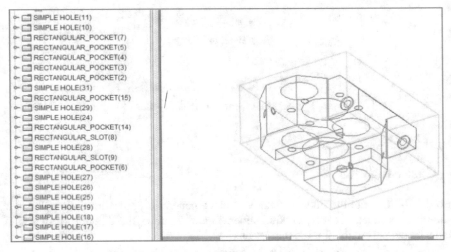

Fig. 8. Model imported to IMPlanner displaying its features

This case study demonstrates the capability to generate the workpiece solid on demand. Any alternate process and feature sequence can be used to generate intermediate workpiece solid for that sequence, which can be used for validation and selection of optimal alternative and sequence. Intermediate solid brings additional validation (in addition to time/cost done in sequence optimization), which is, for example, collision detection for tool, fixture, and machine elements.

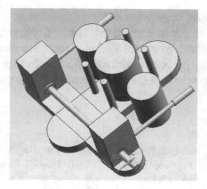

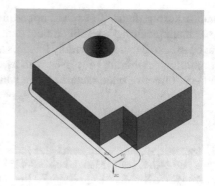

Fig. 9. Generated delta volumes

Fig. 10. Partial workpiece after three processes

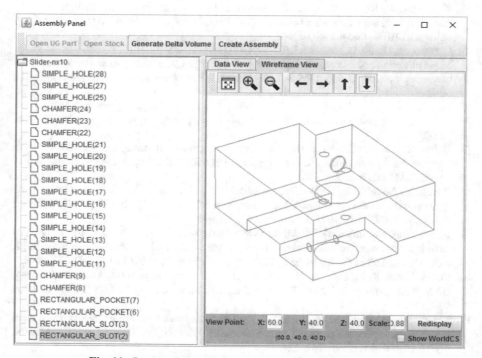

Fig. 11. Intermediate workpiece model in IMPlanner interface

5 Conclusion

The proposed methodology developed a novel procedure for generation of manufacturing digital twin. Intermediate solid models can be used to generate more machining features. The main advantages of this approach are extensive analysis of features and feature interactions which lead to more machining sequences thus offering more flexible manufacturing plans. To expand upon this work, integration of feature modeling with CAD manufacturing assembly model generation can be considered. The same approach can be used to generate assembly features and develop optimal assembly sequence plans. Combining machining processes with assembly processes will lead to a more efficient and cost-effective manufacturing plan which can benefit the manufacturing industry.

References

1. Mokhtar, A., Xu, X., Lazcanotegui, I.: Dealing with feature interactions for prismatic parts in STEP-NC. J. Intell. Manuf. **20**, 431–445 (2008). https://doi.org/10.1007/s10845-008-0144-y
2. Yang, W., Xie, S., Ai, Q., Zhou, Z.: Recent development on product modelling: a review. Int. J. Prod. Res. **46**, 6055–6085 (2008). https://doi.org/10.1080/00207540701343895
3. Ma, Y., Chen, G., Thimm, G.: Change propagation algorithm in a unified feature modeling scheme. Comput. Ind. **59**, 110–118 (2008). https://doi.org/10.1016/j.compind.2007.06.006
4. Kritzinger, W., Karner, M., Traar, G., Henjes, J., Sihn, W.: Digital twin in manufacturing: a categorical literature review and classification. IFAC-Pap. **51**(11), 1016–1022 (2018). https://doi.org/10.1016/j.ifacol.2018.08.474
5. Göppert, A., Grahn, L., Rachner, J., Grunert, D., Hort, S., Schmitt, R.H.: Pipeline for ontology-based modeling and automated deployment of digital twins for planning and control of manufacturing systems. J. Intell. Manuf., 1–20 (2021).https://doi.org/10.1007/s10845-021-01860-6
6. Liu, Z., Wang, L.: Sequencing of interacting prismatic machining features for process planning. Comput. Ind. **58**(4), 295–303 (2007). https://doi.org/10.1016/j.compind.2006.07.003
7. Dipper, T., Xu, X., Klemm, P.: Defining, recognizing and representing feature interactions in a feature-based data model. Robot. Comput.-Integr. Manuf. **27**(1), 101–114 (2011). https://doi.org/10.1016/j.rcim.2010.06.016
8. Zheng, Y., Mohd Taib, J.: Decomposition of interacting machining features based on the reasoning on the design features. Int. J. Adv. Manuf. Technol. **58**, 359–377 (2012). https://doi.org/10.1007/s00170-011-3385-4
9. Houshmand, M., Imani, D., Niaki, S.: Using flower pollinating with artificial bees (FPAB) technique to determine machinable volumes in process planning for prismatic parts. Int. J. Adv. Manuf. Technol. **45**, 944–957 (2009). https://doi.org/10.1007/s00170-009-2023-x
10. Liu, Y.-J., Lai, K.-L., Dai, G., Yuen, M.: A semantic feature model in concurrent engineering. IEEE Trans. Autom. Sci. Eng. **7**, 659–665 (2010). https://doi.org/10.1109/TASE.2009.2039996
11. Khoshnevis, B., Sormaz, D.N., Park, J.Y.: An integrated process planning system using feature reasoning and space search-based optimization. IIE Trans. **31**(7), 597–616 (1999). https://doi.org/10.1080/07408179908969862
12. Sormaz, D.N., Arumugam, J., Ramachandra, H., Patel, C., Neerukonda, N.: Integration of product design, process planning, scheduling, and FMS control using XML data representation I Elsevier Enhanced Reader. https://www.sciencedirect.com/science/article/abs/pii/S0736584510000931. Accessed 01 Feb 2022
13. Sormaz, D.N., Gouveia, R., Sarkar, A.: Rule based process selection of milling processes based on GD&T requirements. Prod. Eng. **21**(2), 26 (2018)
14. Guo, L., Zhou, M., Lu, Y., Yang, T., Yang, F.: A hybrid 3D feature recognition method based on rule and graph. Int. J. Comput. Integr. Manuf. **34**(3), 257–281 (2021). https://doi.org/10.1080/0951192X.2020.1858507

15. Gao, J., Zheng, D.T., Gindy, N.: Extraction of machining features for CAD/CAM integration. Int. J. Adv. Manuf. Technol. **24**(7–8), 573–581 (2004). https://doi.org/10.1007/s00170-003-1882-9

16. Arumugam, J.: Analysis of feature interactions and generation of feature precedence network for automated process planning. Ohio University (2004). https://etd.ohiolink.edu/apexprod/rws_olink/r/1501/10?clear=10&p10_accession_num=ohiou1176142843. Accessed 02 Mar 2022

17. ***: Mill Manufacturing Process, Student Guide, 6 vols. EDS Inc. (2003)

Automating the Generation of MBD-Driven Assembly Work Instruction Documentation for Aircraft Components

Aikaterini R. Papadaki[1], Konstantinos Bacharoudis[1] , David Bainbridge[1],
Nick Burbage[2], Alison Turner[1], David Sanderson[1(✉)] , Atanas A. Popov[1] ,
and Svetan M. Ratchev[1]

[1] Institute for Advanced Manufacturing, Faculty of Engineering, University of Nottingham,
Jubilee Campus, Nottingham NG8 1BB, UK
David.Sanderson@nottingham.ac.uk
[2] Hamble Aerostructures Ltd, Kings Avenue, Hamble-le-Rice, Southampton SO31 4NF, UK

Abstract. The classical approach to the creation of assembly work instructions for high value, complex products is time-consuming and prone to error. It requires a process engineer to write the work instructions step-by-step and manually insert specific technical information, using an encompassing document of manufacturing parameters or life cycle management software. The latter offers synchronisation to design changes through updateable parameters, however major design modifications still require significant manual work to modify the text contents and structure of the work instructions. This leaves the work instruction documentation vulnerable to human error, as well as making the process time-consuming to fully synchronise. A methodology was therefore developed to resolve these issues, utilising JavaScript and VBA for Office to create a simple interface for rapid content generation for work instructions including text, MBD extracted parameters, images and formatting. The overall methodology speeds up the creation of assembly work instructions and reduces errors by implementing automatic insertion of parameters from an MBD model. The implementation and effectiveness of the suggested approach is demonstrated on a case study for the assembly of the joined wing configuration of the RACER helicopter, the latest generation of compound helicopters of Airbus Helicopter.

Keywords: Assembly work instructions · Model-based definition · Joined wing · Digital manufacturing

1 Introduction

Assembly work instructions play a significant role in assembly quality and lead times, capturing the necessary assembly activities and relevant technical information that engineering technicians must read, understand, and apply in order to successfully assemble the product under development. For high value, low volume, complex products such as aircraft components, production of work instructions is particularly elaborate as they

© The Author(s) 2023
K.-Y. Kim et al. (Eds.): FAIM 2022, LNME, pp. 394–403, 2023.
https://doi.org/10.1007/978-3-031-18326-3_38

require a vast number of processes. Many of these processes are also repetitive, occurring at several locations in the aircraft structure, e.g. drilling pilot holes, countersinking, reaming, etc.

The classical approach for creating work instruction documentation first requires a process engineer to retrieve the necessary technical information scattered in CAD models (2D drawings), in spreadsheet documents or through the tacit knowledge of process and design engineers. This technical information must be accumulated and then accurately captured into a document or software. Typical cycle times for the design process in the aerospace field is approximately 5 years [1], therefore retrieving the information accumulated in this period can be a very time-consuming activity, prone to many errors. An inability to create flawless assembly work instructions can jeopardise the entire design activity and confuse or frustrate operators, resulting in poor quality products [2].

Nowadays an increasing number of aerospace companies, such as Boeing and Airbus, are adopting the model-based definition approach [3] to alleviate the problem of managing large amounts of data. Many works discuss and implement MBD technology in the field of manufacturing, e.g. Quintana et al. [4] and Geng et al. [5], however few consider the implementation of MBD to enable product design synchronisation with work instructions. Furthermore, despite fast progression in this field, current research on MBD is limited, particularly for assembly work instructions [3].

A range of modern software packages for product lifecycle management (PLM) are available; one example is Siemens Teamcenter, which offers the extraction of digital work instructions and documentation from 3D models. The specific functionality offered by Teamcenter is the synchronisation of design changes to digital work instructions, and visualisation through interactive 3D assembly sequences. However, the majority of such software available present, including Teamcenter, suffer from the same issues: they can be time consuming to learn and use, and the primary focus is the generation of illustrations for use in technical documentation, not the instructional text contents which are equally vital for shop floor operators.

Alongside these PLM software packages, augmented reality (AR) is one of several technologies that have been devised to optimise the visual representation of these work instructions, thereby improving clarity of their contents (see, for example, Gattullo et al. [6]). This technology is an alternative to traditional documentation and provides a wide range of potential benefits to assembly lines, such as shorter learning time for operators, reduced training costs and overall product quality improvement as seen in various studies [7, 8]. Zauner et al. [9] demonstrates a highly intuitive AR tool that guides the user through simple assembly steps for furniture, with potential for application on more complex assemblies in the future. Although considerable progress has been made in recent years, there are still significant obstacles preventing the incorporation of AR in the production of aircraft structures. These include limitations on object detection using markers for small objects (such as screws, bolts, etc.), high computational requirements, and relatively large upfront investments for equipment installation and training. Furthermore, the majority of AR solutions are offline and therefore any consensus leading to redesign of the product and the assembly processes requires a considerable amount of time and effort to amend the work instruction documentation, since they cannot synchronise to design changes and must be rebuilt again to produce updated instructions [10, 11]. Few works tackle this issue in the field of augmented reality.

Alternate solutions for the automation and improved visual representation of work instructions have also been developed. Geng et al. [5] developed a lightweight tool that is easily accessible by the end user, utilising automatically generated, interactive CAD images embedded in PDFs for improved cognition of work instructions. However, there is no instructional text generated and it is acknowledged that the creation of the work instructions and synchronising design changes are time consuming with this tool, despite its link to MBD, which reaffirms the current gap. Gors et al. [12] also demonstrates a method to determine assembly sequences and generate images to produce work instructions, however there are minimal text instructions for each step. It is therefore clear that few works target the automation of the text content of assembly work instruction documentation, especially when considering the concept of updatable links to manufacturing parameters using MBD.

Based on the presented literature and the everyday industrial practice, there is a clear need for more automatic document generation of assembly work instructions with an updateable link to MBD models, particularly whilst more novel technologies such as AR remain in the experimental stages of implementation. This paper presents a functioning solution to this problem; the methodology is presented in detail in Sect. 2, followed by an exemplary case study in Sect. 3. The selected case study is the joined wing configuration of the RACER (Rapid And Cost Effective Rotorcraft) demonstrator by Airbus [13]. Details of the specific implementation are discussed in Sect. 4. Outcomes of the case study are discussed in Sect. 5, and finally useful conclusions are presented in Sect. 6.

2 Proposed Methodology

This paper suggests a methodology for rapid generation of work instruction documentation, exploiting repetitive assembly processes to automate the process and applying an MBD approach to collate assembly information and error-proof parameter insertion.

Firstly, to enable insertion of manufacturing parameters within the documentation, this methodology presumes the existence of an MBD model in which geometric data and assembly information is captured in a structured approach. The MBD model must be logically structured such that data is extracted and stored in a file format such that data can be identified and retrieved logically. In this way, an updatable and robust link is made between the stored dataset and the MBD. This link ensures major design changes to the product will not compromise the structure of the dataset and therefore the methodology presented can remain synchronised with the product design.

Secondly, a series of process templates for repetitive assembly processes are created. Each template represents the generalised procedure for a particular process, containing user commands that indicate how the instructions will be structured, such as excluding or repeating sections depending on the parameters of the targeted assembly features that the procedure will be carried out on. Placeholders for variables are also used in conjunction with commands to enable MBD parameter insertion during the generation of content. Therefore, a set of chosen process templates must still be created to initiate this methodology, however, the need for manual insertion of parameters is eliminated and the time consumed creating work instructions can be reduced. It is important to note that the process templates only have to be created once when setting up the methodology. Modifications to the process templates are only required if the processes change, which is likely to be infrequent due to the certification requirements in the aerospace industry.

Lastly, two algorithms are used to create the documentation. The first algorithm systematically retrieves process templates based on user selection and links the template placeholders to the MBD parameters of the selected feature. The second algorithm processes the documentation through a series of tasks, such as opening individual process templates, inserting parameters into placeholders, creating repetition or exclusion of text, inserting images, and finally exporting to a final document. The overall structure is summarised in Fig. 1.

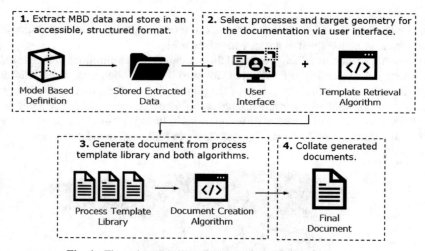

Fig. 1. Flow chart detailing the structure of the methodology.

3 Case Study

The case study selected to demonstrate the developed methodology was the joined wing configuration of the RACER demonstrator [13] shown in Fig. 2.

The specific case study has been selected because sufficient information was present, the assembly processes and assembly sequences associated were relatively complex, and subsequently a successful implementation would mean high applicability to a variety of products. The overall assembly sequence and procedure for this joined wing configuration can be found in [14] and is distinct to a conventional wing assembly, due to the strict and difficult geometric constraints with respect to the location of the wing and the wing interface.

The MBD model of RACER was developed in Dassault's 3DExperience software, using the concept of the joint definition structure. A joint definition is defined by the

(a)   (b)

Fig. 2. Images depicting the RACER demonstrator; (a) shows the conceptual image of the rotorcraft and (b) shows the geometry of the jointed wing configuration [13].

contact of two or more interfacing surfaces (an interface group), where multiple joint definitions may be assigned to a single interface group.

An example of an interface group and the contained joint definitions are captured in Fig. 3. Given the interface group "Upper Cover to Mid Rib", there are three joint definitions associated with the lower wing of the joined wing configuration. Joint definition 14 consists of 11 holes with particular manufacturing parameters, and joint definition 15 and 16 each consist of one hole of differing manufacturing parameters. Together, these three joint definitions make up the "Upper Cover to Mid Rib" interface group.

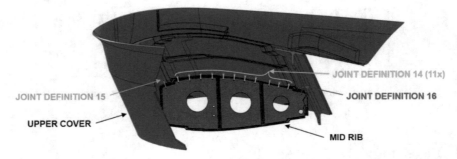

Fig. 3. Joint definition structure displayed within RACER's MBD model

4   Implementation of Methodology

To implement the methodology using the case study of Sect. 3, a select number of process templates were produced by examining the most common processes in the RACER assembly work instructions, namely pilot drill, full size drill, deburr, shim and final fasten.

Pilot drill, full size drill and final fasten were the most repetitive processes in the sample RACER work instructions, and these processes utilised very similar structures such that they can be generalised into the pseudocode shown in Fig. 4. With reference to Fig. 4, italic text corresponds to the printed content of the document, bold text to commands, and plain text to other pseudocode contents.

As can be seen in Fig. 4, a for-loop first iterates through each of the selected interface groups, with following lines defining variables such as the name of each interface within the current interface group. Next, an if-statement determines if the current interface group contains two or three components and inserts text accordingly. Additional logic can be employed according to the needs of the process template format. Finally, embedded within the assembly steps of the defined process, a for-loop is used to iterate through the joint definitions of the current interface group. Parameters extracted from the dataset can be inserted throughout the document, including images.

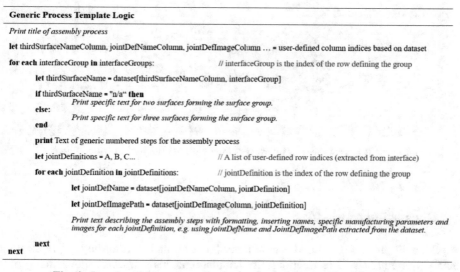

Fig. 4. Pseudocode demonstrating logic of generalised process template.

The deburr and shim processes appeared less frequently within the example RACER work instructions, however they contained elements that would benefit from creating a process template with commands. For example, the deburr process template requires the total number of holes for a target surface (e.g., Upper Cover) and a breakdown table of these holes per joint definition. To complete this manually, the process engineer must check every interface group that includes the target surface, find each joint definition accordingly, create the table and then sum the total number of holes.

Commands within the process template can be used to automatically extract the total number of holes for a target surface and produce a table that lists every associated joint definition with the number of holes to be deburred. This demonstrates the advantage of the methodology even for less repetitive processes, as it reduces the time needed for document creation by also automating the accumulation and manipulation of parameters.

A user interface has been developed using a JavaScript add-in for Office Excel, with embedded VBA coding to enable document generation. The interface contains a layout for the user to input the assembly sequence row by row, using dropdown selection. The dropdown selection is linked and updated with the joint definitions and interface groups using the MBD dataset file. This is shown in Columns D, E, F and G of Fig. 5. The developed tool provides great flexibility; if the assembly strategy changes, it is a matter of few minutes for the user to define a new order of assembly processes, selecting specific interface group and joint definitions for each process. It is assumed that a process template already exists for each process. If not, then the user needs to set up the process template once for each new process and further use them through the interface shown in Fig. 5.

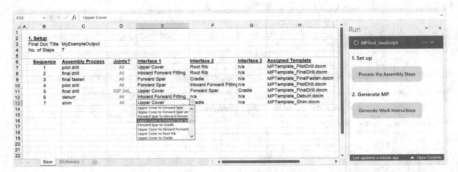

Fig. 5. Image of the user interface (along with the sidebar) in Office Excel.

5 Discussion of Results

A pilot drill process document was produced and formatted according to pre-existing RACER assembly work instructions. The template and final output are shown in Fig. 6(a) and (b) respectively, noting minor post formatting was required such as addition of page numbering and headers. Though these formatting actions can be implemented into the tool, they are typically semi-automated in common word processing programs.

The tool was tested for 10 processes, namely a sequence of four pilot drill processes, two final fasten processes, two final drill processes, one deburr process and one shim process, each with different interface group selections. The approximate time to generate the combined document was 3 min 58 s, containing 23 pages and 6059 words, which included a total of 57 joints. To create approximately 23 pages of work instructions with a more classical approach, an estimate of a few days to a week would be expected. Therefore, it can be seen the timeframe for creating documentation using this methodology is reduced from days, or weeks, to minutes.

Direct comparison to common industrial tools, such as Teamcenter and DELMIA, could not be carried out due to limited access. The literature suggests these tools typically automate image creation, introduce interactive 3D models to the workshop floor and update parameters in real time [15, 16]. However, they require manual text content creation, extensive user training and costly licenses. Therefore, while alternative tools

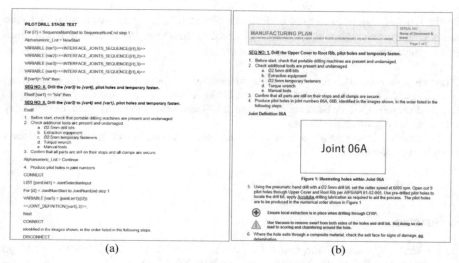

Fig. 6. Images of (a) the pilot drill template, and (b) the resulting document output from the tool.

offer advantages for high quality visuals and cohesive integration with 3D models within the same software package, the demonstrated methodology addresses the shortcomings of these tools. Overall, the methodology provides a low-cost solution for rapid generation of text content for work instructions with updateable, embedded parameters.

6   Conclusions and Future Work

The presented methodology demonstrates the automatic generation of documentation for work instructions with updateable links to MBD data, satisfying the gap identified in relevant literature. Once the methodology is set up then only the user-input instructions are necessary for modifying the assembly sequence or updating the MBD dataset after product design changes. The methodology can be set up by specifying process templates with minimum programming knowledge due to the developed pseudocode whilst still retaining all necessary features and formatting for minimum post-processing. This approach offers the potential to eliminate errors and reduce the time consumed creating documentation by programmatically generating the instructions of assembly processes. This was then verified by the case study implementation, in which the timeframe for instruction creation was shortened significantly.

The methodology also leaves room for future work, such as greater overall automation and collaboration with other methodologies. For example, the implementation of automatic MBD dataset extraction would further speed up the process of assembly work instruction generation. Automatic image extraction and automatic assembly sequence generation from the MBD file could also be implemented for near-complete automation, as demonstrated by Gors et al. [12].

It is also important to note the methodology is not being presented as a complete solution. It generates text for specific processes identified as significantly repetitive, and not the work instructions in their entirety. However, this tool provides a solution to the issues with the conventional method of assembly instruction production in lieu of more novel approaches that are not yet fully matured.

Acknowledgements. This project has received funding from the Clean Sky 2 Joint Undertaking (JU) under grant agreement number CSJU-CS2-GAM-AIR-2014-2015. The JU receives support from the European Union's Horizon 2020 research and innovation programme and the Clean Sky 2 JU members other than the Union.

The tool presented here is available on request; please contact the authors directly to discuss how this can be arranged.

References

1. Shekar, B., Venkataram, R., Satish, B.M.: Managing complexity in aircraft design using design structure matrix. Concurr. Eng. Res. Appl. **19**, 283–294 (2011)
2. Claeys, A., Hoedt, S., Schamp, M., et al.: Intelligent authoring and management system for assembly instructions. Procedia Manuf. **39**, 1921–1928 (2019)
3. Goher, K., Shehab, E., Al-Ashaab, A.: Model-based definition and enterprise: state-of-the-art and future trends. Proc. Inst. Mech. Eng. B. J. Eng. Manuf. **235**(14), 2288–2299 (2020)
4. Quintana, V., Rivest, L., Pellerin, R., et al.: Will model-based definition replace engineering drawings throughout the product lifecycle? A global perspective from aerospace industry. Comp. Ind. **61**, 497–508 (2010)
5. Geng, J., Zhang, S., Yang, B.: A publishing method of lightweight three-dimensional assembly instruction for complex products. J. Comput. Inf. Sci. Eng. **15**(3), 1–12 (2015)
6. Gattullo, M., Scurati, G.W., Fiorentino, M., et al.: Towards augmented reality manuals for industry 4.0: a methodology. Robot Comput. Integr. Manuf. **56**, 276–286 (2019)
7. Frigo, M., da Silva, E.C.C., Barbosa, G.: Augmented reality in aerospace manufacturing: a review. J. Ind. Intell. Inf. **4**(2), 125–130 (2016)
8. Doshi, A., Smith, R.T., Thomas, B.H., Bouras, C.: Use of projector based augmented reality to improve manual spot-welding precision and accuracy for automotive manufacturing. Int. J. Adv. Manuf. Technol. **89**(5–8), 1279–1293 (2016). https://doi.org/10.1007/s00170-016-9164-5
9. Zauner, J., Haller, M., Brandl, A.: Authoring of a mixed reality assembly instructor for hierarchical structures. In: Proceedings of the Second IEEE and ACM International Symposium on Mixed and Augmented Reality, ISMAR, pp. 237–246. IEEE, New York (2003)
10. Goher, K., Shehab, E., Al-Ashaab, A.: Trends in model-based definition based assembly information for high-value manufacturing. In: Hedburg, T. Jr., Carlisle, M. (eds.) Proceedings of the 11th Model-Based Enterprise Summit, MBE, pp. 99–103. NIST, Maryland (2020)
11. Lavric, T., Bricard, E., Preda, M., et al.: An AR work instructions authoring tool for human-operated industrial assembly lines. In: Proceedings of the IEEE International Conference on Artificial Intelligence and Virtual Reality IEEE AIVR, vol. 1, pp. 174–183. IEEE CPS, New Jersey
12. Gors, D., Put, J., Vanherle, B., et al.: Semi-automatic extraction of digital work instructions from CAD models. Procedia CIRP **97**, 39–44 (2021)
13. Blacha, M., Fink, A., Eglin, P., et al.: Clean Sky 2: Exploring New Rotorcraft High Speed Configurations. In: Proceedings of the 43rd European Rotorcraft Forum, ERF, pp. 989–1000. Curran Associates Inc., New York (2017)

14. Bainbridge, D., Bacharoudis, K. C., Cini, A., et al.: Advanced assembly Solutions for the Airbus RACER Joined-Wing Configuration. SAE Technical Paper 2019–01–1884 (2019)
15. Siemens PLM Teamcenter: https://www.plm.automation.siemens.com/global/en/products/teamcenter. Accessed 27 Apr 2022, last accessed 2022/04/27
16. Dassault Systems DELMIA. https://www.3ds.com/products-services/delmia. Accessed 27 Apr 2022

Hyperspectral Imaging for Non-destructive Testing of Composite Materials and Defect Classification

Trunal Patil[1,3]([✉]) [iD], Claudia Pagano[1]([✉]) [iD], Roberto Marani[2]([✉]) [iD],
Tiziana D'Orazio[2]([✉]) [iD], Giacomo Copani[1]([✉]) [iD], and Irene Fassi[1]([✉]) [iD]

[1] STIIMA-CNR, Institute of Intelligent Industrial Technologies and Systems for Advanced Manufacturing, Consiglio Nazionale delle Ricerche, via A. Corti 12, 20133 Milan, Italy
claudia.pagano@stiima.cnr.it
[2] STIIMA-CNR, Institute of Intelligent Industrial Technologies and Systems for Advanced Manufacturing, Consiglio Nazionale delle Ricerche, via G. Amendola 122 D/O, 70126 Bari, Italy
[3] DIMI-Department of Mechanical and Industrial Engineering, University of Brescia, Via Branze 38, 25123 Brescia, Italy

Abstract. Carbon fiber composite materials are intensively used in many manufacturing domains such as aerospace, aviation, marine, automation and civil industries due to their excellent strength, corrosion resistance, and lightweight properties. However, their increased use requires a conscious awareness of their entire life cycle and not only of their manufacturing. Therefore, to reduce waste and increase sustainability, reparation, reuse, or recycling are recommended in case of defects and wear. This can be largely improved with reliable and efficient non-destructive defect detection techniques; those are able to identify damages automatically for quality control inspection, supporting the definition of the best circular economy options. Hyperspectral imaging techniques provide unique features for detecting physical and chemical alterations of any material and, in this study, it is proposed to identify the constitutive material and classify local defects of composite specimens. A Middle Wave Infrared Hyperspectral Imaging (MWIR-HSI) system, able to capture spectral signatures of the specimen surfaces in a range of wavelengths between 2.6757 and 5.5056 μm, has been used. The resulting signatures feed a deep neural network with three convolutional layers that filter the input and isolate data-driven features of high significance. A complete experimental case study is presented to validate the methodology, leading to an average classification accuracy of 93.72%. This opens new potential opportunities to enable sustainable life cycle strategies for carbon fiber composite materials.

Keywords: Hyperspectral imaging · Automatic defect detection · Deep learning · Convolutional neural network · Composite materials · Circular economy

1 Introduction

In the past few decades, carbon fiber composite materials have been largely used in many manufacturing domains such as aerospace, aviation, marine, automation, sports,

K.-Y. Kim et al. (Eds.): FAIM 2022, LNME, pp. 404–412, 2023.
https://doi.org/10.1007/978-3-031-18326-3_39

and civil industries, due to their excellent strength, corrosion resistance, low thermal expansion, and lightweight properties.

It is estimated that by 2025, the annual global CFRP (Carbon fiber reinforced polymers) waste will reach 20 kt; at this year the cumulative amount of CFRC (Carbon fiber reinforced composites) waste which is ready to be recycled in Europe is estimated to reach 144,724 t [1]. This poses great challenges, due to the current environmentally unfriendly CFRC waste management, yet in disagreement with the European Composites Industry Association (EuCIA) directives, which in 2011 has set strict recycling and reuse practices based on the European Union directives on End-of-Life (EOL) vehicles 2000/53/EC and waste 2008/98/EC [2].

1.1 Hyperspectral Imaging (HSI)

The defect detection techniques assist the life cycle assessment, structural performance maintenance, value chain integration, reverse logistic strategy, ecological benefits, and transition into a circular economy model. Several non-destructive techniques (NDTs) are often used for defect detection; however, they cannot usually identify the material and have several disadvantages. Visual inspection and RGB machine vision system methods are extensively affected by the surrounding conditions and unsuitable for hazardous material [3]. Due to the radiation hazard, safety protection tools are required for X-ray imaging and neutron imaging methods [4, 5]. Eddy current tests detect surface and subsurface cracks, but only for conductive materials [6]. The active infrared (IR) thermography is safe and quick with respect to other NDTs [7], but the automatic data processing is challenging despite the recent methodologies such as machine-learning [8], deterministic-differential analyses [9], and deep learning [10]. The resonant inspection method detects the resonant frequency shifts resulting from changes in mass or stiffness of defected areas [11], but the accuracy is highly sensitive to the surrounding noise. Ultrasonic testing is not suitable for the intricate shape of material surfaces [12]. The shearography testing results are difficult to interpret and require a good light source [13].

Along with defect detection, precise material identification is also necessary to address remanufacturing, repair, reuse, and recycling processes in a circular economy perspective. X-ray imaging [14], x-ray computed tomography (XCT) [15], optical microscopy-based imaging [16], and scanning electron microscopy [17] can be utilized for both defect and material identification. However, they require sample extraction, preparation and measurement steps, which are very time-consuming and usually carried out offline; while the online or inline identification of material is essential to set the parameters of manufacturing, repair, remanufacturing, and recycling. This is particularly relevant in the case of composite materials, since the material compositions and, thus, process specifications vary according to different types, designs, manufacturers, and applications.

HSI measures the continuous spectrum of the light for each pixel of the sample with fine wavelength resolution and it can work online or inline directly on parts with no sample preparation nor causing any damage. Therefore, it can perform a very powerful quality control task, playing a vital role in EOL management.

HSI is able to extract a large variety of information on any kind of surface and provides a promising solution to identify the material along with defect classifications.

In compare to other HSI applications [18, 19], very few research studies are reported to identify defects in carbon fiber composite material and still in their infancy [20]. Carbon fiber composites are used in many high precision applications where, in addition to visual and geometric quality control, other quality parameters of performance related defects must be identified in order to comply with the safety of end-users. Therefore, the quality control should be intelligent and robust, to handle the complexities, uncertainties, and variability of the operations.

In this context, the traditional vision systems normally fail, while hyperspectral acquisition system goes far beyond the RGB imaging; it captures the high-resolution spectra at every pixel providing not only physical but also chemical information of the specimen. The large amount of information provided by HSI system comes at the cost of complex data acquisition and a huge volume of data. Indeed, HSI acquires two spatial and one spectral dimension, requiring a complex image analysis for image preprocessing (i.e. calibration, noise removal, data reduction), feature extraction, and classification. The use of HSI integrated with remanufacturing processes will improve the ability of industries to retrieve re-usable CFRC materials from composite components. Consequently, this will reduce the usage of virgin materials, the energy required during manufacturing and logistics operations and, ultimately, the CO_2 footprint along the whole product lifecycle, maintain the waste management legislation, and product emissions regulations.

In this work, a case study shows the possibility of using HSI method to identify material and defects of CFRC. A defect apparatus has been developed to create different defects on carbon fiber composites. A hyperspectral imaging system has been used for acquiring the images and an image processing technique, based on a deep neural network, has been designed to classify the types of defects and identify the different materials.

2   Material and Method

A material and defect identification method is proposed using the middle wave infrared hyperspectral imaging system. Two different types of thermosetting carbon fiber reinforced composites have been studied and two types of defects have been generated on the specimens. The specimens were scanned by the middle wave infrared hyperspectral imaging system, and the spectra have been analyzed with a convolutional neural network (CNN) to identify the unique response from the specimen material and defects.

Two types of thermosetting carbon fiber reinforced composites, FDCA 0.6 (5% CAT 4 layer) and AFD 60 (4 layers) [21], developed by Politecnico di Milano [22], have been investigated. The used samples have square shapes with dimensions of about 10 cm x 10 cm and thickness between 1 and 1.5 mm. Since CFRC material appears black, it has a very limited reflectance in the visible near-infrared and near-infrared spectral region, thus, a MWIR-HSI has been used in this study: a SPECIM broom type camera with spectral range from 2.6757 μm to 5.5056 μm.

The apparatus for defect creation, based on the drop weight impact test procedure, consists of a hollow pipe, a falling body, an impactor holder, and two types of impactors. The two cylindrical shape falling bodies with 0.995 kg and 2.046 kg of mass are used to pass through a hollow pipe at the height of 1 m. The two main defects (*conical* and *hemispherical*) (Fig. 1) have been created using a conical impactor generating an impact energy of 10 J and a pressure of 1.6 bar and a hemispherical impactor generating an impact energy of 20 J and a pressure of 1.8 bar.

Fig. 1. (a) AFD 60 specimen, zoom of (b) a conical defect, and (c) a hemispherical defect

The approximate diameters of conical and hemispherical defects are 2.9 mm and 5.5 mm, respectively. For each specimen, before acquiring the specimen image (I_{raw}), the dark (I_{dark}) and white references (I_{white}) were captured with the camera cover and a frosted aluminum alloy tile, respectively, and used for the calibration procedure to obtain the reflectance values (Eq. 1).

$$I_{corrected} = \frac{I_{raw} - I_{dark}}{I_{white} - I_{dark}} \tag{1}$$

Hyperspectral images of three specimens of each material were acquired before and after the creation of the defects. Figure 2a represents the hypercube of a CFRC-AFD 60 specimen and Fig. 2b three reference spectra, one for each defect type and one for the material of that specimen. Each spectrum was preprocessed to feed a convolutional neural network and used either for the training, the validation or the test. Each spectrum is made of N samples representing the reflection contribution of the specimen at a specific wavelength. Each point of the specimen surface produces a response labeled in one of four classes: one for each material type and one for each defect type.

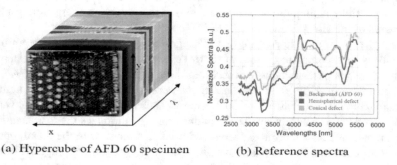

(a) Hypercube of AFD 60 specimen   (b) Reference spectra

Fig. 2. HSI data pattern

The network architecture is shown in Fig. 3. The convolutional layer (Conv 1D) consists of a bank of 35 convolutional filters having 35-entries-length kernels and the classification layer is made of three sublayers: *fully connected layer* - a neural network receives the input features, processed by the previous convolution-based layers, a *softmax layer* - the four outputs of the fully connected layer are arguments of a softmax function and a *classification layer* - the probabilities are compared to select the class with the maximum score (probability).

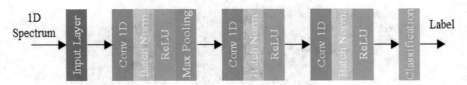

Fig. 3. Representation of the network architecture for data classification

It is worth noting that the network processes each spectral signature alone, without considering further contributions from adjacent pixels. In this way, the defect detection is independent of the shape of the training defects. The classification only depends on the current response of the material for the specific pixel under analysis.

3   Results and Discussions

The acquisition parameters have been optimized to improve the quality of the images, detect small features, and reduce the computational load. A high exposure time value is desirable to increase the signal-to-noise ratio and thus, provide better spectra. However, a trade-off has to be selected to avoid image saturation, in particular for the white reference. Therefore, several tests have been carried out to identify the best exposure time. Moreover, in order to acquire not distorted images of the samples, the correct aspect ratio has to be found varying the frame rate and the scanning speed. Figure 4a and 4b show the effect of an incorrect setting. A proper aspect ratio (Fig. 4c) has been found with a frame rate 21.5 times bigger than the scanning speed. Furthermore, the larger the specimen the better the analysis. Accordingly, the field of view has been reduced as much as possible, taking into account the depth of focus of the camera and the physical constraints of the HSI system.

The classification performance of the proposed network has been proven through the analysis of a complete dataset made of several acquisitions. Six specimens have been used in this case study, which includes three specimens made of FDCA 0.6 and three specimens of AFD 60. All the specimens were scanned by HSI system before and after damaging them so that around 30 defects for each type (hemispherical and conical) have been selected as a train set. The network training minimizes the loss function, i.e. the mean distance between the expected and predicted class, with the Adam optimizer considering each class of the same weight. The stop criterion is computed through a validation procedure. The validation set is used to compute an accuracy value while

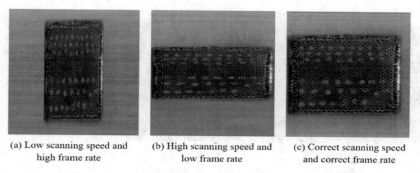

| (a) Low scanning speed and high frame rate | (b) High scanning speed and low frame rate | (c) Correct scanning speed and correct frame rate |

Fig. 4. Aspect ratio to set scanning speed and frame rate parameter

learning a correct prediction ratio, helpful in stating whether the network is converging towards a correct solution.

Due to the reduced dimension of impact areas, in comparison with the area of the specimens, the populations of the classes are unbalanced. Therefore, since the loss optimization is performed considering equally-weighted classes, the most populated classes have been partially reduced but not matched to the least populated ones to induce an implicit bias to the predictions of non-defective regions. The tuning of the network has been completed in 34 epochs achieving an accuracy of 94.32% in label predictions, thus suggesting the exact convergence of the optimization, confirmed by the balanced accuracy, which scales the standard accuracy by the population of the reference class [23]. The testing process confirms the network's capability to classify the input spectra, recognizing the specific constitutive material of the specimens and the defects. Specifically, the global accuracy, considering all correct detections, reaches 93.72%, with an average balanced accuracy of 85.55%. A prediction map of a test specimen is reported in Fig. 5. Conical defects are identified with low sensitivity since they are often confused with defects of the other classes, but not with homogeneous areas, which is very promising for manufacturing quality control. As shown in Fig. 6, the sensitivity can be easily improved, without inducing misclassifications of the hemispherical-shaped defects, whose probabilities are much higher and close to 1 (refer Fig. 6(a)), setting the threshold of classification scores of conical defects to 0.25.

In this work, HSI potential has been exploited with plastic composites with a very limited reflectance. In comparison with another study that used HSI to detect surface damage in carbon fiber reinforced polymer materials [20] no issues related to the background have been observed. Moreover, the low impact energy of the defects (11.9 J) and the complex surface texture, which has been assumed in [20] as the cause of a reduced detection accuracy, are very similar to the conditions of the present study, where good overall classification accuracy is achieved due to a robust computation algorithm. Finally, in this work not only the defects, but also the constituent materials have been detected at the same time as required in industrial product management.

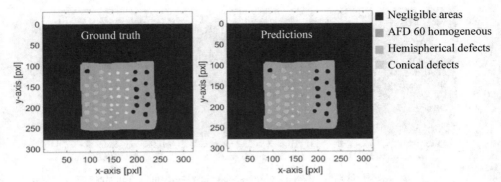

Fig. 5. Prediction map and corresponding ground truth of a test specimen

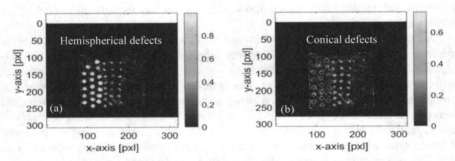

Fig. 6. Classification scores of the defective classes during testing.

4 Conclusions

In this paper, HSI method is proposed for accurate, fast, and reliable quality control of carbon fiber composite parts and demonstrated with a simple case study, in which it was combined with a deep neural network to identify material and defects of thermosetting CFRC samples. The results confirm that HSI is very suitable for non-destructive defect detection and allows not only an efficient classification of various types of defects, but also a reliable identification of material differences in CFRC.

As a practical and managerial implication, this study demonstrated that HSI is a very promising candidate as enabling technology for sustainable manufacturing and the circular economy of composite materials. By systematically adopting HSI as a reliable and accurate solution for in-line or online quality control of manufactured parts, companies will be able to early detect defects and repair parts in order to limit non-quality costs and customers' complaints. In terms of circular economy, the precise identification and characterization of materials and defects through HSI will provide companies the necessary information to select the most appropriate strategies, including repair, reuse, and recycling. This will minimize the production of new parts and, at the same time, will maximize the value of products and materials with significant environmental benefits. This research presents some limitations. First, this methodology can be exploited to all types of materials and, thus, a large variety of product domains, but it is not applicable in the case of internal damage as HSI scans only external surfaces. Second, this research was limited to plain specimens of CFRP, investigating defects bigger than 2.9 mm. Third, from a business point of view, the economic and industrial sustainability of the introduction of HSI was not investigated due to the explorative nature of this research. Accordingly, further work, studies with a larger variety of materials, smaller dimensions of the defect and other geometrical profiles of defects such as wrinkles, cracks, and scratches have to be implemented to verify the sensitivity of the hyperspectral imaging system.

Acknowledgments. This work has been developed in the context of the COMPOSER (Carbon-fiber reinforced composites for Sustainable circular Economy models based on Repair and Remanufacturing for Reuse) project funded by "Fondazione Cariplo" and partially funded by the European Union under the DiManD project (H2020-MSCA-ITN, grant agreement No. 814078).

References

1. Oliveux, G., Dandy, L.O., Leeke, G.A.: Current status of recycling of fibre reinforced polymers: review of technologies, reuse and resulting properties. Prog. Mater. Sci. **72**, 61–99 (2015)
2. Chang, C.-I.: Hyperspectral Imaging: Techniques for Spectral Detection and Classification, vol. 1. Springer Science & Business Media, Boston, MA (2003)
3. Shanmugamani, R., Sadique, M., Ramamoorthy, B.: Detection and classification of surface defects of gun barrels using computer vision and machine learning. Measurement **60**, 222–230 (2015)
4. Garcea, S.C., Wang, Y., Withers, P.J.: X-ray computed tomography of polymer composites. Compos. Sci. Technol. **156**, 305–319 (2018)
5. Banhart, J., et al.: X-ray and neutron imaging – complementary techniques for materials science and engineering: dedicated to Professor Dr. H.-P. Degischer on the occasion of his 65th birthday. Int. J. Mater. Res. **101**, 1069–1079 (2010)
6. García-Martín, J., Gomez-Gil, J., Vázquez-Sánchez, E.: Non-destructive techniques based on Eddy current testing. Sensors (Basel). **11**, 2525–2565 (2011)
7. Balageas, D., et al.: Thermal (IR) and other NDT techniques for improved material inspection. J. Nondestr. Eval. **35**(1), 1–17 (2016). https://doi.org/10.1007/s10921-015-0331-7
8. Marani, R., et al.: Automatic detection of subsurface defects in composite materials using thermography and unsupervised machine learning. In: Proceedings of the IEEE 8th International Conference on Intelligent Systems, p. 516 (2016)
9. Marani, R., et al.: Two-dimensional cross-correlation for defect detection in composite materials inspected by lock-in thermography. In: Proceedings of the 22nd International Conference on Digital Signal Processing, pp. 1–5 (2017)
10. Marani, R., Palumbo, D., Galietti, U., D'Orazio, T.: Deep learning for defect characterization in composite laminates inspected by step-heating thermography. Opt. Lasers Eng. **145**, 106679 (2021)
11. Stultz, G., Bono, R., Schiefer, M.: Fundamentals of resonant acoustic method NDT. Adv. Powder. Metall. Part. Mater. **3**, 11 (2005)
12. Felice, M.V., Fan, Z.: Sizing of flaws using ultrasonic bulk wave testing: a review. Ultrasonics **88**, 26–42 (2018)

13. Hung, Y., Ho, H.P.: Shearography: an optical measurement technique and applications. Mater. Sci. Eng. R-Reports **49**, 61–87 (2005)
14. Bunaciu, A.A., Udriştioiu, E.G., Aboul-Enein, H.Y.: X-Ray diffraction: instrumentation and applications. Crit. Rev. Anal. Chem. **45**, 289–299 (2015)
15. Maire, E., Withers, P.J.: Quantitative X-ray tomography. Int. Mater. Rev. **59**, 1–43 (2014)
16. Shen, Y., Wan, W., Zhang, L., Yong, L., Lu, H., Ding, W.: Multidirectional image sensing for microscopy based on a rotatable robot. Sensors (Basel). **15**, 31566–31580 (2015)
17. Goldstein, J.I., Newbury, D.E., Michael, J.R., Ritchie, N.W.M., Scott, J.H.J., Joy, D.C.: Scanning electron microscope (SEM) instrumentation. In: Scanning Electron Microscopy and X-Ray Microanalysis, pp. 65–91. Springer, New York (2018). https://doi.org/10.1007/978-1-4939-6676-9_5
18. Wu, D., Sun, D.-W.: Advanced applications of hyperspectral imaging technology for food quality and safety analysis and assessment: a review — part I: fundamentals. Innov. Food Sci. Emerg. Technol. **19**, 1–14 (2013)
19. Candiani, G., et al.: Characterization of fine metal particles derived from shredded WEEE using a hyperspectral image system: preliminary results. Sensors (Basel). **17**, 1117 (2017)
20. Yan, Y., et al.: Nondestructive Testing of Composite Fibre Materials with Hyperspectral Imaging : Evaluative Studies in the {EU} {H2020} FibreEUse Project, CoRR. abs/2111.0 (2021)
21. Jin, Y., et al.: Recent advances in dynamic covalent chemistry. Chem. Soc. Rev. **42**, 6634 (2013)
22. Fortunato, G., et al.: Simultaneous recovery of matrix and fiber in carbon reinforced composites through a diels-alder solvolysis process. Polymers (Basel) **11**, 1007 (2019)
23. Marani, R., Palumbo, D., Galietti, U., Stella, E., D'Orazio, T.: Enhancing defects characterization in pulsed thermography by noise reduction. NDT E Int. **102**, 226–233 (2019)

Correction to: Development Process for Information Security Concepts in IIoT-Based Manufacturing

Julian Koch, Kolja Eggers, Jan-Erik Rath, and Thorsten Schüppstuhl

Correction to:
Chapter "Development Process for Information Security Concepts in IIoT-Based Manufacturing" in: K.-Y. Kim et al. (Eds.): *Flexible Automation and Intelligent Manufacturing: The Human-Data-Technology Nexus*, **LNME, https://doi.org/10.1007/978-3-031-18326-3_31**

The book has been inadvertently published with the incorrect affiliation of all the authors in Chapter 31, which has now been corrected. The book and the chapter have been updated with the change.

The updated original version of this chapter can be found at
https://doi.org/10.1007/978-3-031-18326-3_31

K.-Y. Kim et al. (Eds.): FAIM 2022, LNME, p. C1, 2023.
https://doi.org/10.1007/978-3-031-18326-3_40

Author Index

© The Editor(s) (if applicable) and The Author(s) 2023
K.-Y. Kim et al. (Eds.): FAIM 2022, LNME, pp. 413–414, 2023.
https://doi.org/10.1007/978-3-031-18326-3

Printed in the United States
by Baker & Taylor Publisher Services